6th AIChE/SPE Joint Workshop 2011

Challenges in Flow Assurance

Houston, Texas, USA
26 – 28 September 2011

ISBN: 978-1-5108-1775-3

Printed from e-media with permission by:

Curran Associates, Inc.
57 Morehouse Lane
Red Hook, NY 12571

Some format issues inherent in the e-media version may also appear in this print version.

Printed by Curran Associates, Inc. (2016)

AIChE
120 Wall Street, FL 23
New York, NY 10005-4020

Phone: (800) 242-4363
Fax: (203) 775-5177

www.aiche.org

Additional copies of this publication are available from:

Curran Associates, Inc.
57 Morehouse Lane
Red Hook, NY 12571 USA
Phone: 845-758-0400
Fax: 845-758-2634
Email: curran@proceedings.com
Web: www.proceedings.com

TABLE OF CONTENTS

SESSION SIX: THE ROLE OF CHEMICAL ADDITIVES

SESSION SEVEN: MULTIPHASE FLOW II

SESSION EIGHT: THE ROLE OF FA ENGINEERS IN PROJECTS AND SUBSEA OIL SPILL FA

Hydrate Flow Assurance

Dendy Sloan, Carolyn Koh, Amadeu Sum

and the Hydrate Busters

SPE/AIChE Flow Assurance Workshop

Houston, Texas
September 27, 2011

COLORADOSCHOOLOFMINES
engineering the way

FOR RESEARCH CENTER HYDRATE

Hydrate Busters

Associate Director
Prof. David Wu

Affiliated Faculty
Prof. Matt Liberatore

Postdoctoral

Eric Grzelak | Prasad Kerkar | Jonathan Levine | Sergey Skiba | Somendra Chakraborty

UG Students

Nathan Welch | Kyle Olcott | Kyle Springer | Michaela Wagar | Samantha Hawkins

Graduate Students

Matt Walsh | Sanjeev Joshi | Luis Zerpa | Ishan Rao | Patrick Lafond | Eric Webb | Zach Aman | Giovanny Grasso | Gary Grim | Erika Horning | Mike Jones

Hongfei Xu | Zach Ward | Chris Cabusao | Martin Braniff | Matt Gilmer

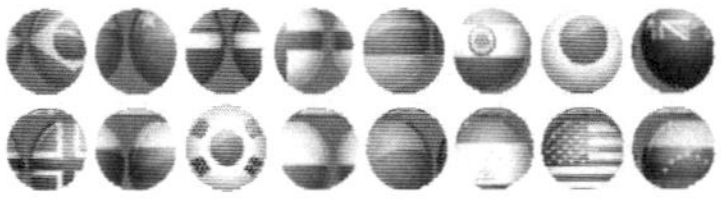

COLORADOSCHOOLOFMINES
engineering the way

2

FOR RESEARCH CENTER HYDRATE

Three Principal Ideas

1. Flow assurance has changed from:
   - (a) apprehension to (b) avoidance to (c) management
   - Flowing solids are acceptable

2. Model predicts hydrate plug formation wrt:
   - (a) GOR, (b) Salinity, (c) Insulation, (d) Restart

3. Imminent Future:
   - Extensions to all phases flow conditions and
   - Interactions (e.g. hydrates and waxes)

Major (58 BBOE Total) Deepwater (>500m) Giants are Oil Reserves

Region	Oil (BBOE)	Gas (BBOE)	% Gas
Gulf of Mexico	14	3	17.5
African West Coast	15	0.5	3.2
Brazil	13	0.9	6.5
North Sea	0.6	3.5	85
Asia/Pacific	0.8	11	93

Flow Assurance (FA) Cooperation

(FA: largest technical issue for 110 oil companies)*

*Welling & Associates Survey, 1999

1. Large Investment = Share Cost,Technology
   - US$1B/Platform = Economic Partners
   - Drilling 39% of investment; wells/flowlines 38%
   - Engineering Companies = Full Technical Partners

2. One Problem Affects the Entire Industry
   - BOEMRE concern for company accidents
   - Experience gives comfort & best practices
     - End of the deepwater beginning ?

COLORADOSCHOOLOFMINES engineering the way

CENTER FOR HYDRATE RESEARCH

Three Flow Assurance Challenges:

Hydrates, Waxes, Asphaltenes

Ayres, R.R., Thermally Managed Flowline Engineering, DeepStar CTR 5307, (2003)

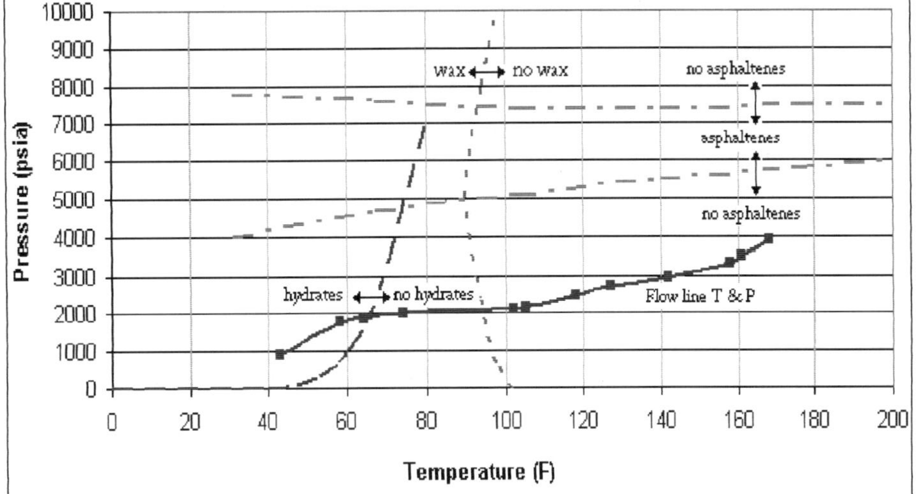

COLORADOSCHOOLOFMINES engineering the way

6

CENTER FOR HYDRATE RESEARCH

Solid deposition examples (from Fogler)

Wax deposition

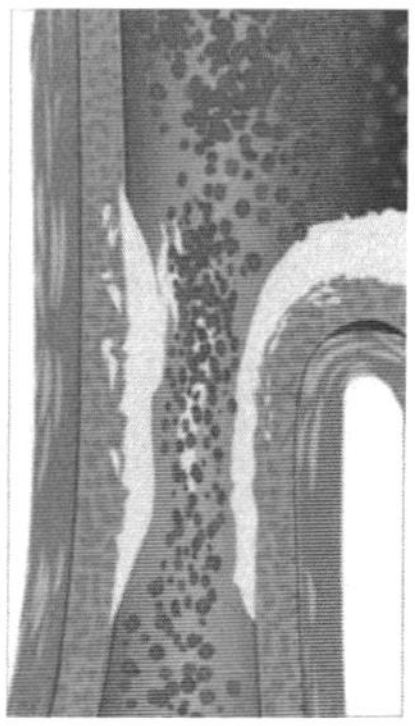

Cholesterol deposition

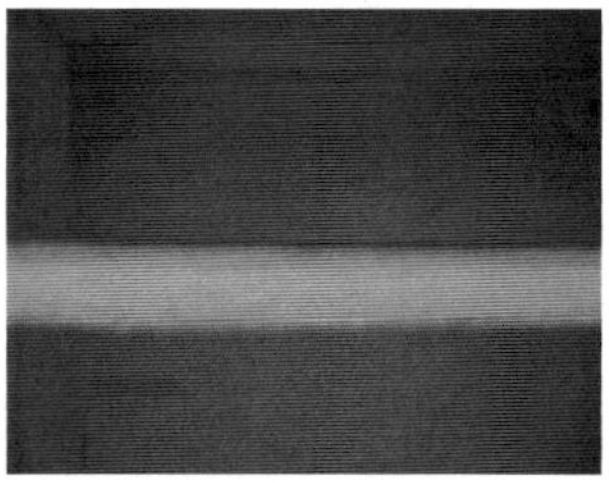

Hydrate formation

COLORADOSCHOOLOFMINES engineering the way

CENTER FOR HYDRATE RESEARCH

Some Flow Assurance Heuristics

1. 10s Rule of Thumb:
   – Hydrates > waxes > asphaltenes > napthenates
     - In both (a) magnitude and (b) formation speed
     - Scale between waxes & asphaltenes
2. Avoidance (thermo) ➔Risk management(dynamics)
   – Progress requires (a) definition (b) short exptl. time
3. Experience provides comfort with risk
   – Hydrate plugs rare, manageable, & recoverable
   – Wax plugs due to 2 errors: design or operating
   – 94% of downtime < 10 hours

COLORADOSCHOOLOFMINES engineering the way

CENTER FOR HYDRATE RESEARCH

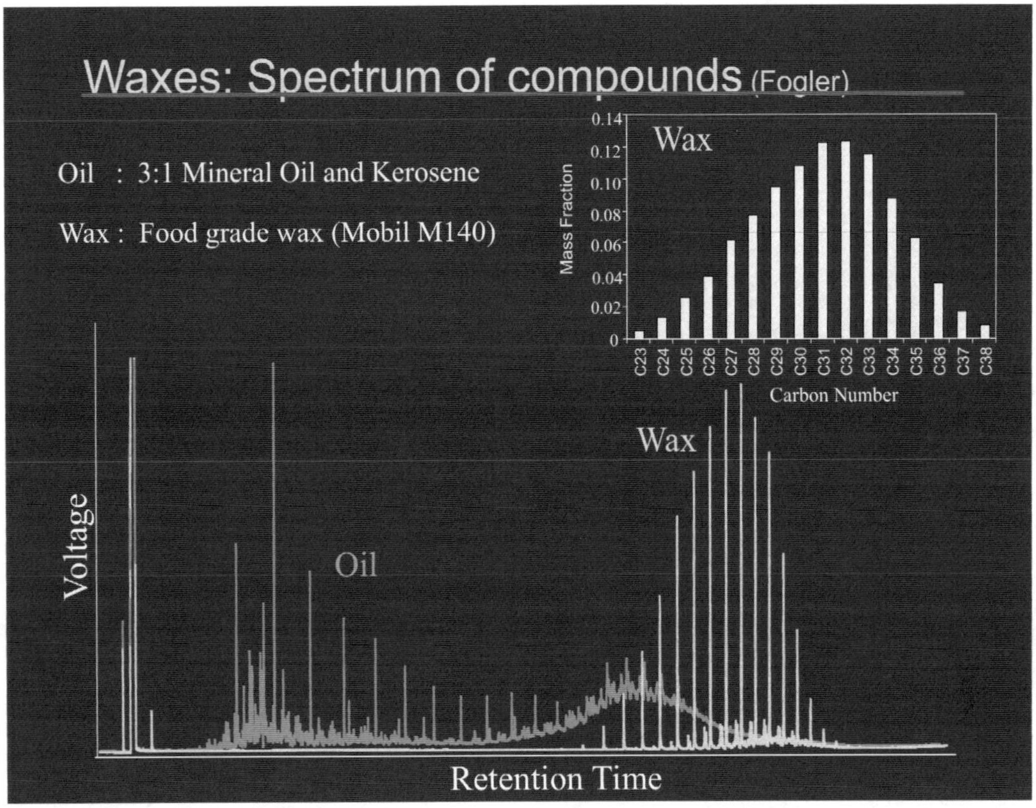

Hydrate Plug Formation

1. **What Are Hydrates?**
    a. Crystal Structures Dictate Formation Conditions
    b. The Interface is More Important than the Bulk
2. **How Do Hydrate Plugs Form?**
    a. Inside and Outside the Pipeline
    b. How can we use this to enable flow assurance

COLORADO SCHOOL OF MINES engineering the way

10

CENTER FOR HYDRATE RESEARCH

Three Principal Hydrate Structures

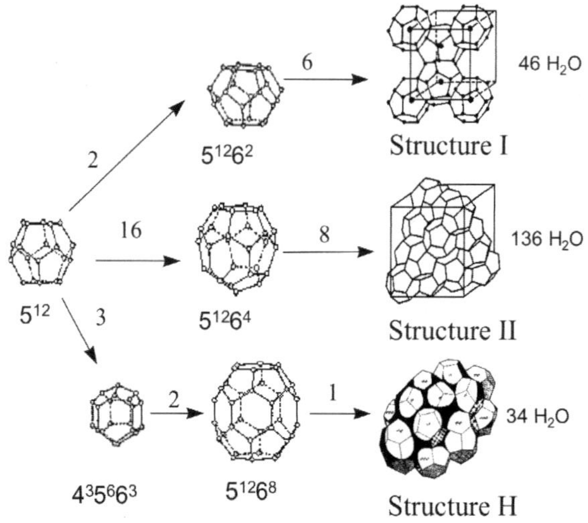

- ≤ 1 guest / cavity
- Ideal solid solution
- All three structures
  - 85 mole% H_2O
  - 15 mole% guests

Molecular Growth Visualization

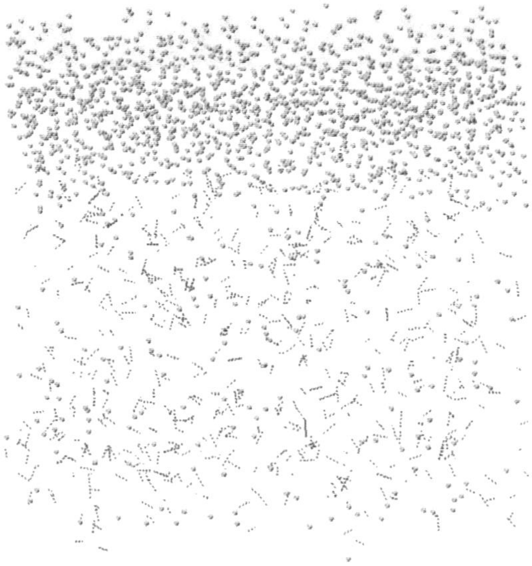

Methane Hydrate Growth Occurs Predominately at the L_w-V Interface

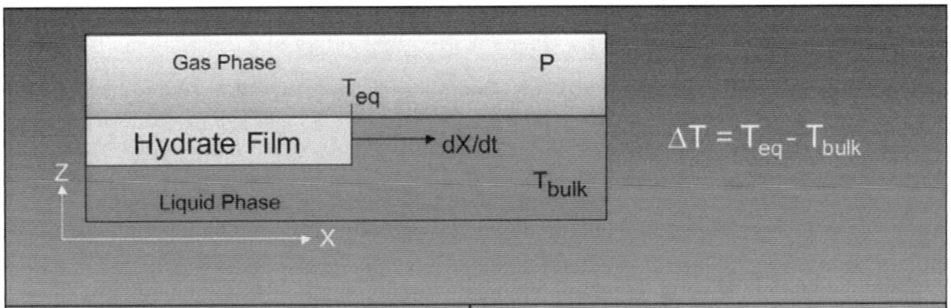

Methane Solubility at T = 25°C			Water Solubility at T = 25°C		
X_{CH4} in H_2O[1]	<<	X_{CH4} in Hydrate	X_{H2O} in CH_4[2]	<<	X_{H2O} in Hydrate
0.000839		0.15	0.001		0.85

[1] Culberson and McKetta , 1951

[2] Kobayashi and Katz, 1953

COLORADOSCHOOLOFMINES engineering the way

13

FOR RESEARCH CENTER HYDRATE

Determination of Film Growth Rates

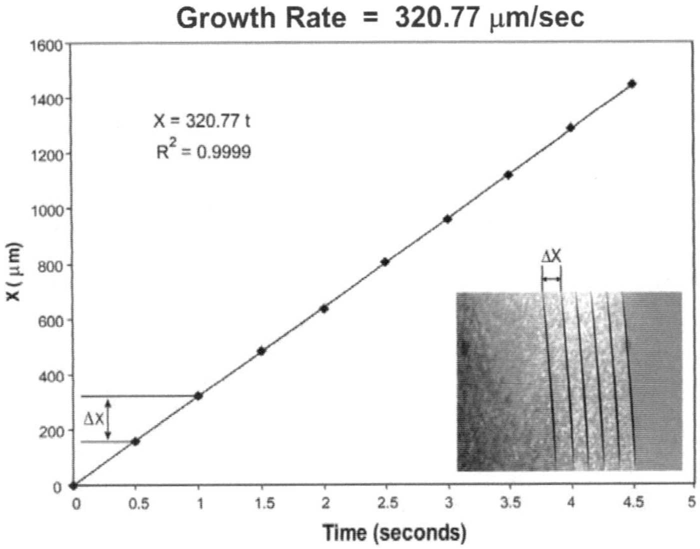

COLORADOSCHOOLOFMINES engineering the way

14

FOR RESEARCH CENTER HYDRATE

Concept: Emulsified Droplet Conversion

- Initial shell formation limited by heat transfer **(Freer 2001)**
- If droplets are small they convert completely **(Lachance 2008)**
- Large droplets convert slowly limited by mass transfer **(Turner 2005)**

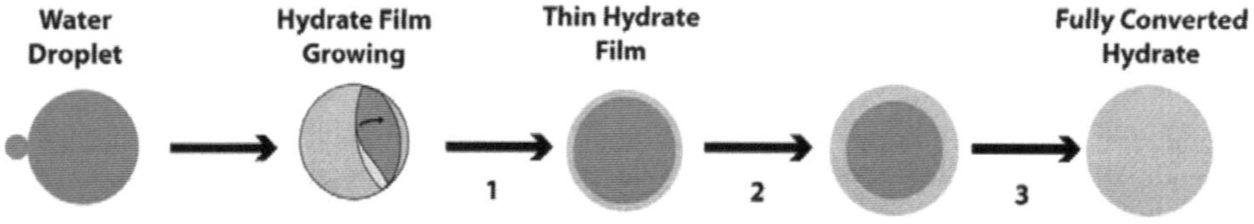

COLORADOSCHOOLOFMINES engineering the way

CENTER FOR HYDRATE RESEARCH

Model of Hydrate Formation in Oil-Dominated System

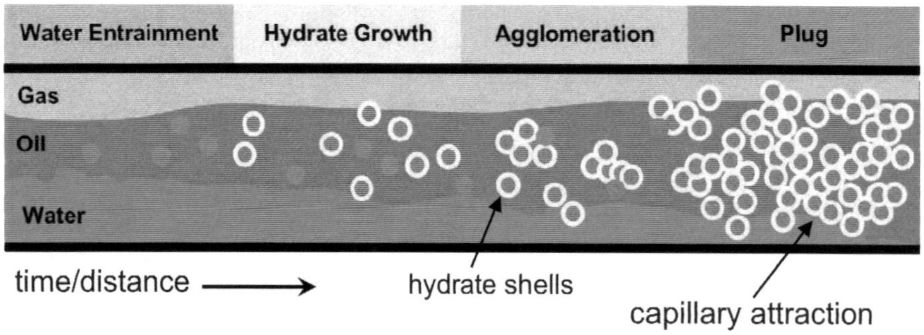

Agglomeration Prevention
=
Plug Prevention

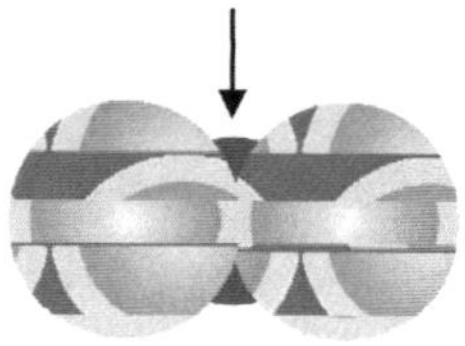

COLORADOSCHOOLOFMINES engineering the way

CENTER FOR HYDRATE RESEARCH

Hydrate Plugging in Pipelines

Hydrates Cause Major Economic & Safety Risks During Energy Production & Transportation

- **Hydrate formation in oil/gas flow lines**
- **#1 problem in flow assurance**
- **Costly to prevent**
- **Costly to remove**
- **Safety concern**

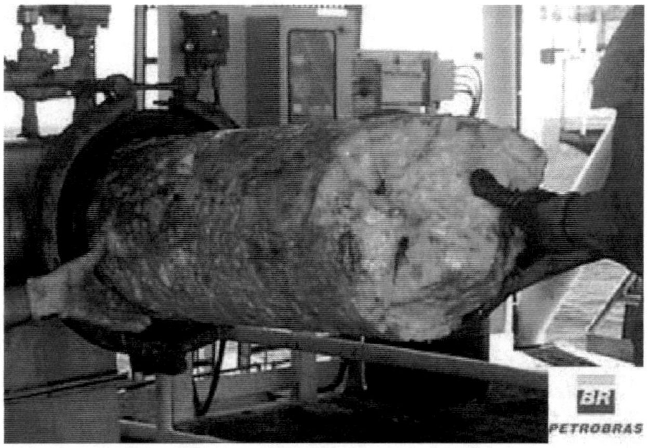

Hydrate plug

COLORADOSCHOOLOFMINES engineering the way | 17 | CENTER FOR HYDRATE RESEARCH

Flow Assurance

Risk Management Requires More

than Thermodynamics:

Dynamics or Time-Dependence

COLORADOSCHOOLOFMINES engineering the way | 18 | CENTER FOR HYDRATE RESEARCH

Paradigm Shift in Flow Assurance

Avoidance

- Avoid Hydrate *T* & *P*
  - Thermodynamics
  - Time Independent
  - Accuracy:1.5°F; 10%*P*
- Avoid Hydrates *T* & *P*
  - Insulation
  - Inhibitor Injection
  - Salt

Risk Management

- Prevent Blockages
  - Dynamics
  - Time Dependent
  - No Agglomeration
- Examples
  - Kinetic Inhibitors
  - Anti-Agglomerants
  - Slurry Flow
  - Direct Elect. Heating

COLORADOSCHOOLOFMINES engineering the way 19 CENTER FOR HYDRATE RESEARCH

COLORADOSCHOOLOFMINES engineering the way 20 CENTER FOR HYDRATE RESEARCH

Managing Risks in Oil-Dominated Systems

not solely arterio-stenosis as in wax lines - but prevention of particle aggregation

COLORADOSCHOOLOFMINES
engineering the way

21

CENTER FOR HYDRATE RESEARCH

Hydrate Formation in Multiphase Pipeline Flow

Gas, Oil, Free Water (W/O and O/W Emulsions)

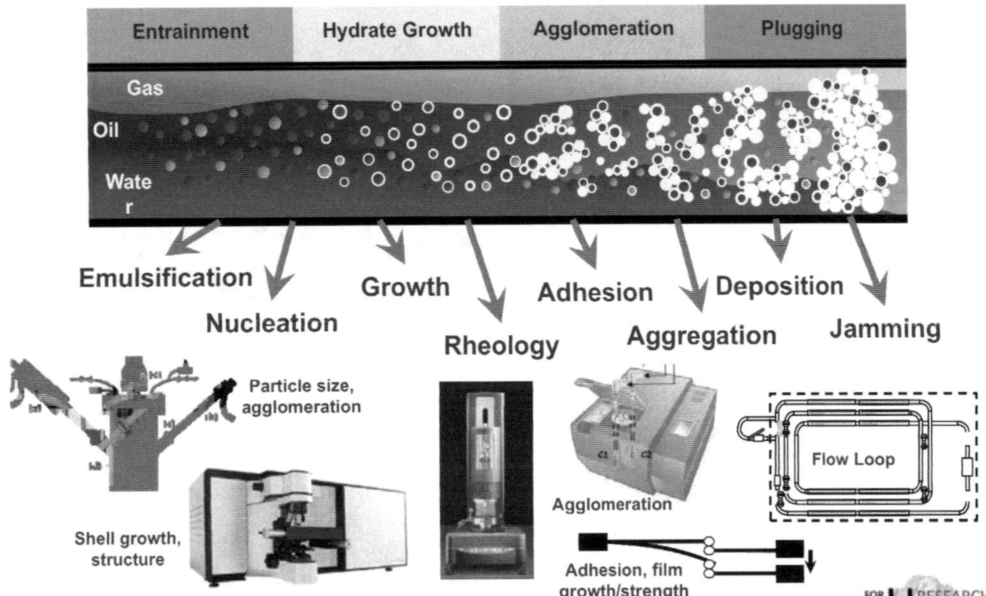

COLORADOSCHOOLOFMINES
engineering the way

22

CENTER FOR HYDRATE RESEARCH

Particles Aggregate:
Hydrate Forces Have been Measured

- Ice inter-particle ~10 mN/m
- C_5H_{12} hydrates ~ 1 mN/m
- Forces cause aggregates
  - Chemistry weakens
  - Shear breaks

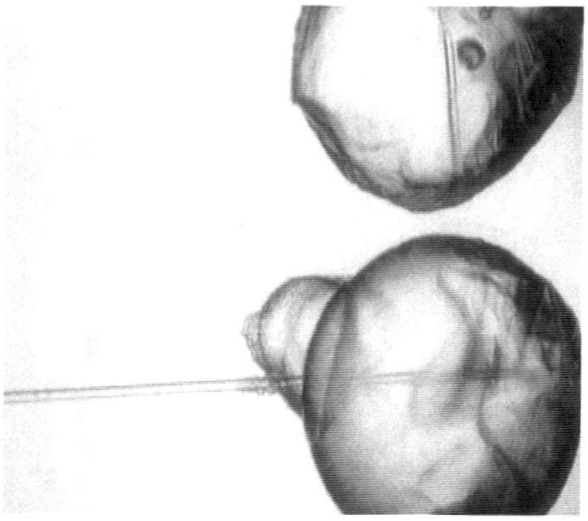

Jamming and Plugging:
Particles form an arch/dome when jamming occurs

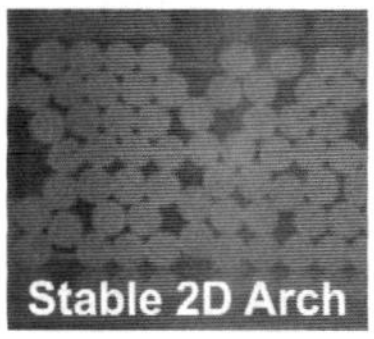

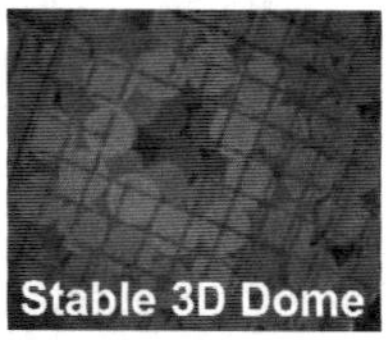

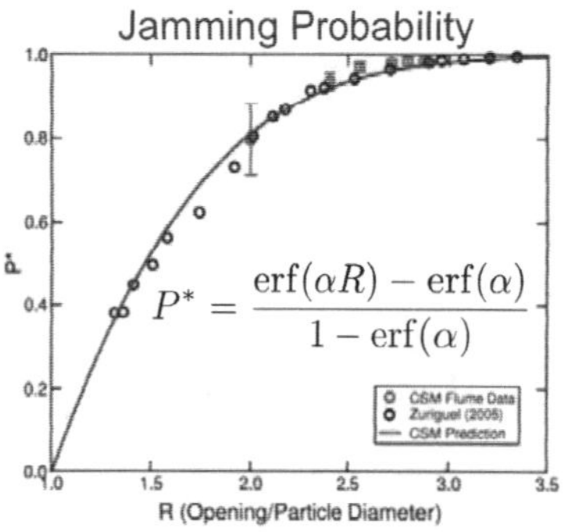

2,500 polystyrene spheres (5/8" dia.)

Hydrate Formation Model: CSMHyK/OLGA

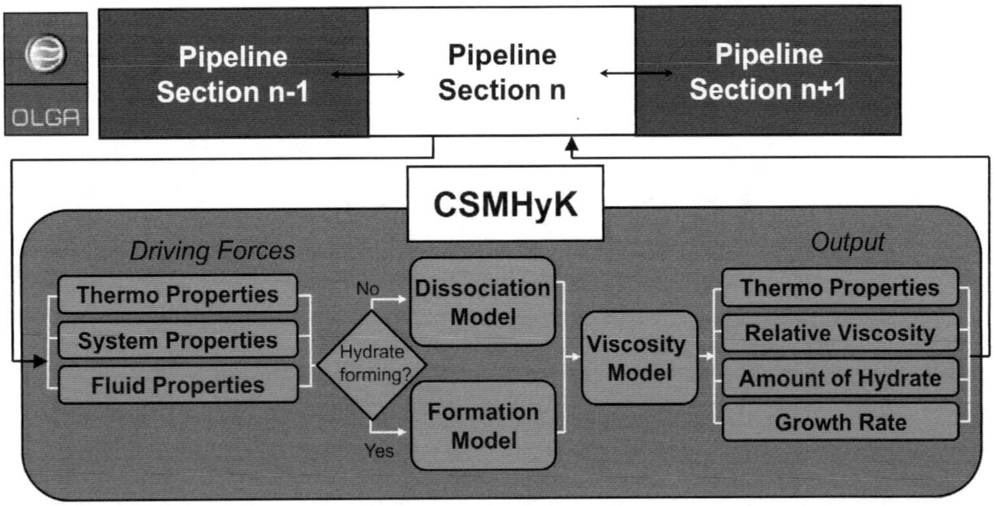

COLORADOSCHOOLOFMINES
engineering the way

25

CENTER FOR HYDRATE RESEARCH

Results from Flow Loop Experiments

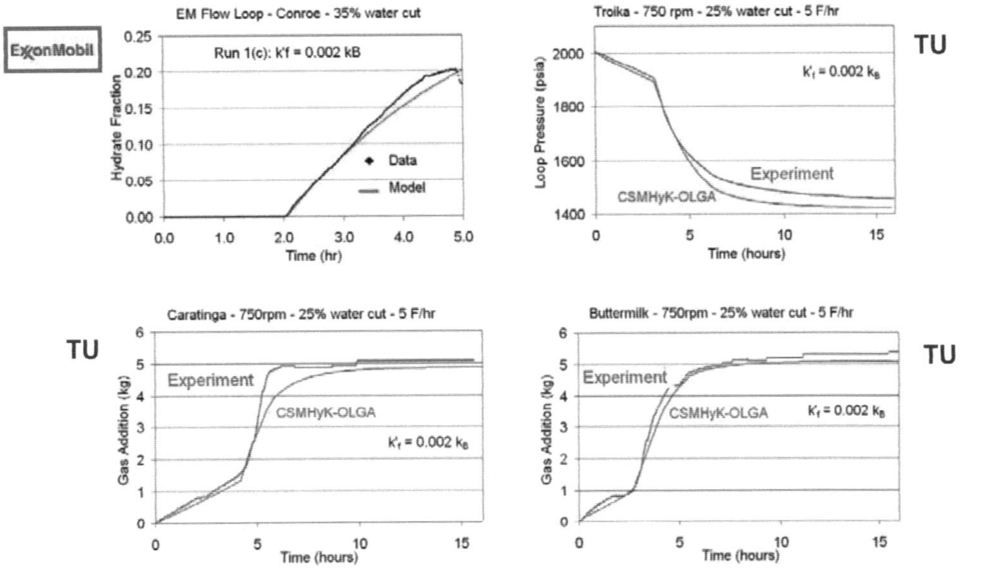

COLORADOSCHOOLOFMINES
engineering the way

26

CENTER FOR HYDRATE RESEARCH

Creek et al. (0TC22017) use model to show

- Hydrate formation limited by
  - (a) Insulation (b) high salinity (c) low GOR (gas: oil ratio)
- Example: Insulation (a) preserves reservoir heat (b) limits hydrate amount

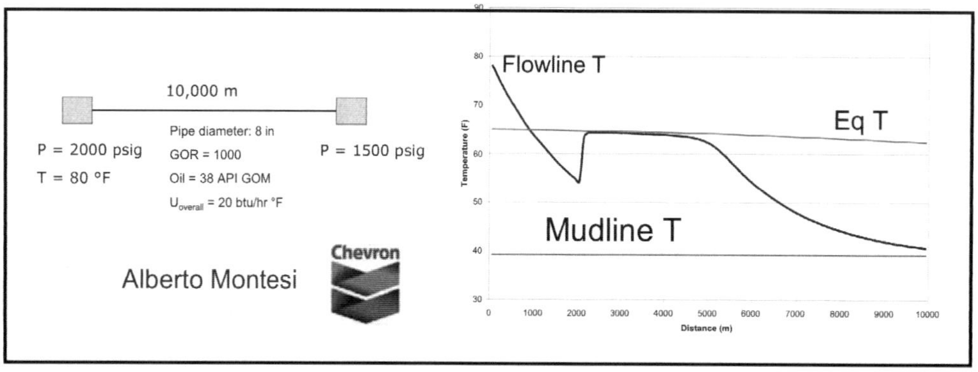

Pioneering Use of CSMHyK-OLGA®

Chevron	Leon field condensate line (D. Estanga, 2009)
ExxonMobil	(XoM Vrsn) Cold-flow & flowloop experiments
Statoil	Tommeliten field subsea tieback and cold-flow
BR PETROBRAS	Case study: Caratinga field offshore well
TOTAL	Case study: design of subsea flowline network

Creek et al. (2011 0TC22017) suggest for W/O Systems without free water and gas:

1. Flowline may tolerate 22 vol% hydrates
   - If particles interact, limit can be as low as 10 vol%
   - Hi salinity (e.g. 130,000 mg/L) can tolerate 50% WC
2. Can sometimes “Do Nothing”
3. Can reduce some typical inhibitor costs
   - e.g. $750,000/d for MeOH
4. Model is useful for risk evaluation

This work will be presented in the upcoming OTC Brazil, 4-6 October, 2011

OTC 22406

Hydrate Risk Assessment and Restart Procedure Optimization of an Offshore Well Using a Transient Hydrate Prediction Model

L.E. Zerpa, E.D. Sloan, C.A. Koh, and A.K. Sum, Colorado School of Mines

Well restart procedure sensitivity

- Considering intermediate risk case, transient shut-in/restart
- THI concentration varied from 0 to 40 wt% Ethanol
- Two valves control shut-in & restart operations

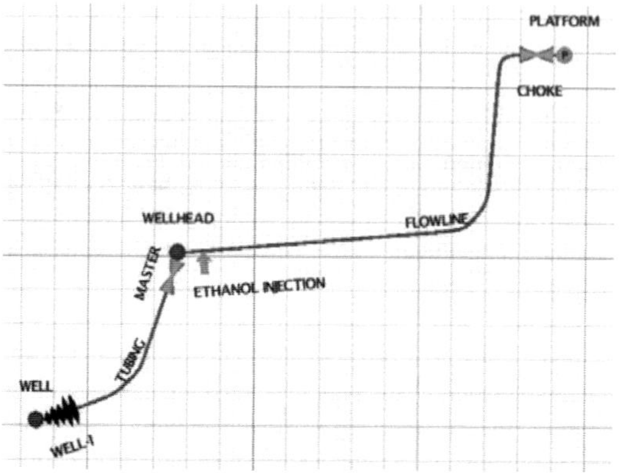

Production restart generates an accumulation of hydrates that travels downstream

1. Moving hydrate accumulation shown by hydrate volume peaks
2. Bank of fluids without hydrates follow hydrate accumulation
3. With time system equilibrates at stable hydrate fraction
4. Hydrate accumulations are critical condition in flowline

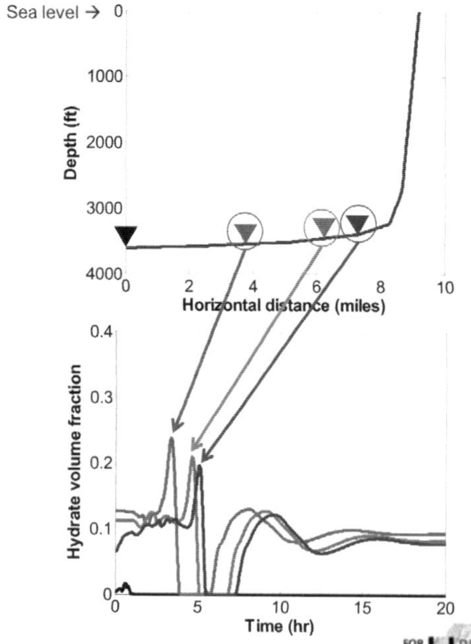

Future Directions

1. Better multi-phase flow understanding
   - Free water
   - Free gas
   - Combine all in comprehensive model

2. Better understanding of FA interactions
   - Example: Hydrates and wax co-formation
   - Example: Corrosion inhibitors affecting hydrate inhibitors

Three Flow Assurance Challenges: Hydrates, Waxes, Asphaltenes

Ayres, R.R., Thermally Managed Flowline Engineering, DeepStar CTR 5307, (2003)

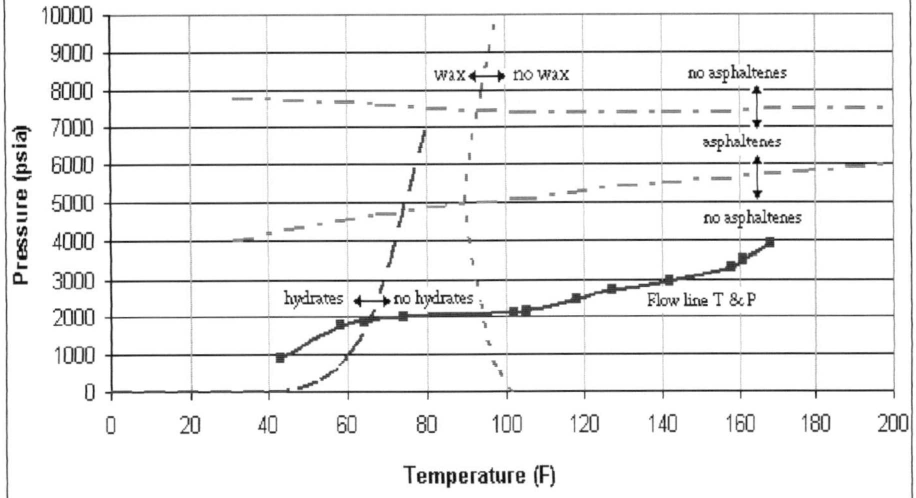

Fundamental Principle:
Hydrate Forces: ↑ H2O-wet; ↓ HC-wet

Høiland, Aspenes, Barth (Bergen) & Aman (CSM)

- 10 yrs of water-wet interactions
  - Top: particle-particle (Aggregation)
  - Lower: particle-steel (Deposition)
- H_2O-wet = Hi force (plug?)
  - These two videos
- HC-wet = Low forces (no plug?)
  - See next slide

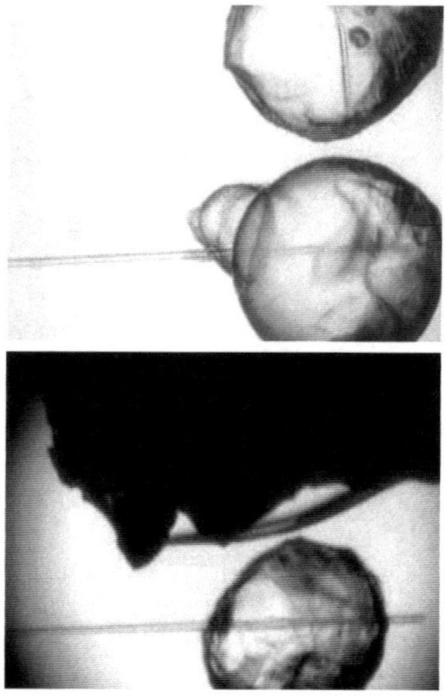

COLORADOSCHOOLOFMINES engineering the way   35   CENTER FOR HYDRATE RESEARCH

Wax Deposition = No Hydrate Deposition?

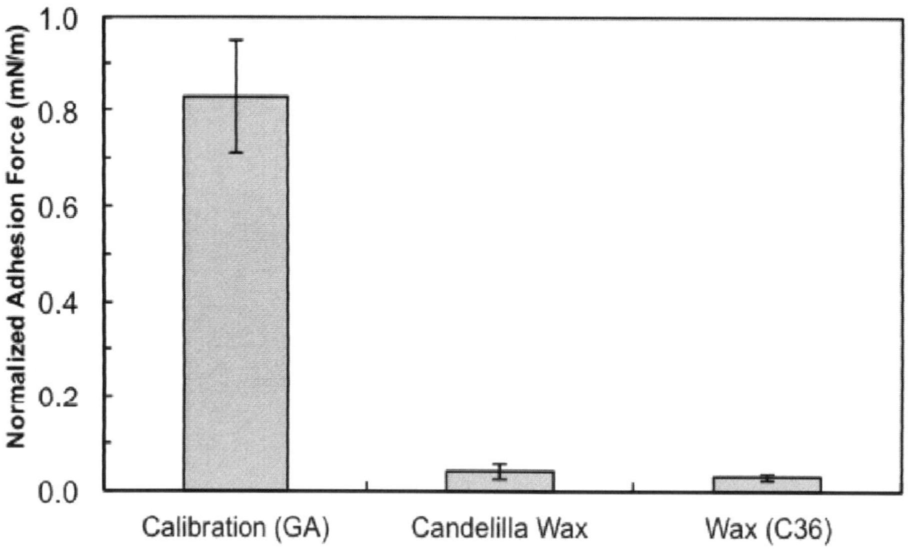

COLORADOSCHOOLOFMINES engineering the way   36   CENTER FOR HYDRATE RESEARCH

Future: Study Problem Interactions

- Why do Corrosion Inhibitors affect Hydrate inhibitors?
  - Related to the hydrate/wall wetting principle
- Does wax deposition affect hydrate deposition?
  - WAT > HAT (Hydrate Appearance Temperature)
    - Study history of hydrate problems in waxy crudes
- Consider the following Case Study of Staffa Field
  - Is it possible that hydrate formation caused wax plug?

Case: Hydrates/Waxes in Staffa N. Sea Field (1993)

1. Un-insulated, single 8 inch flowline, 6.3 miles long
   - no capability for round-trip pigging
2. 6000 BOPD; GOR = 1600 scf/bbl; 1% WC
   - WAT = 79°F; wax content = 5%; arrival at 44°F
3. Wax filled line after MeOH Injection lost
   - 1.2 miles of flowline replaced, filled with wax
   - Line again plugged with wax
4. 1995: multiple problems (hydrates & wax)
   - Field abandoned

- Did Hydrate nucleation cause Wax plugging?

Conclusions

1. Flow assurance has changed from:
   - (a) apprehension to (b) avoidance to (c) management

2. W/O model: hydrate plugs prevented by
   - (a) Low GOR
   - (b) Hi Salinity
   - (c) Hi Insulation
   - (d) Specifying Restart Conditions

3. Future: FA Interactions e.g. Hydrates & Waxes

COLORADOSCHOOLOFMINES engineering the way

CENTER FOR HYDRATE RESEARCH

ExxonMobil
Upstream Research

Causes of Hydrate Blockages

Larry D. Talley

ExxonMobil Upstream Research Company

September 27, 2011
AIChE/SPE Flow Assurance Workshop

Multiple Mechanisms Known

- Jamming
- Film Growth
- Slurry Viscosity
- Bedding
- Agglomeration

ExxonMobil 4” Flow Loop

Sticky hydrates in gas/condensate

12vol% water, 60% liquid loading

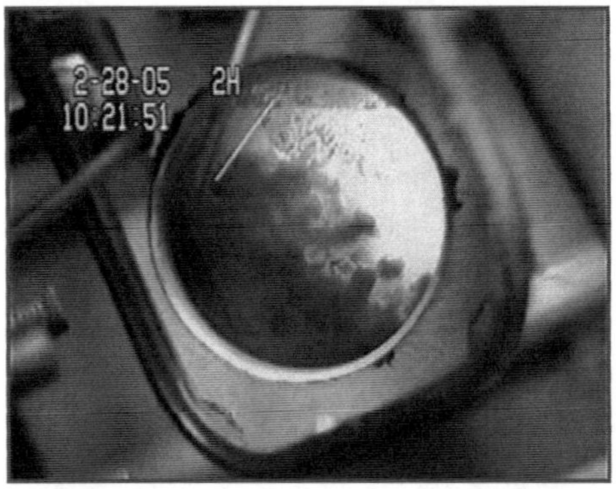

Jamming

- Statistical Approach modeled by CSM
  - **Geometric distribution of single-size, spherical non-interacting particles**
  - **Models advancing to include**
    - **hydrate particle size distribution, shape and hydrate particle interactions**
  - Measure jamming in field by dP in multiple locations

- **Prevention: Generate small hydrate particles and prevent agglomeration; eliminate sloughing; use anti-agglomerants. Otherwise, prevent hydrate formation with KHI's or THI's**

Critical number of particles (n_c) through normalized restriction size (R).

(From Patrick G. Lafond et al, 7th ICGH, 2011)

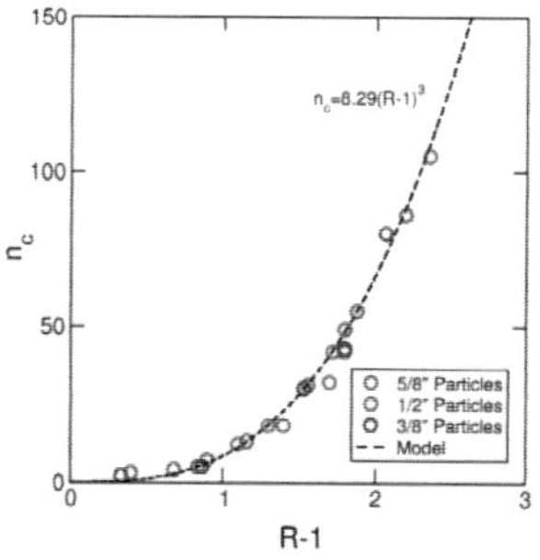

Bedding

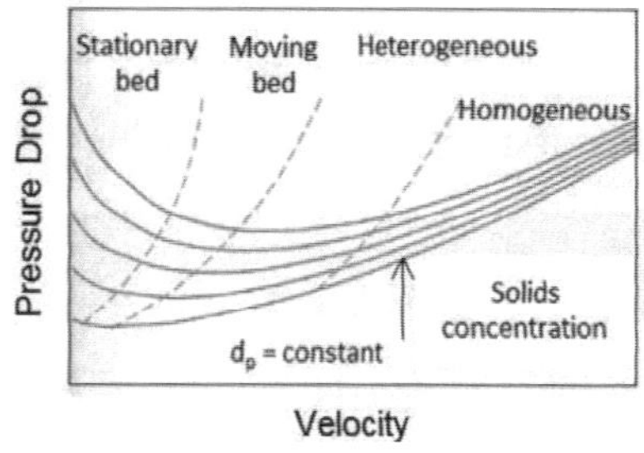

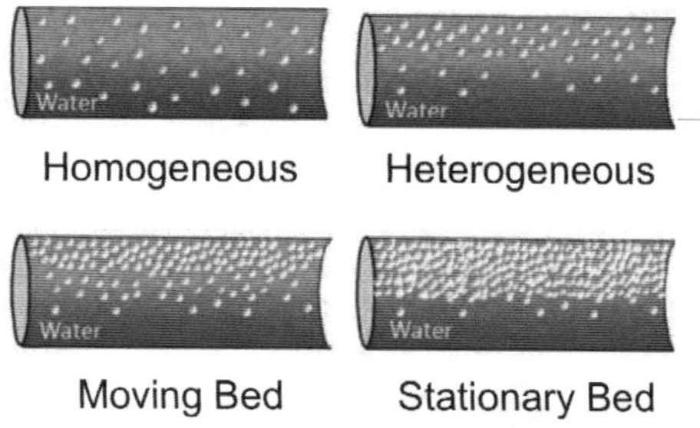

- Floating bed modeled by CSM
  - **Models advancing to include**
    - **hydrate particle size/shape distribution and interaction forces**
    - **Sanjeev Joshi, et al., 7th ICGH, 2011**
  - Measure by time evolution of local dP

- **Prevention: Generate small hydrate particles using AA's or mechanically; optimize flow rate; Otherwise, prevent hydrate formation with KHI's or THI's**

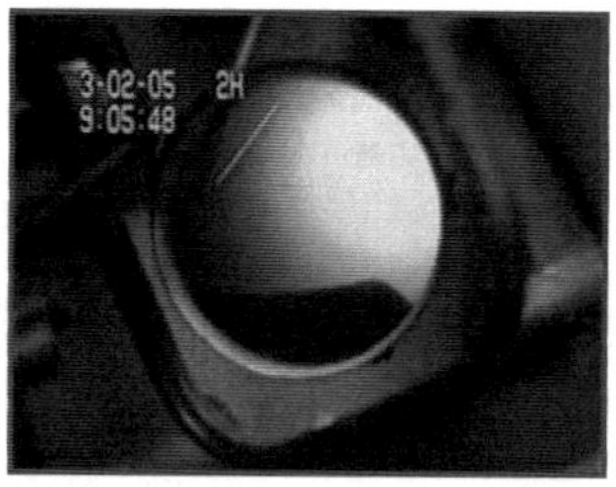

Dry hydrates in a gas condensate 16vol% water, 80% liquid loading in ExxonMobil 4" Flow Loop

Film Growth

Mechanisms include

- Water wetting followed by hydrate nucleation on wall
- Frost crystallization followed by
  - water wetting and growth
  - adsorption of wet hydrate from bulk

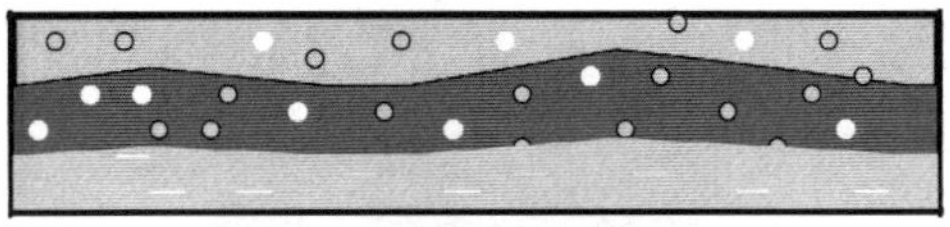

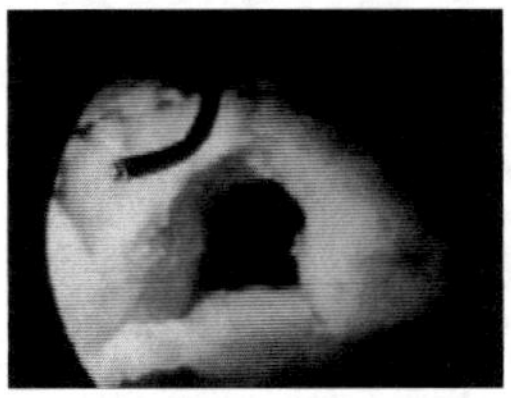

Stenosis buildup in the SWRI flowloop (Hatton and Kruka, 2002)

Film Growth, cont.

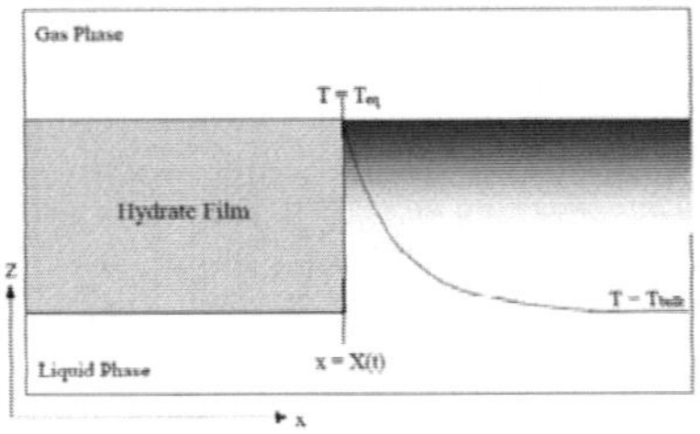

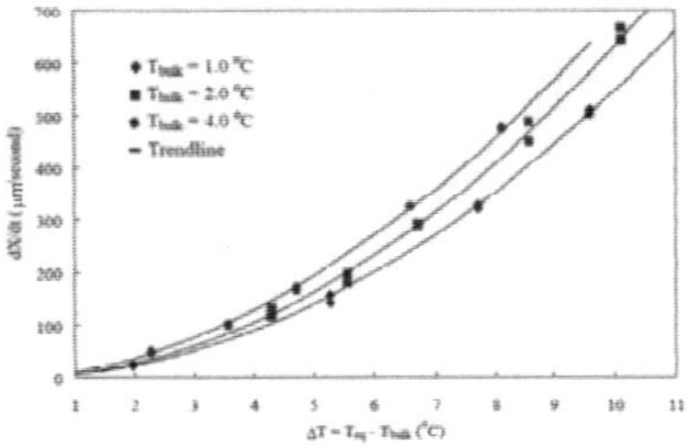

- CSM modeled interfacial hydrate growth in x-, y- and z-directions
  - **Methane hydrate by Erik Freer, M.S. Thesis, Colorado School of Mines, 2000**
  - **CO_2 hydrate by Craig Taylor, M.S. Thesis, Colorado School of Mines, 2005**
  - **Models need to include**
    - **heat transfer rates, mass transfer rates, bulk hydrate interactions – adsorption, sloughing, oil and condensate systems, gas systems**
  - **Film growth promotes jamming**
    - **by adsorption of large/small hydrate particles from bulk**
    - **by sloughing**
  - **Hydrate film growth may occur even with AA's**

Film Growth, cont.

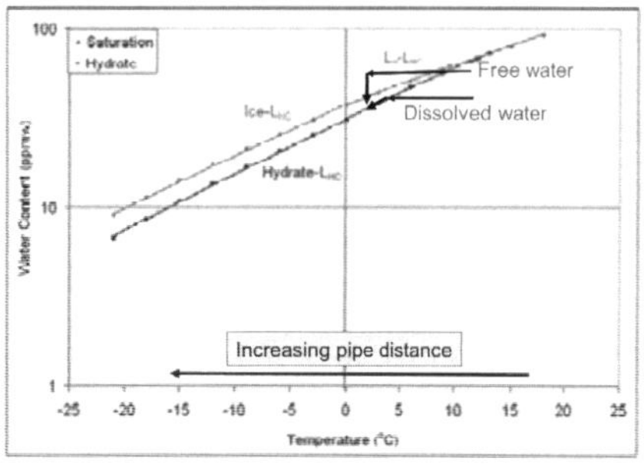

It was hypothesized that in the presence of a water saturated hydrocarbon, hydrates would grow on the pipe wall as the fluid cooled below its equilibrium temperature. This hypothesis was confirmed using a single pass condensate flowloop. Water was continuously dissolved into the flowloop inlet stream as water deposited in the flowloop test section, resulting in both a pressure drop and fluid temperature increase. This work illustrates the need for a hydrate wall growth model.

Joe Nicholas, et al., 6th ICGH, 2008

- CSM/Westport Intertek/ExxonMobil measured film growth in once-through apparatus (Joe Nicholas, 6th ICGH 2008)
  - **Models advancing to include**
    - **Effects of subcooling, liquid composition, gas composition, liquid loading, flow regime, heat transfer rates, mass transfer rates, bulk hydrate interactions – adsorption, sloughing, chemicals**
- Measure in field by X-ray imaging, dP in multi locations, videoscope
- **Prevention: Prevent hydrate formation with KHI's or THI's, dehydrate gas/liquid**

Film Growth, cont.

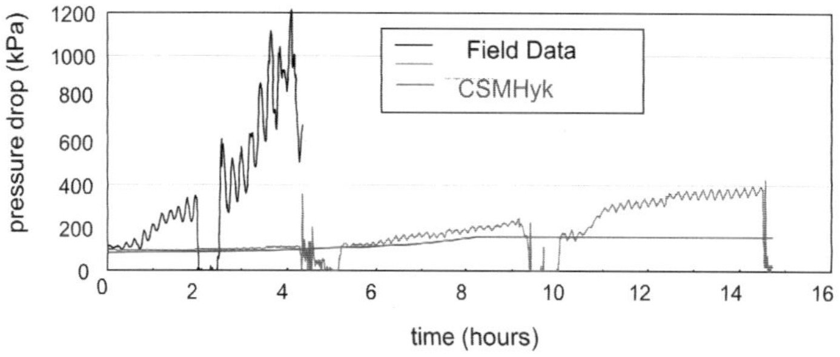

ExxonMobil Once-through Field Test

- ExxonMobil measured film growth in once-through flowline
  - **First run with agglomeration dominating**
  - **Second run without agglomeration; film growth observed**
  - **Models needs to include film growth**

Agglomeration

CSM modeling hydrate agglomeration

- Hydrate agglomeration affected by all fluid components, wetting properties, capillary forces, shear forces, flow regime, phase volumes
- Affects jamming, bedding, slurry viscosity, film growth/sloughing
- Measure agglomeration in field by PVM/FBRM

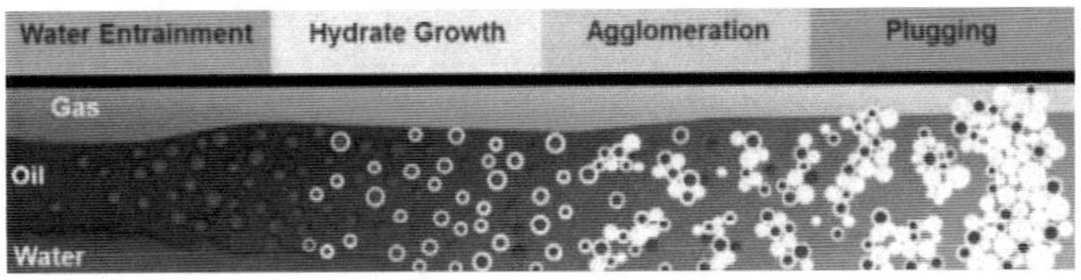

Low water cut agglomeration

(From Sanjeev Joshi et al., 7th ICGH, 2011)

- **Prevention: Generate small hydrate particles and prevent agglomeration; eliminate sloughing potential; use Cold Flow methods or anti-agglomerants; prevent hydrate formation with KHI's or THI's**

Slurry Viscosification

- Prediction of steady state hydrate slurry viscosity available
  - Model by IFP (Martin Gainville, et al., 7th ICGH, 2011) depends on
    - Applied wall shear stress which reduces size of transient agglomerated hydrate particles
    - Effective hydrodynamic volume of hydrates at steady state
  - Steady state models only applicable far downstream of hydrate formation location

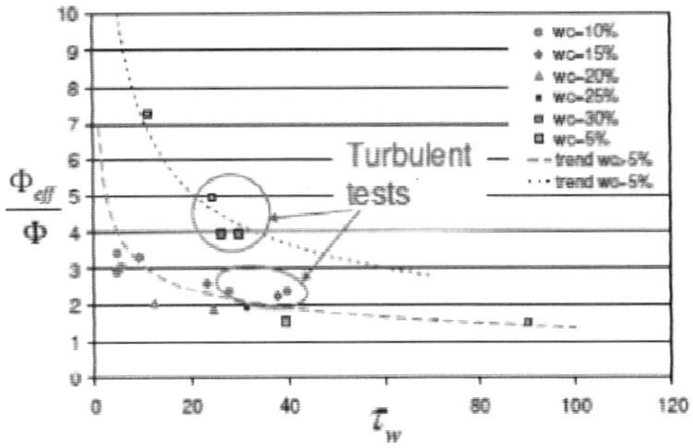

Agglomeration viscosity (from effective hydrate volume) versus Wall Shear Stress at Steady State

Slurry Viscosification, cont.

- Transient hydrate slurry viscosity at hydrate formation location depends on many factors
- Factors that favor formation of low-viscosity hydrate slurries include:
  - High Reynolds Number
  - High Capillary Number
  - High mass transfer and heat transfer rates
  - High liquid loading
  - Low watercut
  - High superficial fluid velocities
  - Dispersed bubble flow/small gas bubbles/ small water droplets
- Measure viscosity by dP vs flow rate
- Prevention of viscous blockage achievable through controlling hydrate formation step, using anti-agglomerants or preventing hydrate formation with THI's or KHI's

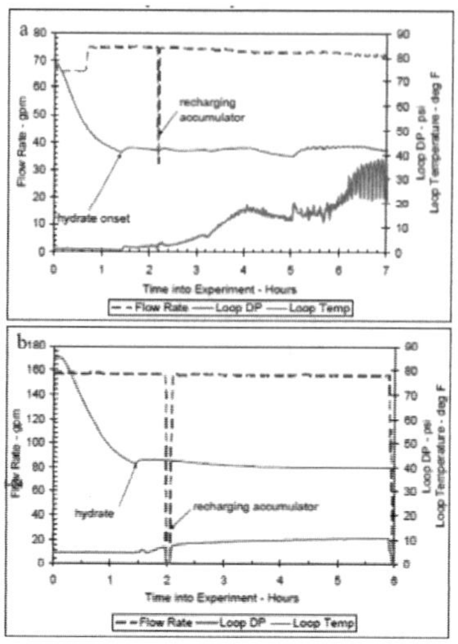

Figures show that higher flow rate resulted in lower slurry viscosity.

Doug Turner, et al. (6th ICGH, 2008)

Causes of Hydrate Blockages

- **Jamming**
- **Film Growth**
- **Slurry Viscosity**
- **Bedding**
- **Agglomeration**
- **Other**

Morphology and Rheology of Cyclopentane-Hydrate Emulsions

Morris, J. F., Benjamin Levich Institute, City College of CUNY
Karanjkar, P. U., Levich Institute, City College of New York
Lee, J. W., The City College of New York

Clathrate hydrates generate great interest and concern in the petroleum industry because of the potentially severe consequences of pipeline blockage by hydrate plug formation. In this study, the mechanical properties of cyclopentane-hydrate forming emulsions will be examined, seeking primarily to relate the morphology of the hydrate formed for different parameters to the bulk rheological and solid mechanical properties. Cyclopentane forms hydrates under atmospheric conditions, and these structure II hydrate formed is stable in a temperature window above the melting point. These features allow ready experimentation under standard laboratory conditions with clear connection of hydrate crystal morphology to the observed properties.

A combination of single-drop visualizations and rheometric studies using density matched emulsions (to avoid sedimentation/creaming effects) will be described which show that surfactant plays a key role in the development of the hydrate layer formed on water drops. The work focuses on the oil-soluble surfactant sorbitan monooleate (Span 80), and will illustrate that depending on its loading it can have a range of effects; the hydrate layer can vary from a robust polycrystalline layer at small loadings of surfactant to a mushy, hairy structure at large loadings. The impact of this morphology on the rheology of the mixture includes cases of large viscosification and jamming, although the influence of the surfactant is found to be dependent on the water fraction.

Statoil

Do we control hydrates or do they control us?

Upstream Engineering & Flow Assurance Forum
Houston September 27th, 2011

Hydrate control: There is no general solution!

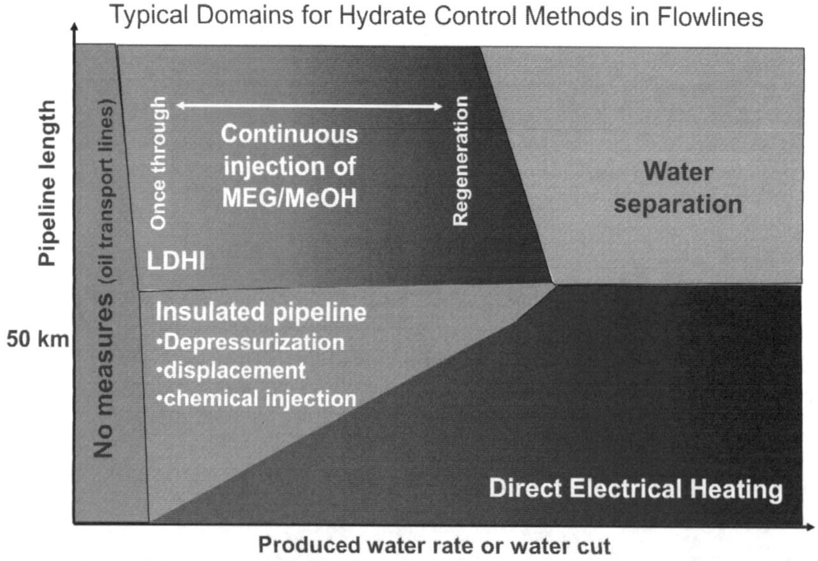

Statoil

Interface with other disciplines

- Flow assurance is a key driver in concept selection
- Hydrate control requires a strong interface with several disciplines
- Important to ensure understanding of technical requirements from Flow Assurance with their economic consequences

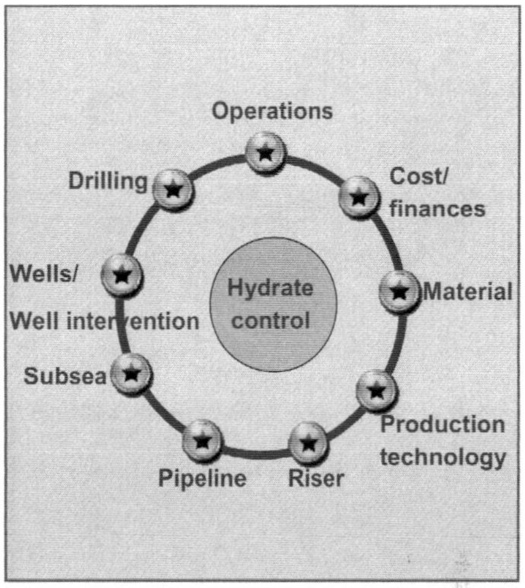

Functional and operational requirements (1)
Well completion

- Downhole chemical injection
- Artificial lift (gas lift, ESP)
- Sand control

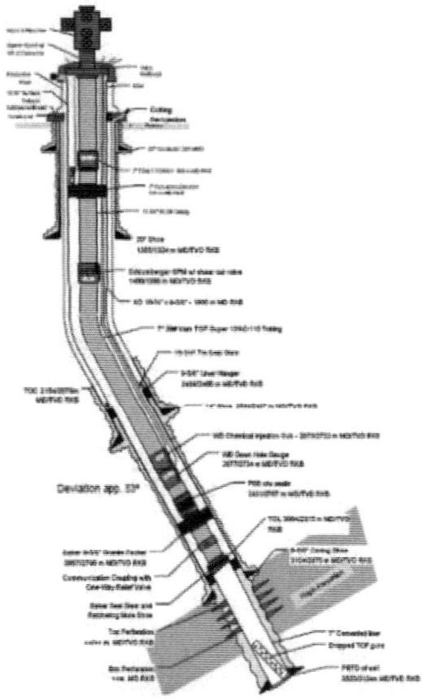

Functional and operational requirements (2)
Pipelines and risers

- Configuration (single, dual, bundle, etc.)
- Dimensions
- Insulation
- Burial
- Active heating (DEH, bundle etc.)
- Scraper pigging for wax control (pigging loop, subsea pig launcher)
- Minimum / maximum design temperature
- Maximum fluid velocities (erosion)
- Gas lift (riser base)

Statoil

Functional and operational requirements (3)
Subsea

- Multiphase metering
- Insulation of subsea structures
- Boosting / artificial lift
- Piping design (dead legs, low points, etc.)
- Chemical injection (points, number of lines, capacity)
- Access for mechanical tools (e.g. coiled tubing for hydrate plug removal)

Statoil

Functional and operational requirements (4)
Receiving facilities

- Slug catcher size
- Liquid handling capacity
- Flare capacity
- Fluid heating requirements
- Active slug control
- Chemical injection
- Chemical storage

7 - Classification: Internal 2011-07-07

Statoil

Statoil's pioneer endeavour
Direct Electrical Heating of Pipelines

- Primarily for hydrate control purposes
- Installed on 7 Statoil operated fields on multiple lines
  - Operational flexibility – "push the button"
  - Rapid start-up after shutdowns
  - No need for depressurisation
  - End zones still need additional hydrate control measures
- Heat input uniform along the line
- Enables "single pipeline" tie-back
  - Tyrihans: 4 templates on a 43 km pipeline
  - Morvin: 2 templates on a 20 km pipeline
  - Fossekall-Dompap: 3 templates on a 26 km pipeline
- Development of new wet insulation materials very promising
- DEH also selected by other operators recently

8 - Classification: Internal 2011-07-07

Statoil

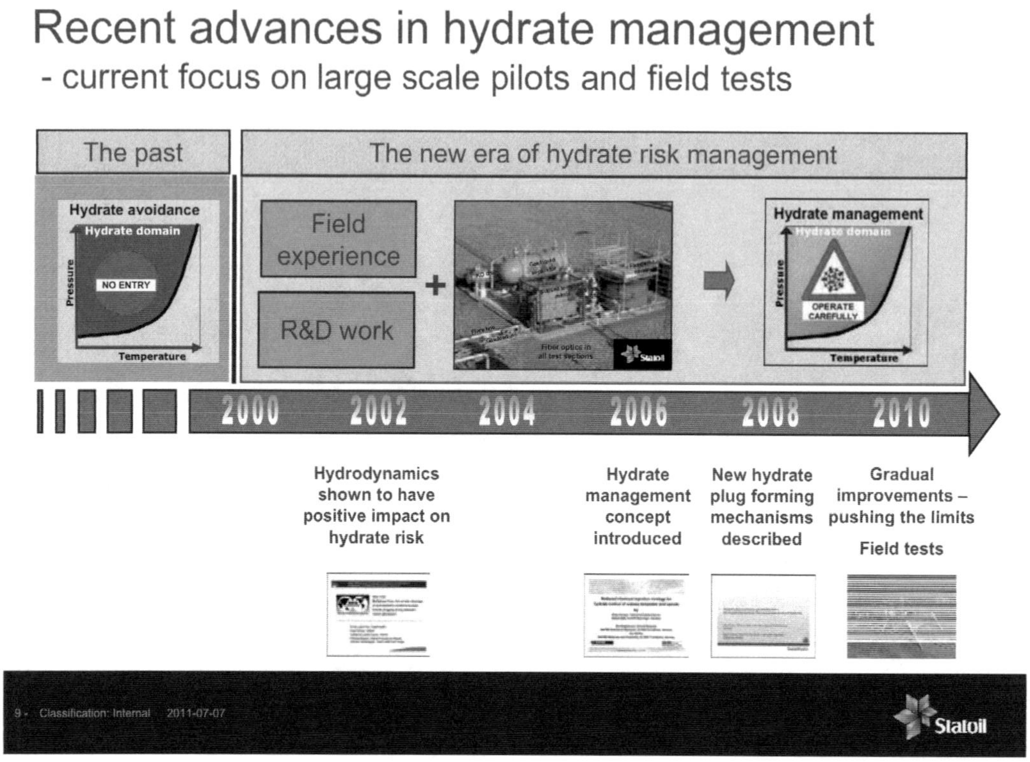
Recent advances in hydrate management
- current focus on large scale pilots and field tests
The past
The new era of hydrate risk management
Hydrate avoidance
Hydrate domain
NO ENTRY
Pressure
Temperature
Field experience
R&D work
+
Hydrate management
Hydrate domain
OPERATE CAREFULLY
Pressure
Temperature
2000
2002
2004
2006
2008
2010
Hydrodynamics shown to have positive impact on hydrate risk
Hydrate management concept introduced
New hydrate plug forming mechanisms described
Gradual improvements – pushing the limits
Field tests
9 - Classification: Internal 2011-07-07
Statoil

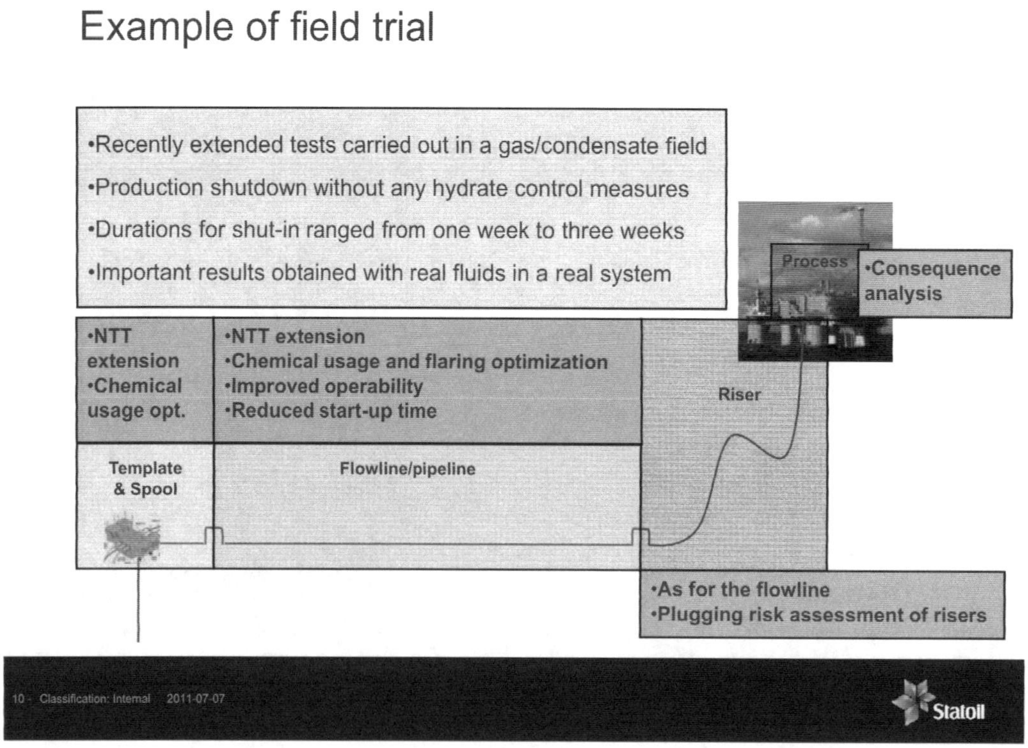
Example of field trial
•Recently extended tests carried out in a gas/condensate field
•Production shutdown without any hydrate control measures
•Durations for shut-in ranged from one week to three weeks
•Important results obtained with real fluids in a real system
Process
•Consequence analysis
•NTT extension
•Chemical usage opt.
•NTT extension
•Chemical usage and flaring optimization
•Improved operability
•Reduced start-up time
Riser
Template & Spool
Flowline/pipeline
•As for the flowline
•Plugging risk assessment of risers
10 - Classification: Internal 2011-07-07
Statoil

Operational support

- Flow Assurance work extends beyond the limits of early phase design
- FA Duty Team is well established and actively used by operational units in Statoil:
  - 24/7 support provided throughout the year
- Large savings obtained through:
  - reduced downtime and problem solving for fields in operation
  - Lessons learned implemented in new fields

Importance of the Fluid

- Implementation of new advances requires an understanding of the risks
  - Not available through software simulations
  - Requires adequate test programs
- Sophisticated hydrate control methods based on risk management require sound and detailed fluid behavior testing
  - Lack of information is costly!
  - Sufficient volumes of fluid should be sampled during appraisal phase
  - Advances in sampling techniques and cost reduction should be conducted in parallel

The future ….

Longer distance

Arctic/harsh environment

Increased complexity

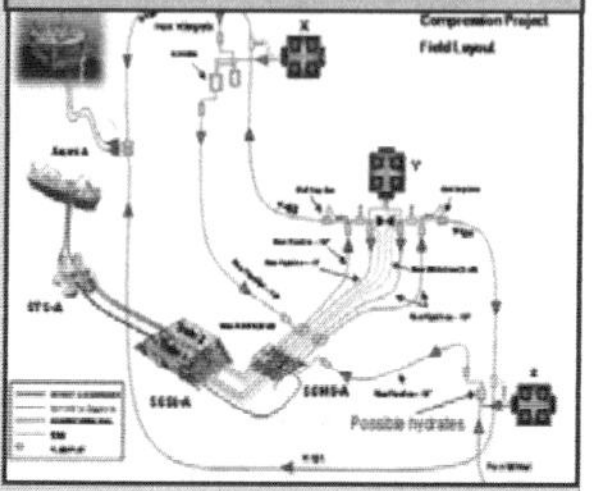

Hydrate control lessons learned in the past pave the way for the future

Deeper water

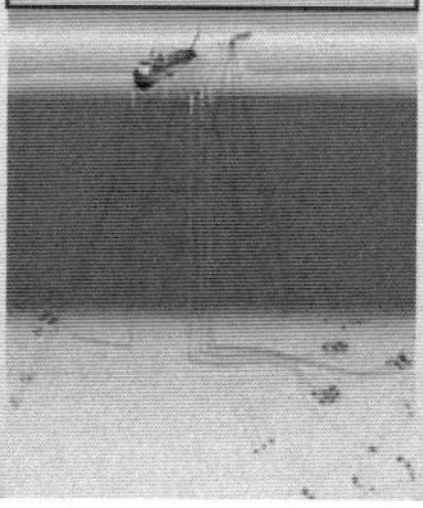

More difficult fluids

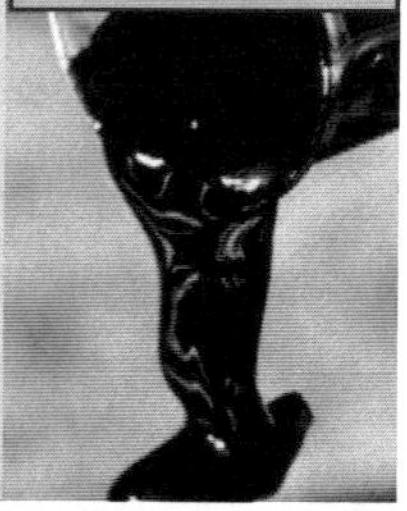

Statoil

Key messages

- Hydrate control has many ***functional requirements*** to secure successful design and operations.
  - Large impact on the design complexity and cost.
  - Safety hazards and remediation costs call for robust but adequate prevention strategy
- Hydrate Control plays an important role in ***interface management*** along the entire value chain (from reservoir to point of delivery).

Hydrate management with a risk based approach is the new trend setter in the industry for safe and cost effective hydrate control. strategies.

Statoil

Use of Equations of State in Fluid Property Simulation

Creek, J., Chevron Petroleum Technology Corporation

One of the key uses of equations of state in E&P is for the evaluation of key fluid properties related to molar volume and viscosity. For reservoir these two variables are important for conversion of simulation tools that operate in moles and the petroleum industry that operates on volume. Viscosity is important for resistance to fluid flow and pressure performance in reservoirs. Even correlations for properties like IFT depend on density prediction. This talk will discuss how properties are calculated and the correlations that are typically used. The talk will also discuss the shortcomings of current practices and make recommendations for the future.

Challenges in Fluid Sampling for Flow Assurance Testing

Hani Elshahawi

Fluid Evaluation and Sampling Technologies (FEAST)

Shell International E&P Inc.

AIChE-SPE Joint Workshop

Challenges in Flow Assurance

September 26-28, 2011, Houston, TX, USA

Impact of Fluid Properties on E&P Business

- Project Design
  - Concept selection and well completions
- Process Requirements
  - Flow assurance, separation, treatment
- Operations
  - Start-up, Shut down, Surveillance
- HC processing, refining
  - LNG, Gas to liquids
- Marketing
  - HC specification, Crude quality and Valuation

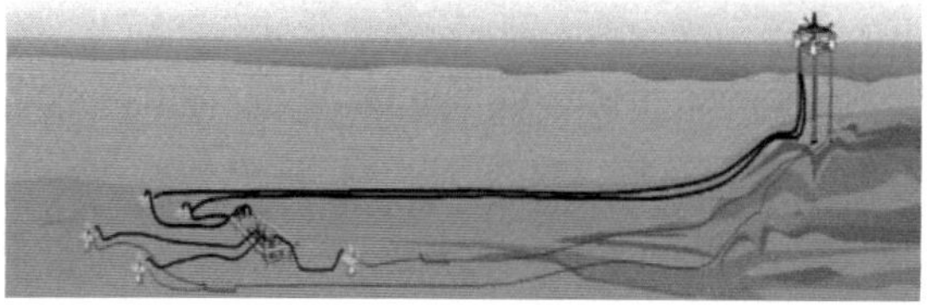

Impact of Fluid Properties on Flow assurance

Asphaltenes

Paraffins

Hydrates

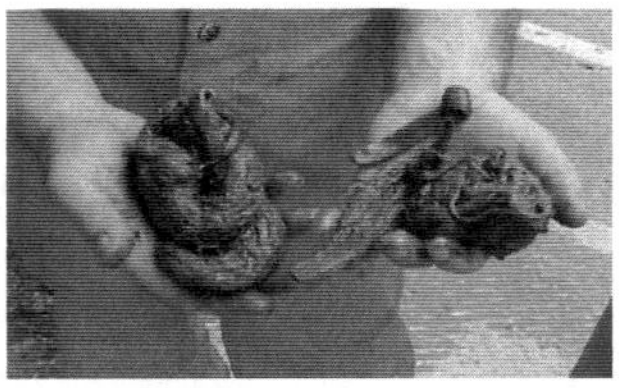

Naphthenates

Formation Testing Applications

- Pressures
- Gradients and Contacts
- Sampling
- Downhole fluid analysis
- Permeability
- Insitu stress testing

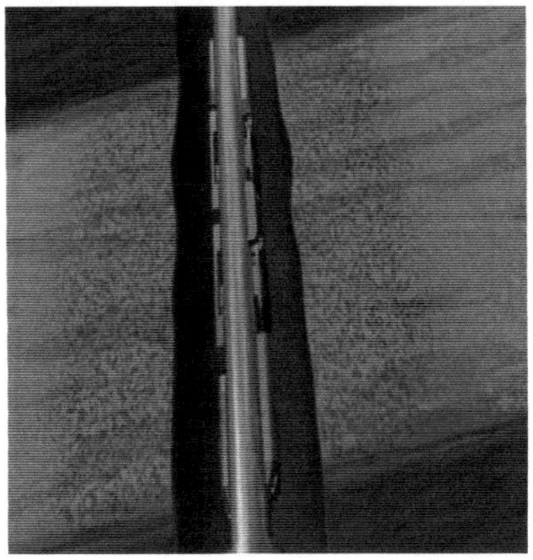

Animation courtesy of Halliburton

WFT Planning

WFT objectives

- Pressures, gradients, samples, permeability, miniDST, microfrac, etc.
- Prioritization: Critical success factor, nice to have, intermediate priority, low priority, etc.

Well conditions

- Expected rock and fluid properties
- Mud type – WBM/OBM/silicate mud, LCM, salinity
- Temperature
- Absolute and differential pressures
- Formation stability, consolidation, wash outs, rugosity

Well configuration

- Deviation – method of conveyance: wireline/pipe
- Borehole Size – (open hole logs, esp. caliper, rock strength)

WFT Execution

Filtrate Clean-up

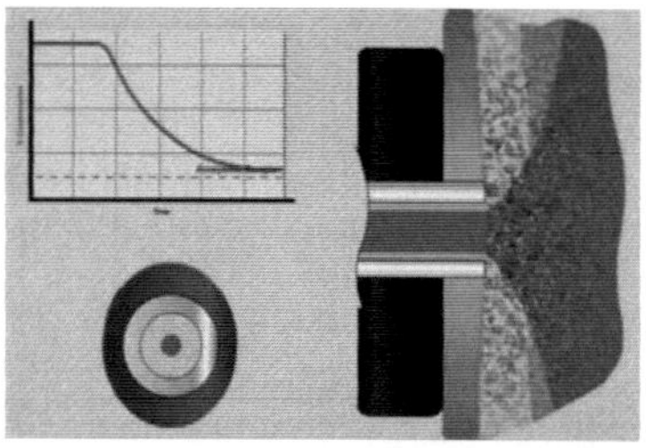

- Downhole pump and downhole fluid analysis
- Contamination and drawdown control
- Conventional or focused probes

Sample Capture

- Low/zero Shock: no disturbance of contamination or drawdown
- Over-pressure sample to limits of pump/system or using N2 cushion

Sample Preservation

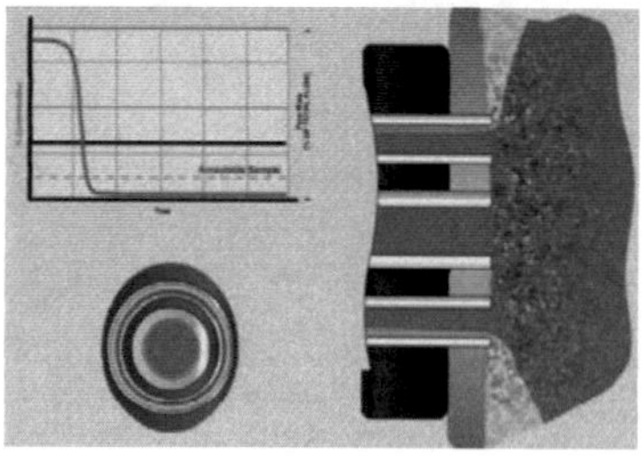

- Maintain pressure on sample to the lab or otherwise minimize time in multi-phase region
- Proper restoration prior to transfers

Real-Time Operations

Real-Time operations enable operating and service personnel to share information, decisions, and solutions

ANYWHERE, ANYTIME

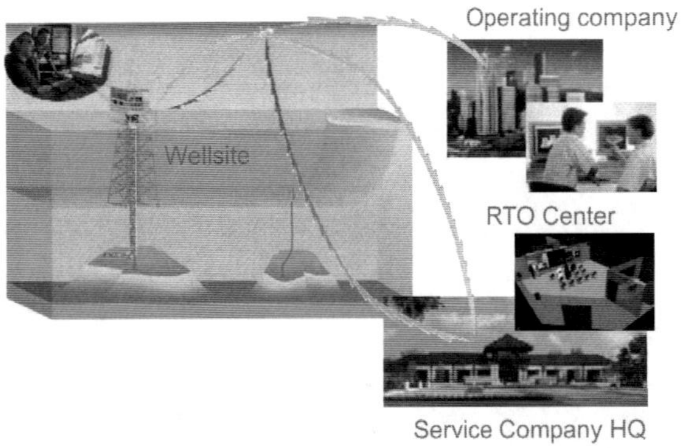

Requirements

- Secure web link
- Minimal delay
- Bi-directional

Benefits

- Travel, logistics
- HSE, personnel security
- Decision making when it counts
- Interdisciplinary collaboration/knowledge transfer

Representative Samples for Flow Assurance

- No phase separation or plugging during cleanup and sampling. Asphaletene stability impacted by changes in oil composition
- Low contamination: Asphaletene stability impacted by OBM contamination as well as by water content, pH, etc.
- Multiple samples per depth; consistency checks on each.
- Minimal dead volume and lost gas/liquid.
- Minimal phase changes, unnecessary sample transfer or handling, and long term storage of sampling bottles. Asphaletene stability impacted by temperature and pressure changes above Psat.
- Heated transfers, for recombining waxes and liquid drop-out.

Effect of OBM Contamination

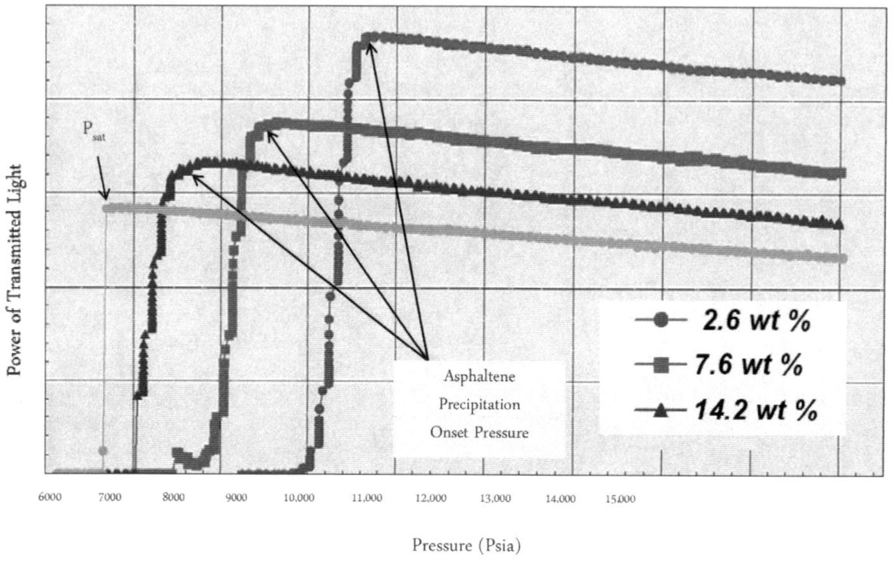

Effect of Sample Capture and Handling

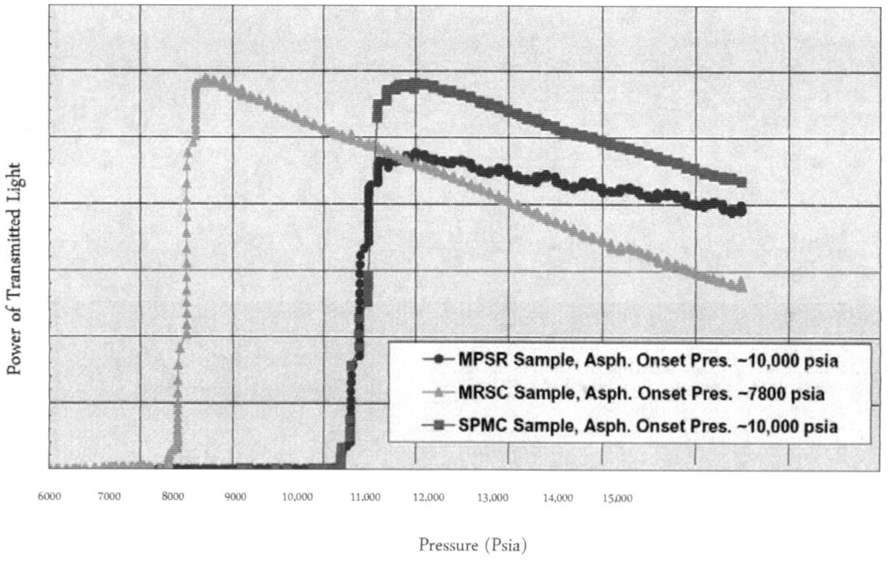

Impact of Water content in HC samples

- Stable emulsions have a significant impact on viscosity and IFT
- Entrainment of gas molecules in liquid produce foamy oil

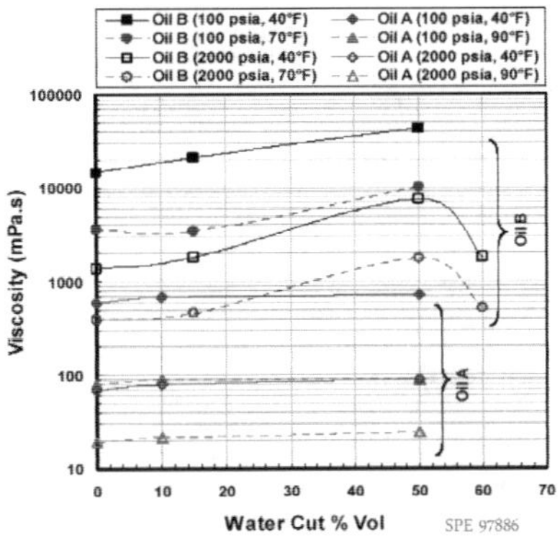

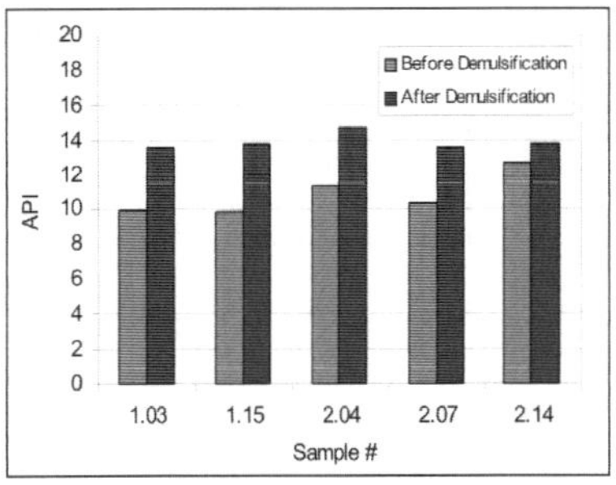

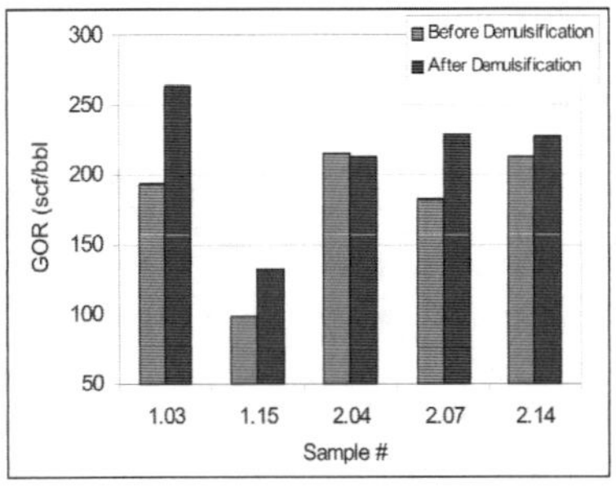

Example 1 – Water sampling in OBM

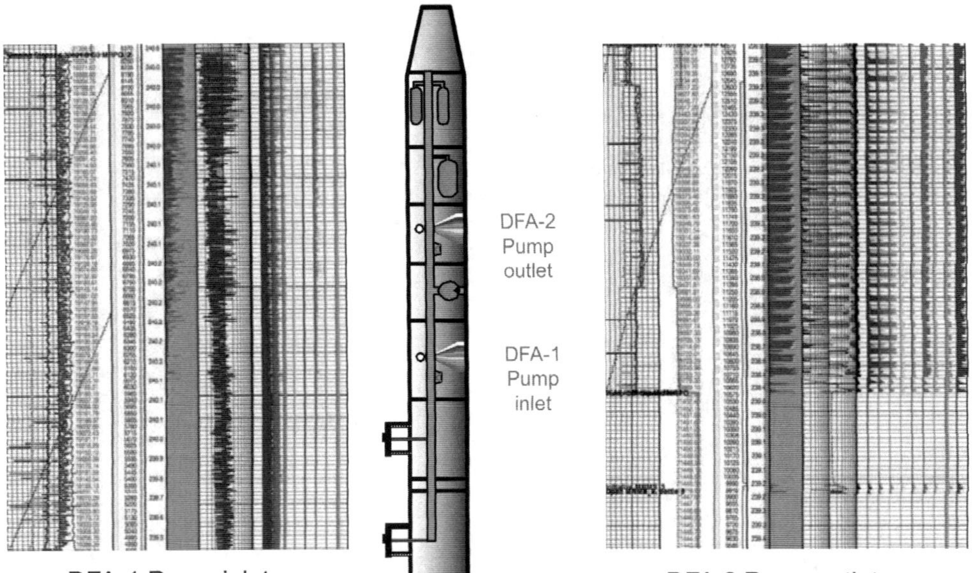

DFA-1 Pump inlet

DFA-2 Pump outlet

Example 2 – Oil Sampling in OBM

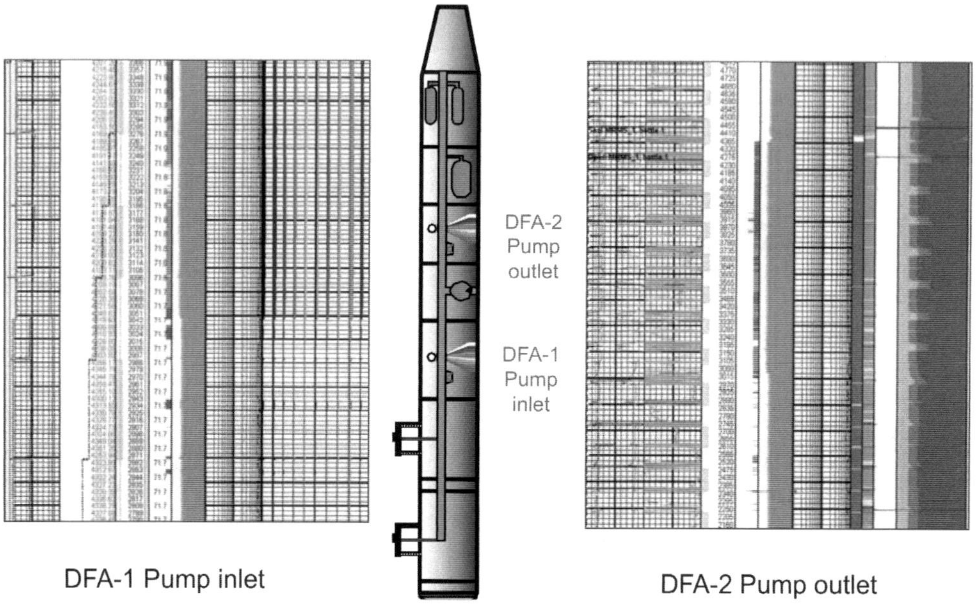

Example 3 - Heavy Oil Sampling

Successful, representative heavy oil samples

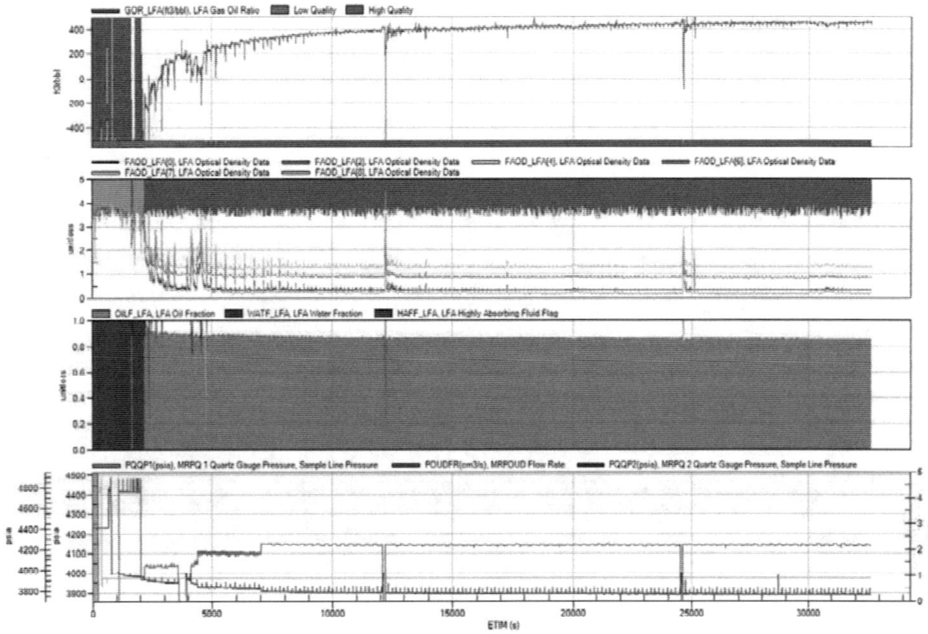

Unrepresentative heavy oil samples

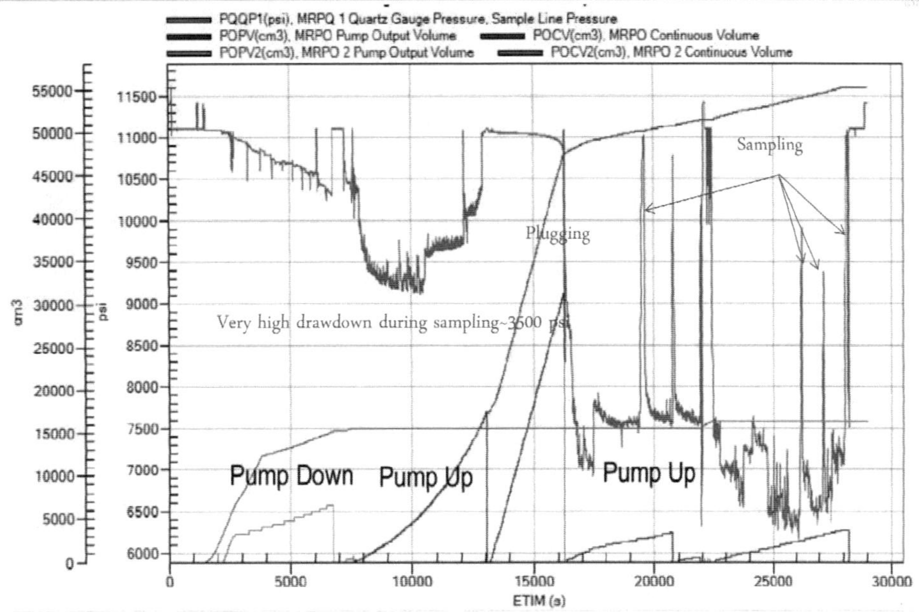

Unrepresentative heavy oil samples

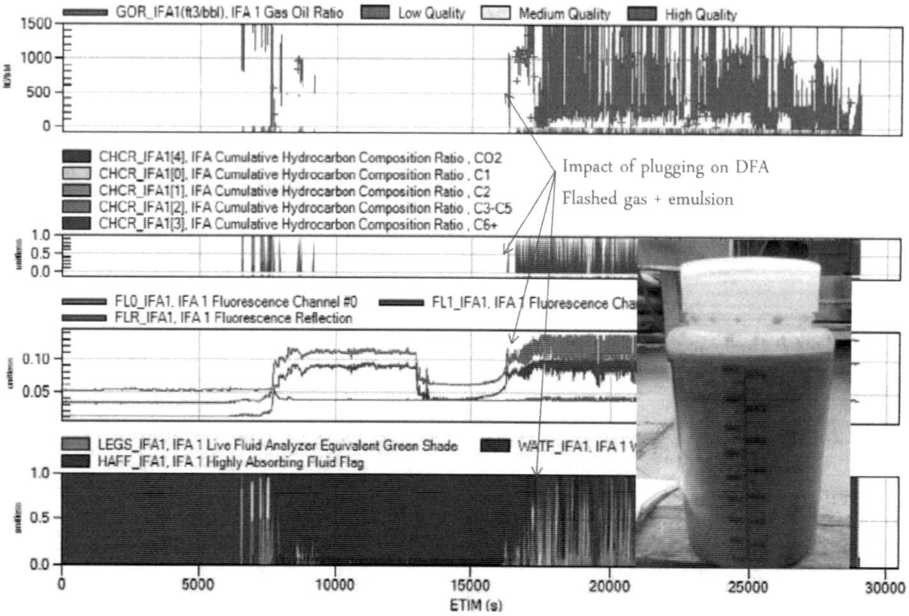

Example 4 – DFA for Compositional Grading

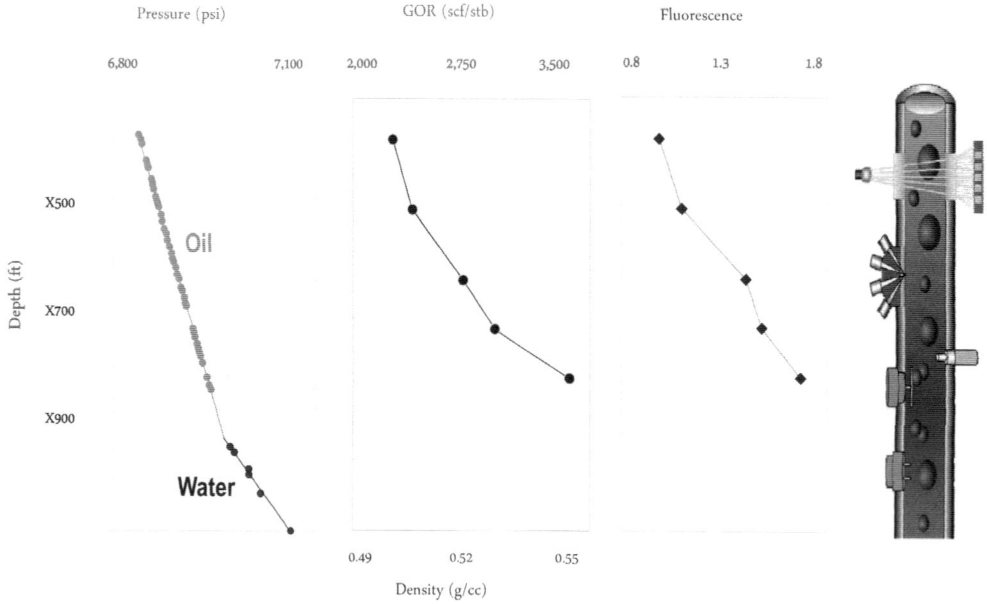

Example 5 – DFA for fluid fingerprinting

Depth A: Little gas

Depth B: Much gas

Depth C: Medium gas

Depth D: Medium gas

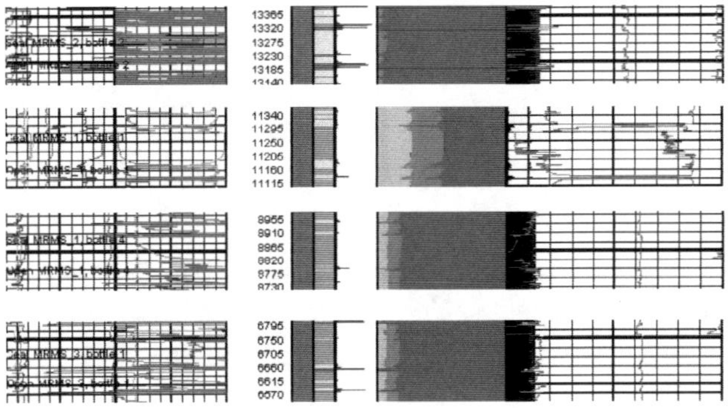

DFA analysis (left) and **PVT lab analysis (right)**

Depth	C1 (wt%)	C2-C5 (wt%)	C6+ (wt%)	GOR (scf/bbl)	Color
Depth A	1%, 2%	7%, 9%	92%, 89%	230, 400	
Depth B	27%, 25%	25%, 31%	48%, 44%	4900, 6200	
Depth C	8%, 7%	11%, 14%	81%, 79%	940, 1030	
Depth D	7%, 6%	11%, 13%	82%, 81%	860, 875	

Learnings and Recommendations

- Identify main objectives and challenges
- Plan the execution - execute the plan
- Experience matters
- Integrate...integrate...integrate
- Never walk away with insufficient data
- Real time monitoring and control is key

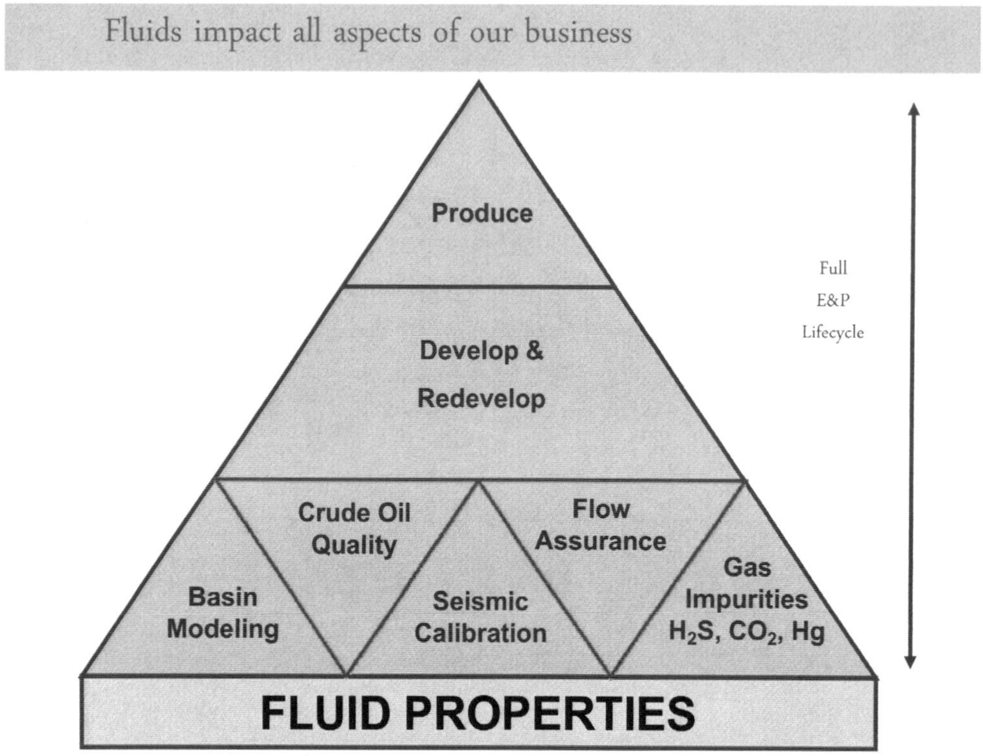
Fluids impact all aspects of our business
Produce
Develop &
Redevelop
Crude Oil
Quality
Flow
Assurance
Basin
Modeling
Seismic
Calibration
Gas
Impurities
H_2S, CO_2, Hg
FLUID PROPERTIES
Full
E&P
Lifecycle

Fluid Property Characterization for Flow Assurance

John Nighswander
AIChE/SPE Joint Workshop – Challenges in Flow Assurance
September 2011

Schlumberger

Presentation Outline

- Introduction to Flow Assurance
- Role of Production Chemistry
- Characterization of solid deposits
  - Composition
  - Phase Boundaries
  - Transient/Kinetic Behavior
- Case Study – Wax Deposition

What is Flow Assurance?

Flow assurance is an engineering analysis process used to assure hydrocarbon fluids are economically transmitted from a reservoir to the end user over the lifetime of a project

Excellence in Execution

Flow Assurance Disciplines

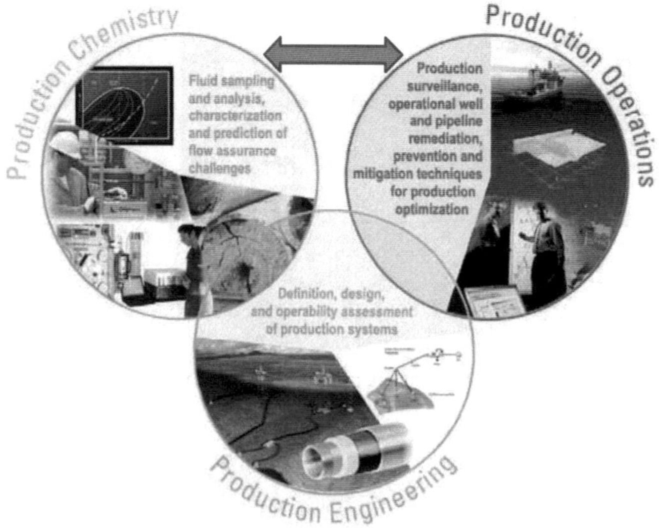

Excellence in Execution

Deepwater Production Scope of Activity

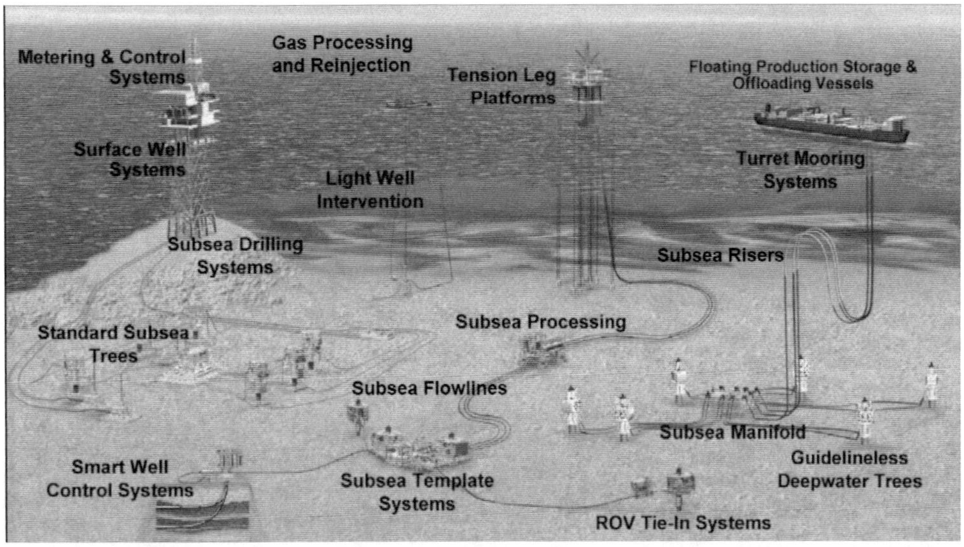

Excellence in Execution

Schlumberger

Production Chemistry – Major Issues

Knowledge of produced fluid chemistry and dynamic phase behavior over the full range of production conditions are required to develop mitigation strategies for controlling potential solid phase formation and deposition in the production system

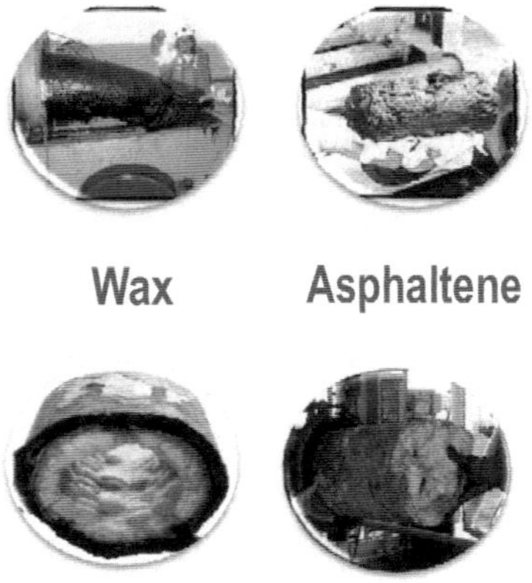

Wax | Asphaltene

Scale | Hydrate

Excellence in Execution

Schlumberger

Production Chemistry Discipline

- Sampling, Sample Handling/Conditioning
- Compositional Analysis
- Transport Properties
- Thermodynamic Phase Boundaries
- Transient Phase Measurements
  - Solid Formation/Dissolution Kinetics
  - Deposition rates
  - Chemical Inhibitor screening (Thermodynamic/Kinetic)
  - Formation Damage

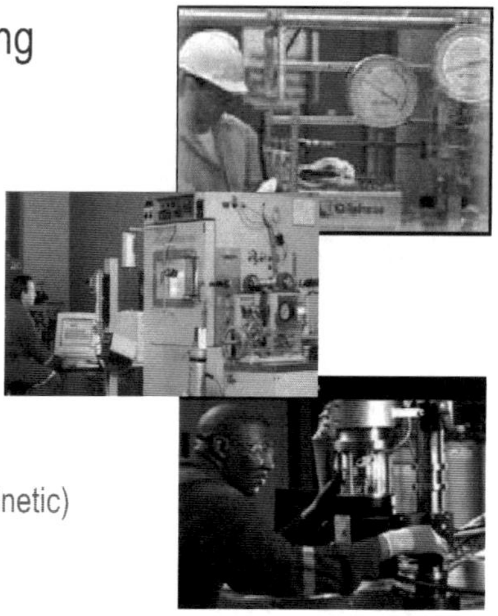

Modeling & Simulation

Fluid Composition

- Reservoir Fluid Oil and Gas
  - Standard PVT report – C15+ on gas – C30/36+ on oil
- Formation Water
  - Ion Chromatography, pH
- Crude Oil Detailed Characterization
  - Wax - HTGC (C70+), UOP wax content
  - Asphaltene – SARA – IP143 C7 Asphaltene
- Potential Injection Fluids
  - Seawater for waterflood
  - Gas lift gas
  - Commingled reservoir fluid streams

- Measurement uncertainty in compositional analysis is both fluid and test specific.
– Must assess on a case by case basis

Fluid Composition

- Necessary input for fluid modeling applications
  - Requires additional data for model tuning
  - May be used directly in empirical correlations to predict likelihood of potential issues based analogous fluids and their observed behavior

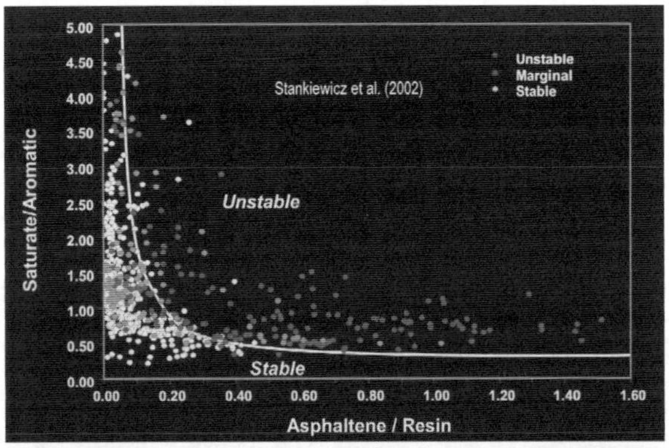

Transport Properties - Viscosity

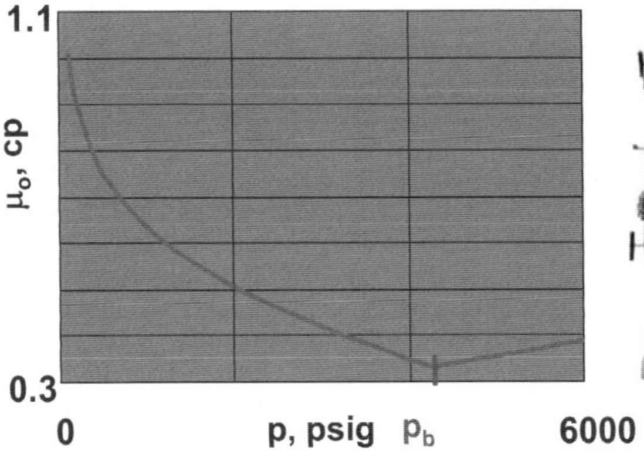

- High Uncertainty in live fluid viscosity (+/- 20%)

Emulsion Apparent Viscosity

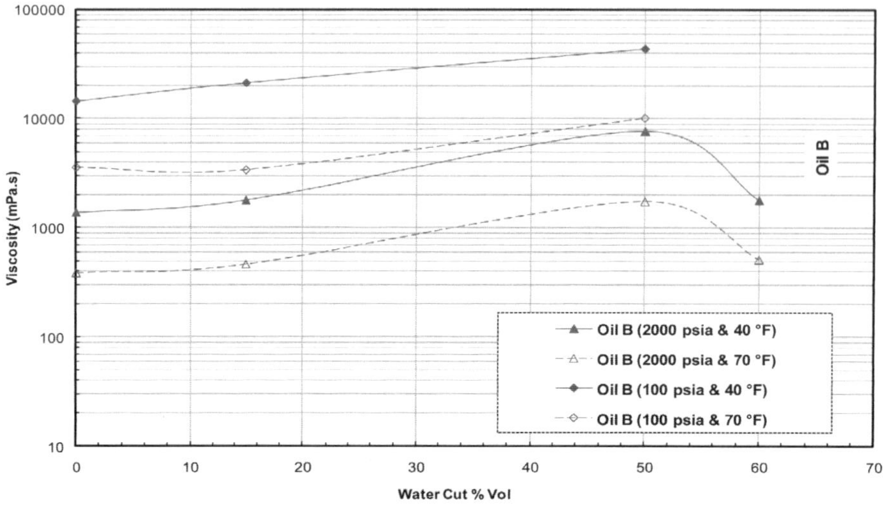

Density

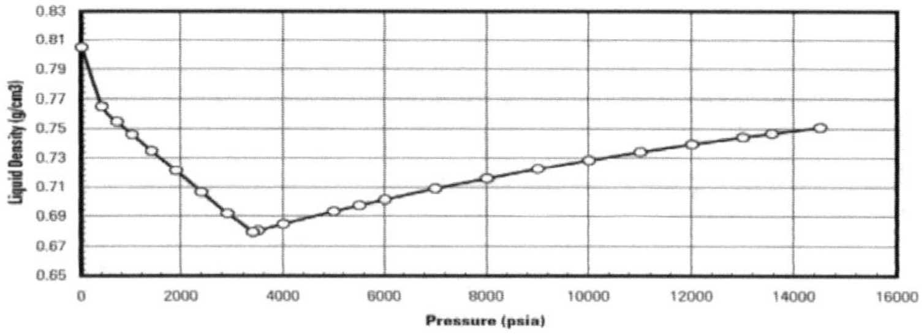

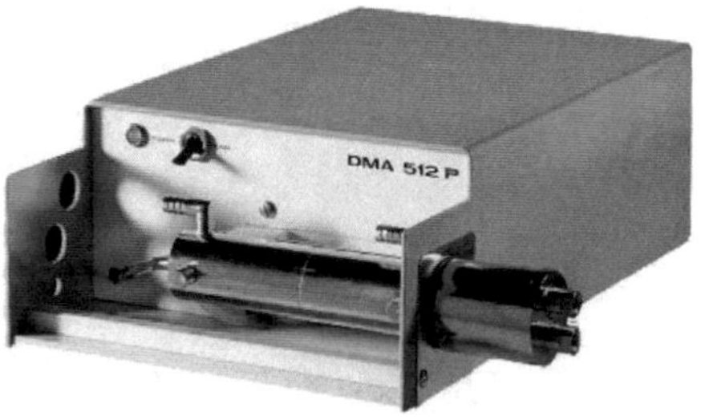

Density Measurements accurate to +/- 2%

Phase Boundaries

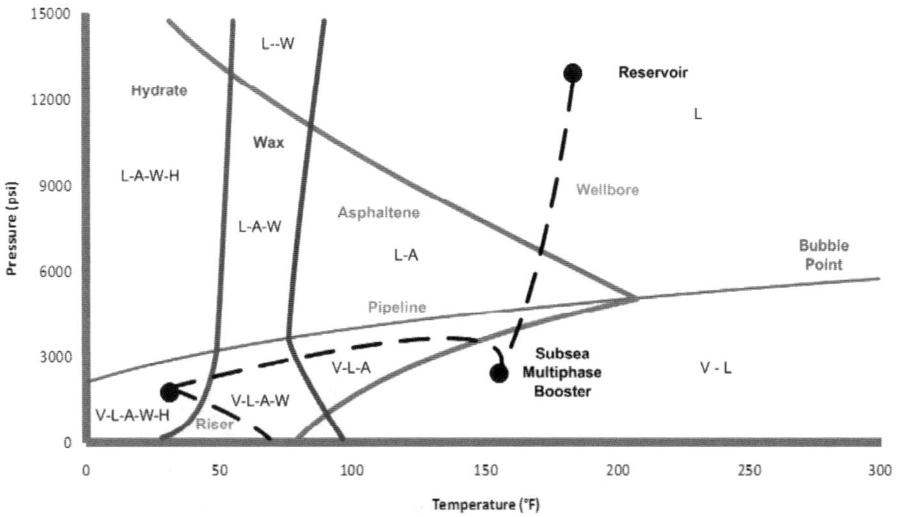

SDS and HP-CPM Systems

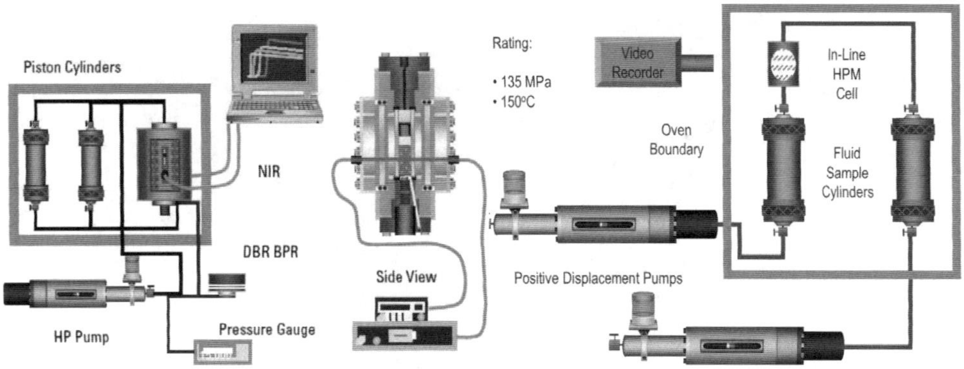

HP-CPM Images and SDS Plot For Wax

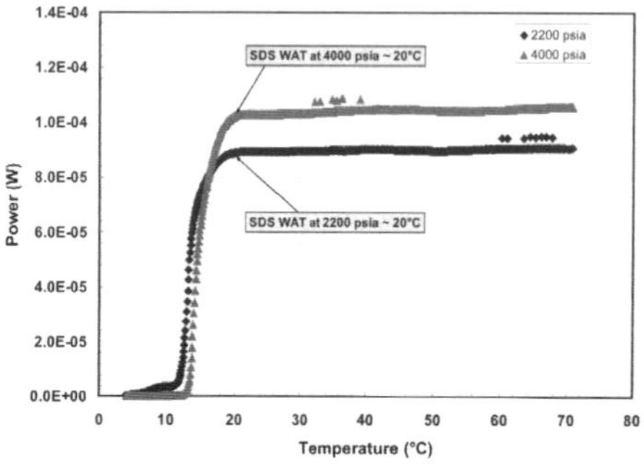

HTGC wax content: 11 wt%
STO WAT = 27°C
Uncertainty usually <5C

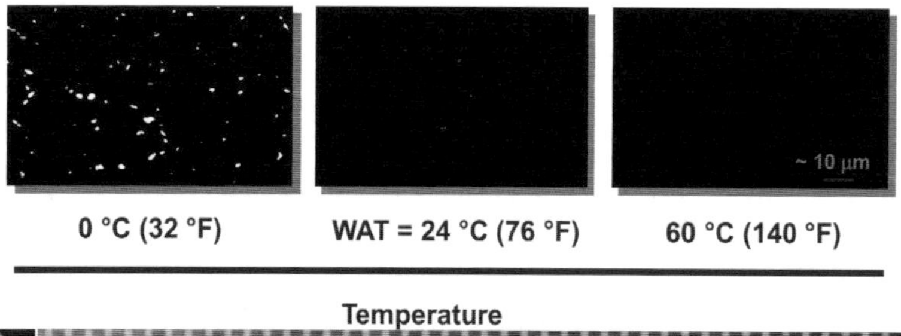

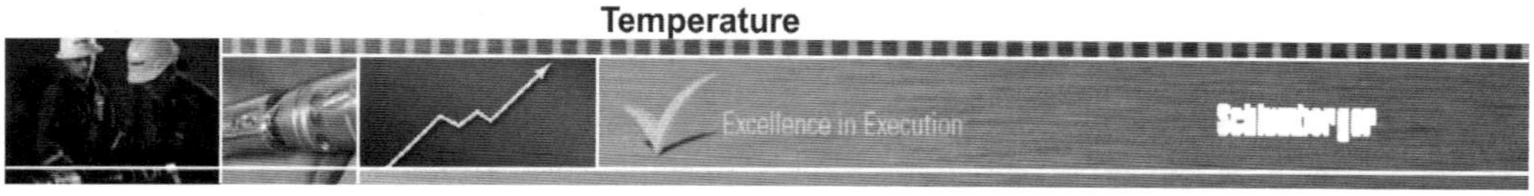

HP-CPM Images and SDS Plot For Asphaltene

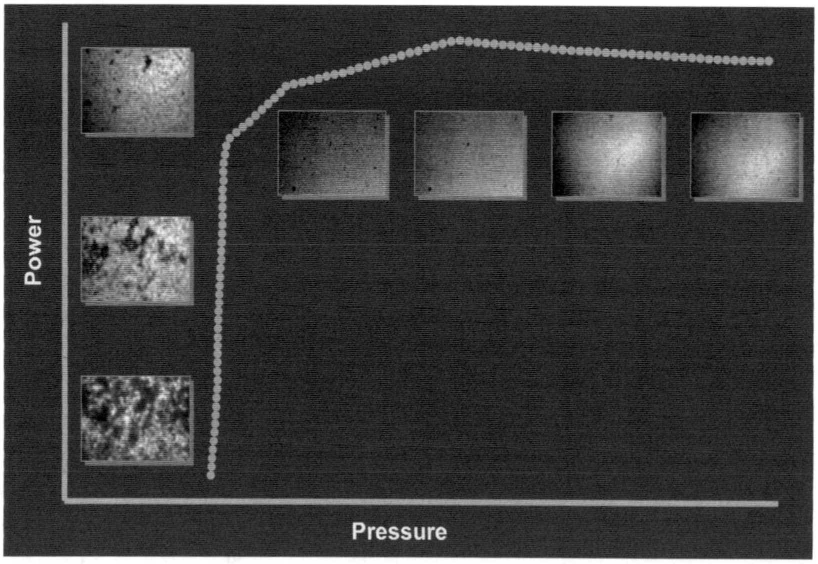

Transient Solid Deposition

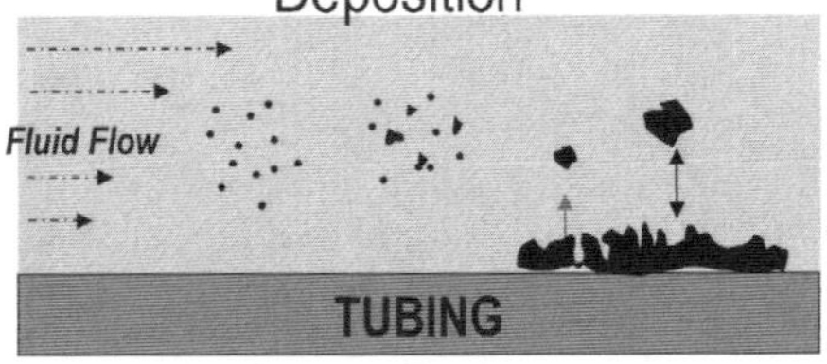

- **Precipitation** is the formation of a solid phase out of a liquid phase.
- **Deposition** is the formation of a solid layer on a surface.
- **Impairment** is the formation of solid layers and/or bridging of pore throats in the reservoir or completion
- Precipitation is a necessary but not a sufficient condition for deposition and impairment.

Deposition Mechanisms

- Molecular Diffusion
- Shear Dispersion
- Brownian Diffusion
- Gravity Settling
- Sloughing/Shear Stripping
- Crystal growth/kinetics
- Particle Diffusion and Inertia
- Adsorption/adhesion-cohesion
- Aging
- Gelation
- ???

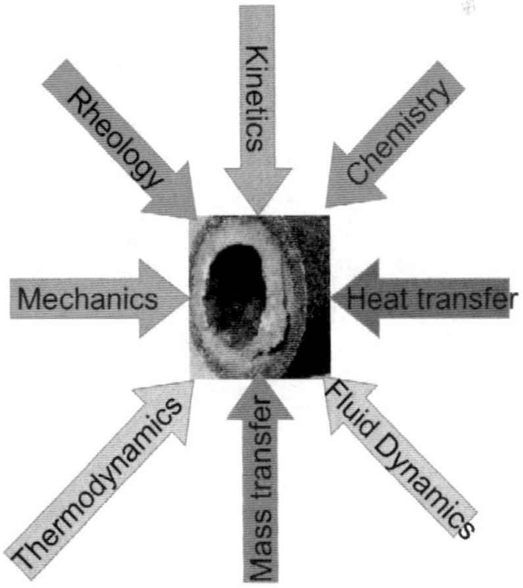

Transient Solid Deposition Measurements

Wax

- Shut in Conditions (transient bulk fluid temperature cooling)
  - Pour Point, Viscosity = f(T), gel strength (Model Pipeline Test)
- Producing Conditions – (transient deposition on cold surface)
  - Flow Loop Tests, Benchtop Testing (Real View)

Asphaltene

- Injection Fluid Compatibility (transient fluid mixing)
  - gas lift gas titration, miscible solvent titration
- Producing Conditions - (transient deposition due to pressure reduction)
  - Benchtop Testing (Real View) , STO capillary tube ΔP (correlated to live oil deposition rate)

Wax Deposition Measurement Techniques

(Producing Conditions)

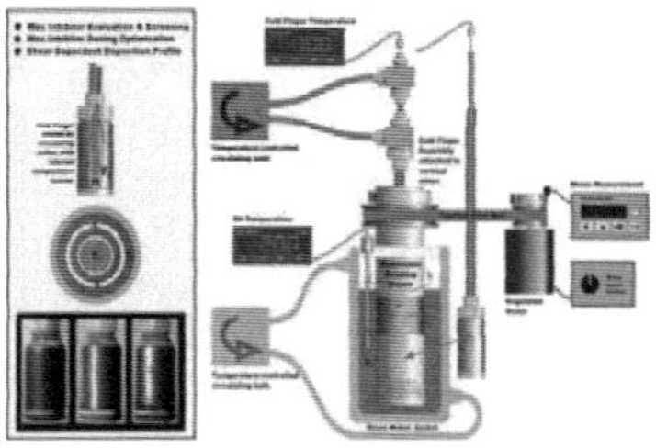

RealView Deposition Cell: Configuration

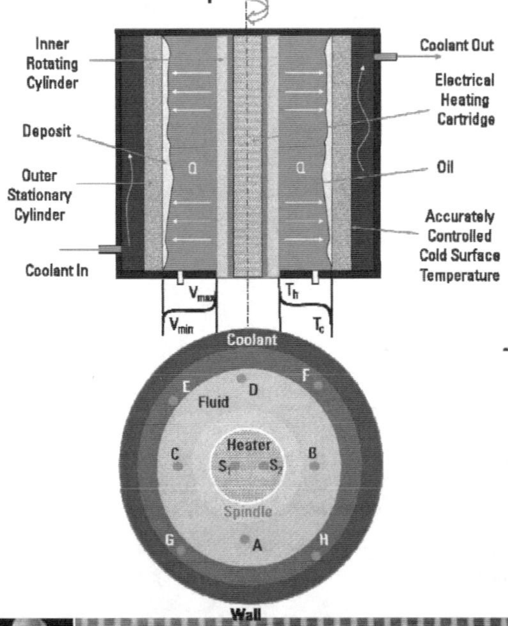

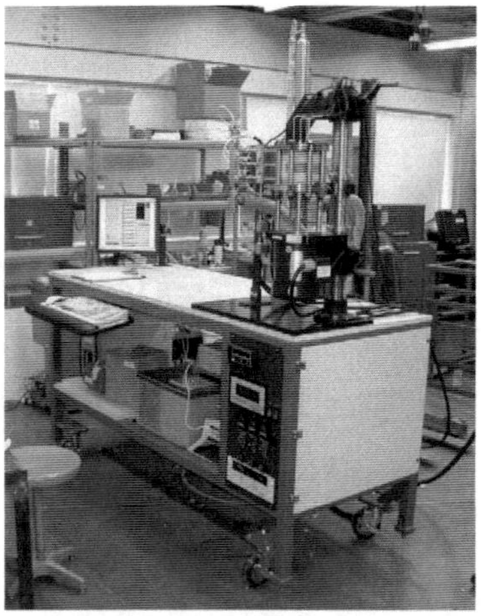

Equivalent Pipe Segment

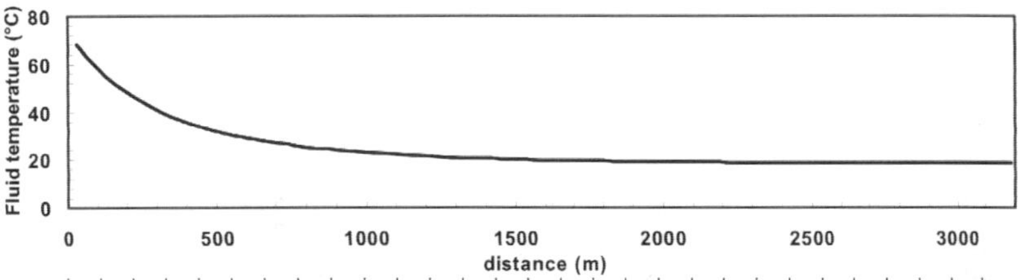

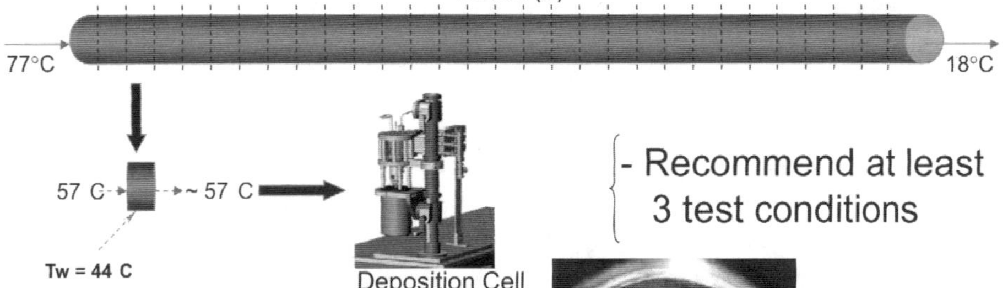

Deposition Cell

- Recommend at least 3 test conditions

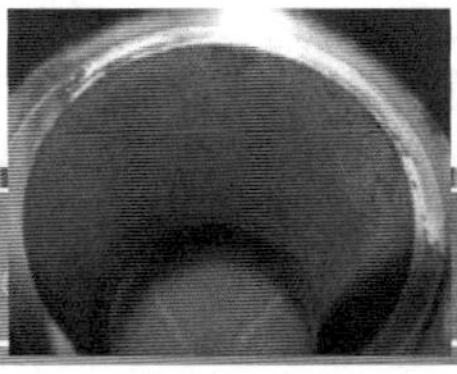

Live vs. STO:

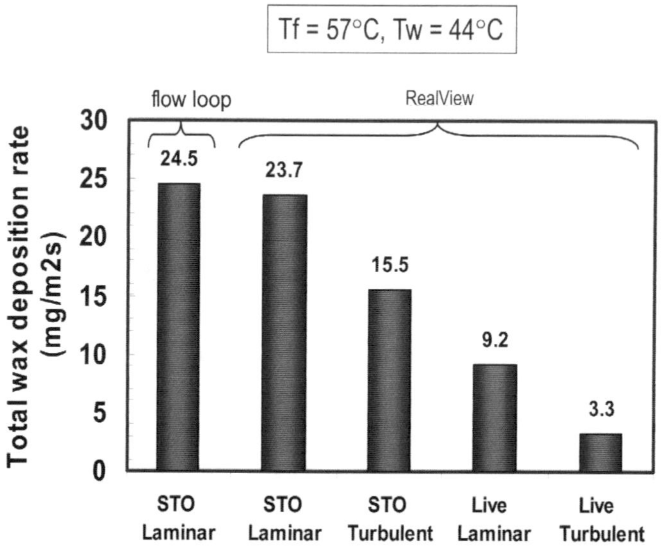

Wax inhibitor effect

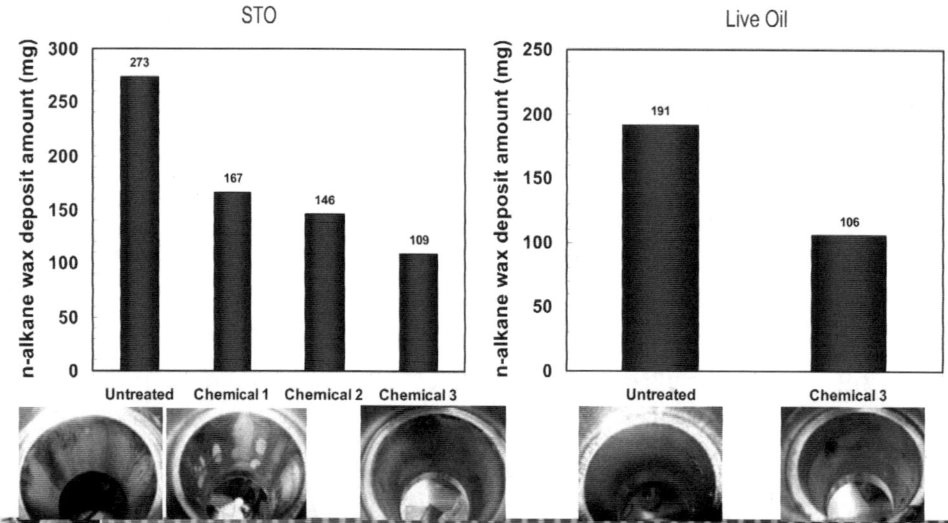

Case Study : Waxy Crude from West Africa

SPE 115131 Akbarzadeh et. al. (2008)

Reservoir Fluid Properties	
Tres (°C)	77
Pres (psia)	3180
Psat @ Tres (psia)	800
GOR (SCF/STB)	230
°API	36
Wax Content (wt%)	13.1
STO WAT at Pres (°C)	60
Live Oil WAT at Pres (°C)	47
STO Pour Point (°C)	29
Live Oil Pour Point (°C)	18*
Hydrate Formation T (°C)	12

* Measured at 700 psia

Production System Properties	
Pipe ID (in)	6
Pipe wall thickness (in)	0.625
Tieback Length (km)	16
Water Depth (m)	300
Seabed Temperature (°C)	13
Production Rate (B/D)	5000
Fluid Entry P & T	Pres & Tres

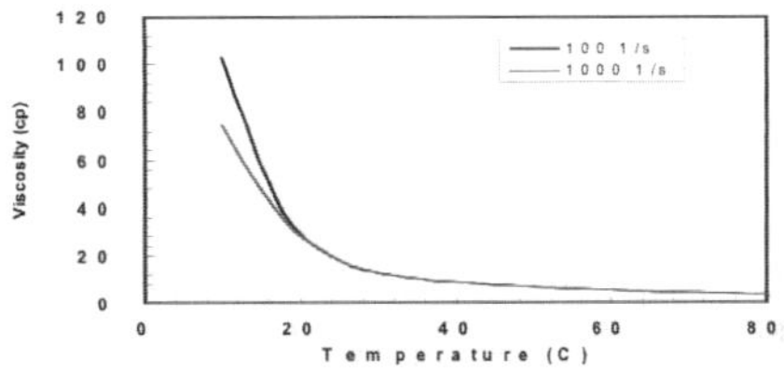

Step 1 – Complete Lab-scale Real View Tests

Recommend minimum of 3 Tests over a range of shear rates determined from expected pipeline production rates

- Assume shear stress equivalence between Real View and Pipeline
- Number of tests matches number of adjustable parameters in deposition model

Test #	Rotational Speed (rpm)	Shear Stress (Pa)	Deposit Mass (mg)	Entrapped Oil (wt%)	Deposition Rate (g/m2/hr)
1	180	0.1	419	83	33.3
2	3600	12.0	149	69	11.8
3	6000	28.7	47	70	3.7

Operating Conditions: Tf = 57°C, Tw= 44°C, P=1000 psia, t=1 hr

Step 2 - Scale Real View Data to a 6" ID 1m Long Equivalent Pipe Segment (EPS)

	EPS #1	EPS#2	EPS#3
Velocity (m/s)	0.19	2.46	4.00
Deposit mass (g)	15.9	5.6	1.8
Deposit volume (cc)	17.7	6.2	2.0
Deposit Thickness (mm)	0.037	0.013	0.004

Excellence in Execution

Schlumberger

Step 3 Tune OLGA Wax Deposition Model

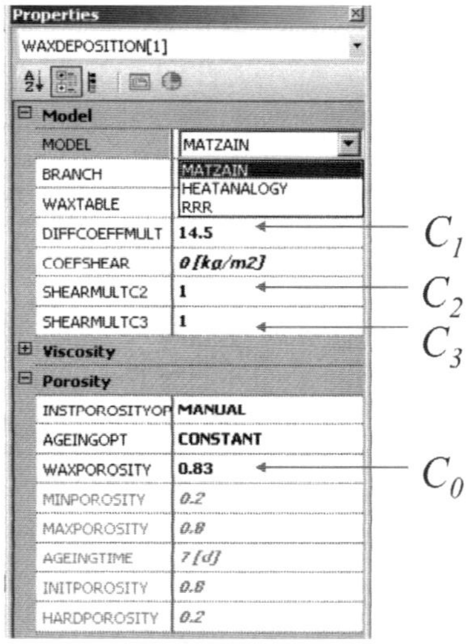

Matzain Model

$$\frac{d\delta}{dt} = \frac{\Pi_1}{1+\Pi_2} D_{ow} \left(\frac{dC_w}{dr} \right)_{\text{interface}}$$

$$\Pi_1 = \frac{15C_1}{1-C_0}$$

$$\Pi_2 = 0.055 C_2 N_{SR}^{1.4C_3}$$

Excellence in Execution

Schlumberger

Tuned Model Predictions vs. Experimental Data

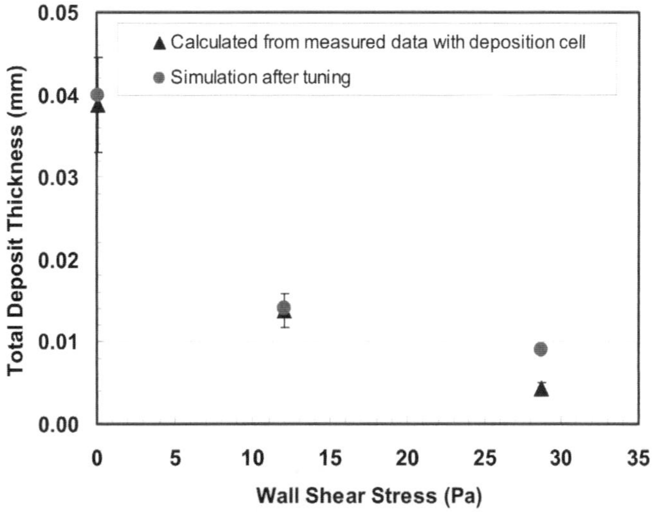

C0 = 0.7
C1 = 3.7
C2 = 2.0
C3 = 4.0

Predicted deposit growth over time along pipeline

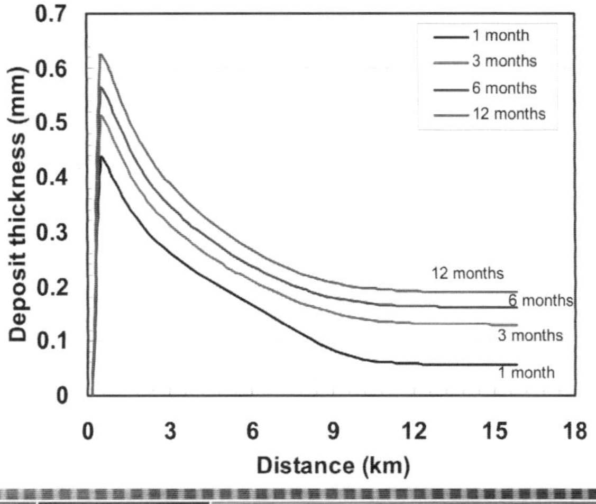

Beware Using Unrealistic Input Data for Model Tuning

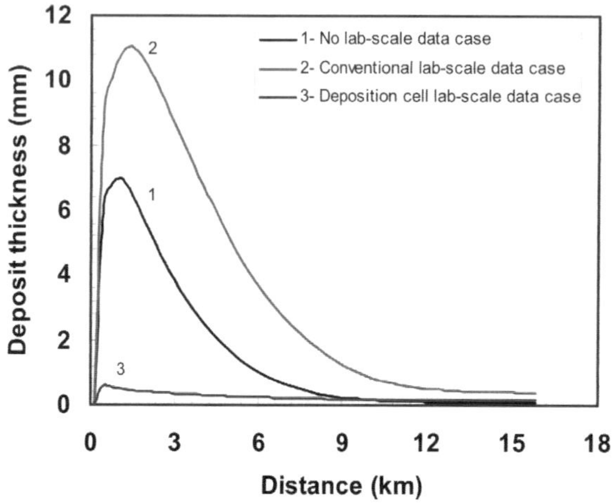

Conclusions

- production chemistry information is necessary input to production engineering discipline for production facility design
  - Fluid Composition and Transport Properties
  - Phase Boundaries
  - Thermal and Solid Deposition Transient Behavior
- All lab measurements are subject to measurement uncertainty
  - Actual uncertainty is both fluid and test specific
  - Good judgment must be used in assessing which data is most reliable and suitable for modeling
    - Example - Care must be taken to insure lab scale solid deposition data is properly scaled to pipeline conditions prior to inclusion in pipe deposition models

Issues in Equation of State Modeling for Flow Assurance Applications

Walter G. Chapman
Sai R. Panuganti
Anju Kurup
Rice University, Houston, TX

RICE

Outline

- Introduce Flow Assurance Models
- Introduce Asphaltenes
- Equation of State Models for Asphaltenes
- Characterization Procedure and Data Issues
- Predictions from the Equation of State Model

2

Use of Thermodynamic Models in Upstream

Uses of Thermodynamic Models Include

- Density
- Phase Behavior
- Enthalpy
- Compositional Grading
- Properties for Viscosity Models
- Interfacial Properties

RICE

3

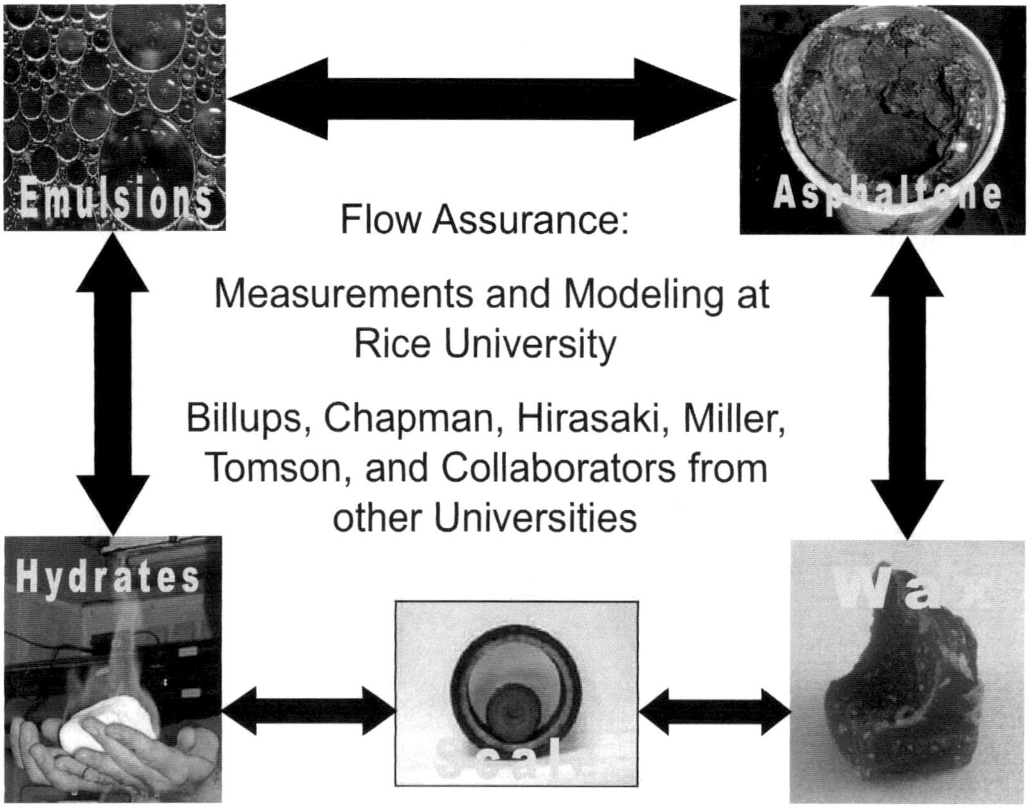

Phase Behavior Models for Flow Assurance

- Gas Hydrates
  - van der Waals–Platteeuw (e.g., CSM-GEM)
- Asphaltenes
  - Oil Compatibility Model (Wiehe, Soluble Solutions)
  - Solution Model (e.g., PC-SAFT Eqn. of State, Rice Univ.)
- Wax
  - Solid solution model (e.g., Coutinho Wax Model)
- Scale
  - Brine model (e.g., ScaleSoftPitzer™ from Rice Univ.)

5

RICE

Thermodynamic Models used in Petroleum Industries

Method	Application	Operating Conditions
Soave-Redlich-Kwong	General hydrocarbon	Pressure > 1 bar
API SRK	General hydrocarbon	Pressure > 1 bar
Peng-Robinson	Cryogenics < -70 °C	Pressure > 10 bar
Benedict-Webb-Ruben-Starling	Single species	Pressure > 1 bar
Grayson-Streed	Heavy end hydrocarbon	Moderate P >7bar <200bar Temperature –18 °C to 430 °C
ESSO	Heavy end hydrocarbon	Pressure < 7bar Temperature 90 °C to 200 °C
SAFT (PC-SAFT, SAFT-VR, SAFT1, Soft-SAFT, CPA)	General hydrocarbon with methanol and water; Hydrocarbon – gases	-

Adapted from Edwards JE. Process Modeling Selection of Thermodynamic Methods. P & I Design Ltd, MNL 031B. November 2008

6

RICE

Phase Behavior Models for Flow Assurance

Approach in Modeling Phase Behavior

- Choose Appropriate Model
  - Understand the physical phenomena
  - Understand model approximations and limitations
- Obtain Data for Model Correlation and Validation
  - Understand how data was obtained
  - Verify laboratory data
- Correlate the Model to Available Data
- Predict Flow Assurance Issues
- Refine Model with any Additional Data

RICE

Asphaltenes in Flow Assurance

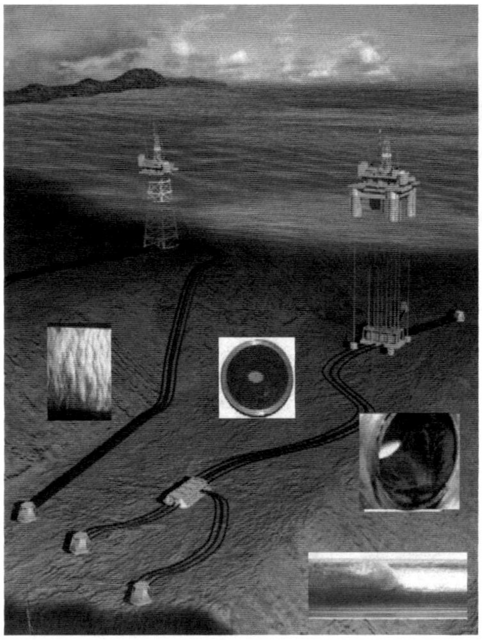

Flow Assurance Prediction – Operator's Savings:

- Intervention cost to remove solids
  ~ $300K/well-dry tree $3,500K / well – wet tree
  - Loosing the well
  ~ $50,000K to replace the well with a side track
  - Losses due to downtime
  ~ $ 700K /day (for prod. of 7,000bbls/day)

8

RICE

Fast Facts about Asphaltenes

- Polydisperse mixture of the heaviest and most polarizable fraction of the oil
- Defined in terms of its solubility
  - ✓ Miscible in aromatic solvents, but insoluble in light paraffin solvents
- Deposition mechanism and molecular structure are not completely understood
- Behavior depends strongly on *P*, *T* and $\{x_i\}$

(a) n-C_5 asphaltenes

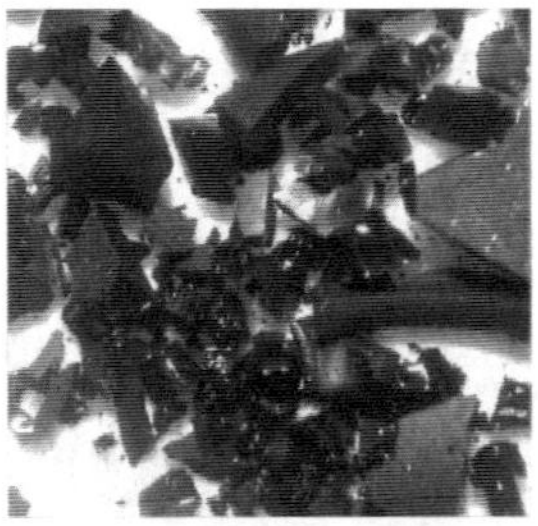

(b) n-C_7 asphaltenes

Buckley, J.
To be published.

Francisco Vargas, fvargas@rice.edu

9

RICE

Modeling using Cubic EoS (Crude A)

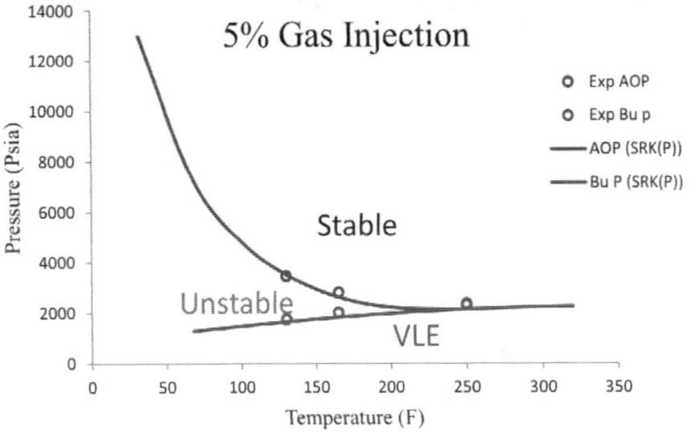

Crude A

- Characterized the crude oil system using PVT-Sim of Calsep
- The Cubic EoS employed was SRK-P

10

RICE

Modeling using Cubic EoS (Crude A)

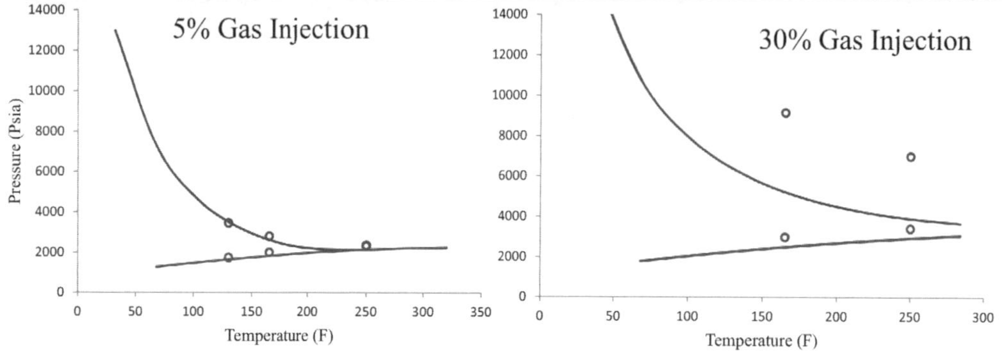

The optimized Cubic EoS parameters from 5% were used to predict the phase behavior for 30% injected gas

Limitations of cubic equation of state

- Cubic EoS are best for small molecules and least accurate for liquid phase properties
- For asphaltenes, results are very sensitive to parameters

Larry GC et al. Advances in Thermodynamics (Volume 1): C7+ Fraction Characterization. Taylor & Francis; 1989

RICE

11

A Little History of Equation of State Models

Begins with the van der Waals equation of state

$$P = \frac{RT}{v-b} - \frac{a}{v^2}$$

- The first term, RT / (v-b), is only accurate for hard spheres in one dimension.
- An accurate hard sphere equation of state was not available until the work of Norman Carnahan
- Parameters of modern cubic (a, b, omega)

Carnahan, N. F.; Starling, K. E. *J. Chem. Phys.* **51**, 635 (1969)

RICE

12

Cubics are Commonly fit to the Critical Point

- Classical equations of state perform well far from the critical region but fail near the critical point

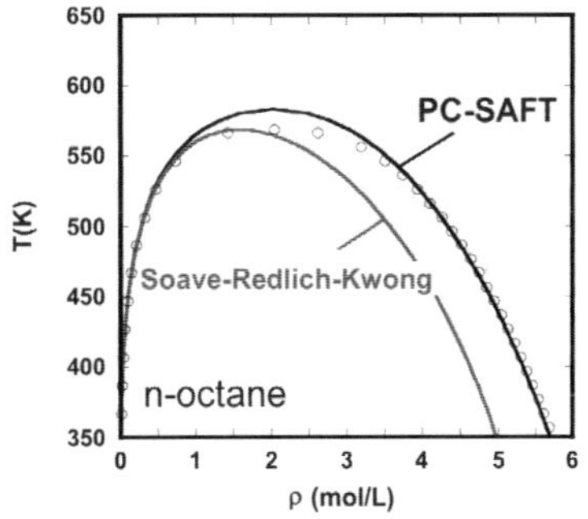

- Critical parameters are not known for most components above C20.
- Better to fit to liquid density and vapor pressure.
- Crossover corrections based on renormalization-group theory can be developed

13

Introduction to the SAFT Eqn. of State

Chapman et al., *Mol. Phys.* (1988); *I&EC Res.* (1990)
Wertheim, *J. Stat. Phys* (1986)

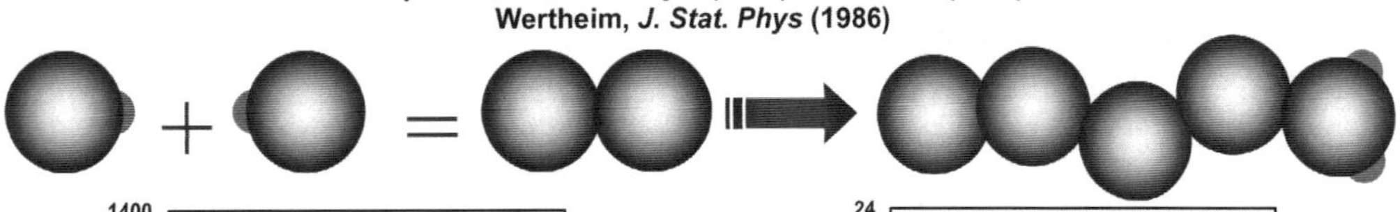

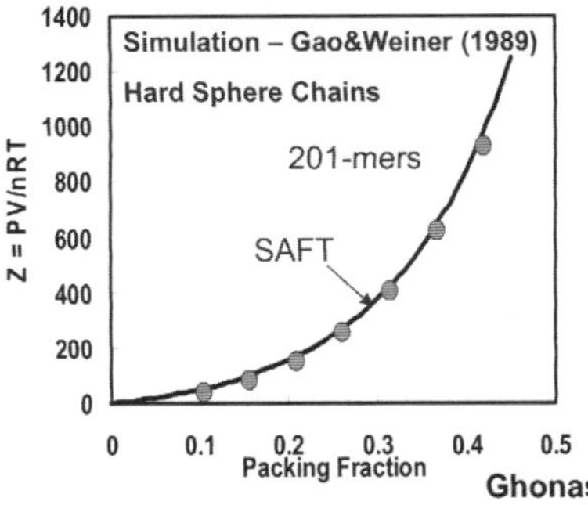

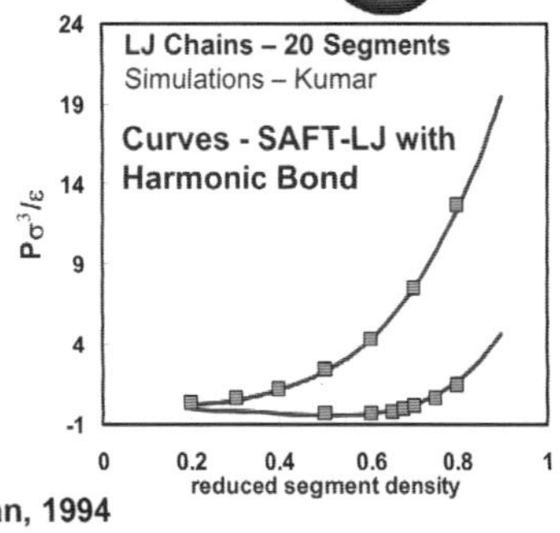

Ghonasgi and Chapman, 1994

Introduction to the SAFT Eqn. of State

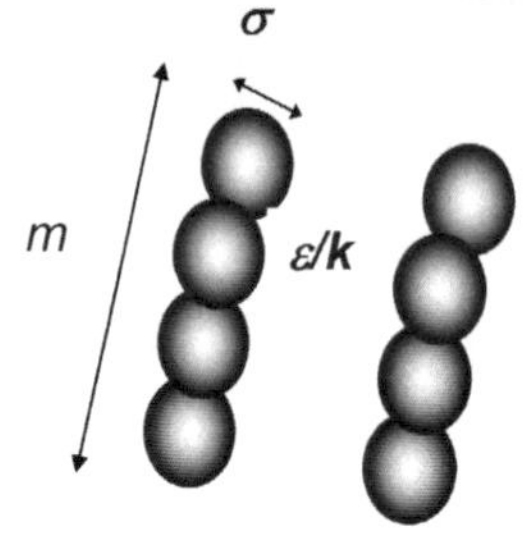

$$\frac{A^{res}}{RT} = \frac{A^{seg}}{RT} + \frac{\Delta A^{chain}}{RT} + \frac{\Delta A^{assoc}}{RT}$$

Chapman, Jackson, and Gubbins, Mol. Phys. 65, 1057 (1988)
Chapman, Jackson, Gubbins, and Radosz, I&EC Res, (1990)
Gross & Sadowski, Ind. Eng. Chem. Res., 40, 1244-1260 (2001)
Gonzalez, et al. Energy & Fuels 19, 1230-1234 (2005)

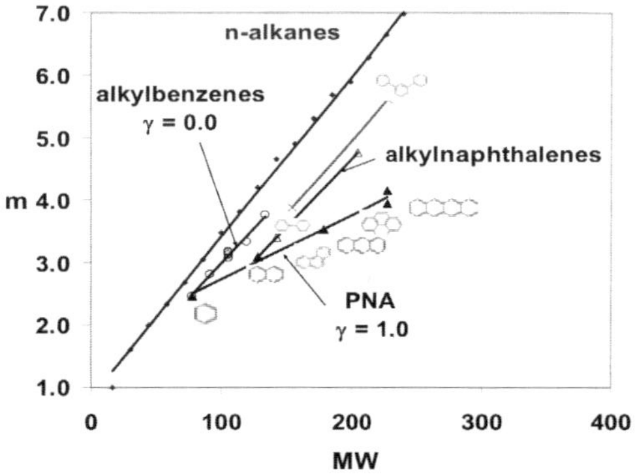

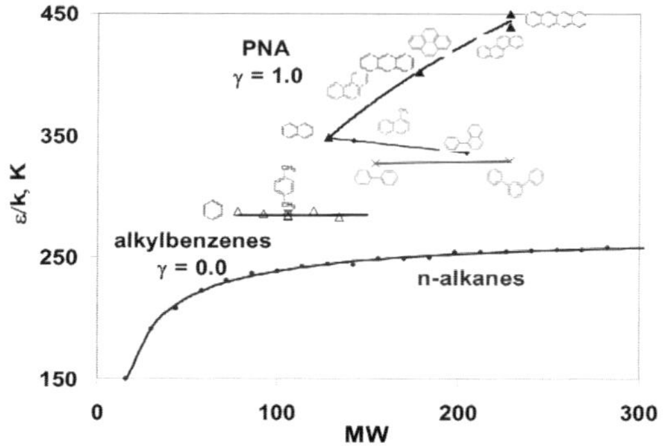

SAFT EoS available in InfoChem UK, VLXE and Calsep
SAFT parameters are well correlated to molecular weight

RICE

Characterized Stock Tank Oil

Crude B

	MW	Basis 100 g of STO	Saturates		Aromatics + Resins	
Component	g/mol	mass%	mass (g)	mol%	mass (g)	mol%
N_2	28.04	0	0	0	0	0
CO_2	44.01	0	0	0	0	0
H_2S	34.08	0	0	0	0	0
C_1	16.04	0	0	0	0	0
C_2	30.07	0.04	0.04	0.37	0	0
C_3	44.10	0.30	0.30	1.70	0	0
iC_4	58.12	0.25	0.25	1.10	0	0
nC_4	58.12	0.92	0.92	4.03	0	0
iC_5	72.15	1.00	1.00	3.52	0	0
nC_5	72.15	1.59	1.59	5.60	0	0
C_6	84.00	3.59	3.59	10.87	0	0
Mcyclo-C_5	84.16	0.45	0.45	1.35	0	0
Benzene	78.11	0.14	0	0	0.14	1.50
Cyclo-C_6	84.16	0.32	0.32	0.97	0	0
C_7	96.00	3.60	3.60	9.54	0	0
Mcyclo-C_6	98.19	0.62	0.62	1.60	0	0
Toluene	92.14	0.70	0	0	0.70	6.20
C_8	107	3.80	3.80	9.04	0	0
C_2-Benzene	106.00	0.22	0	0	0.22	1.72
m&p Xylene	106.17	0.64	0	0	0.64	4.94
o Xylene	106.17	0.04	0	0	0.04	0.29
C_9+	268.4	81.72	49.5	50.31	29.42	85.36
			C_9+ Sat MW 250		C_9+ Sat MW 280.6	

Asphaltenes			
Component	MW	mass	mole
Asphaltenes	1700	2.8	0.0016

Modeled STO molefractions

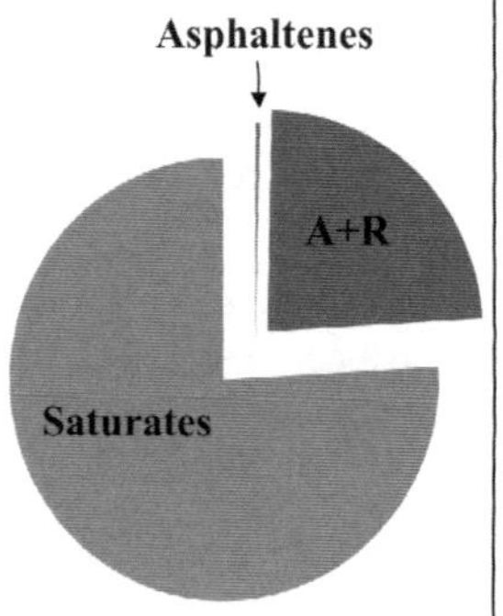

Current Characterization Procedures

- Katz and Firoozabadi (1978)
  - n-paraffin's boiling temperature was used for separating the carbon number fraction
  - Does not consider aromatic compounds
- Whitson (1983)
  - Average boiling point correlations (Riazi-Daubert 1980)
  - Correlations not applicable for heavier compounds (Tarek 1989)
- Leelavanichkul (2004)
  - Paraffins-naphthenes-aromatics based on refractive index (Riazi 1996)
  - The maximum refractive index does not reach the expected 1.7 value that has been estimated for asphaltenes

17

RICE

SAFT Characterization

Developed a standardized characterization procedure based on:

SARA analysis

Flashed Liquid and Gas compositions

Liquid Density

Bubble Pressure

Asphaltene Onset Pressure

Gas-to-oil Ratio

Methodology:

- Requires three parameters per compound similar to a cubic EoS
- Temperature independent binary interaction parameters for all compounds are small and well established

Panuganti SR et al. "PC-SAFT Characterization of Crude Oils and Modeling of Asphaltene Phase Behavior" Fuel – In Press

18

RICE

Data Issues

- Lab PVT report
  - Ask how experiments were run and consistency checks used (e.g., Vargas, et al "One-third Rule")
  - Ask for raw-unprocessed data
  - Submit duplicates to check reproducibility

- Was your sample clean?
  - Samples contaminated with oil based mud or inhibitors are different from clean oil

- Check your crude oil characterization versus other laboratory data

19

RICE

SARA Issues

- SARA of crude C reported by thin-layer chromatography with flame ionization detection (TLC-FID)

SARA Analysis	Saturates	Aromatics	Resins	Aspaltenes
Percentage (wt/wt)	49.5	40.2	7.2	3.1

- SARA of crude C reported by high-pressure liquid chromatography (HPLC)

SARA Analysis	Saturates	Aromatics	Resins	Aspaltenes
Percentage (wt/wt)	73.42	19.32	7.05	0.17

- Crude C – Light oil (API ~ 40)
- Models fit what they are fitted to – inaccurate SARA will produce inaccurate predictions of phase behavior

Fan and Buckley, Rapid and Accurate SARA Analysis of Medium Gravity Crude Oils. *Energy & Fuels*, **2002**, 16, 1571-1575
Fan, Wang, Buckley, Evaluating Crude Oils by SARA Analysis. SPE 75228
Francisco Vargas, Private Communication (2011).

20

RICE

Pseudo-component Scheme

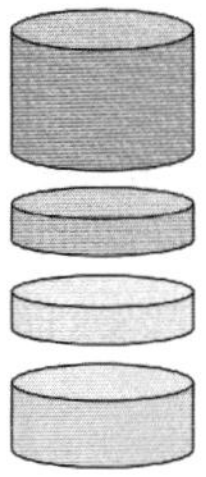

Methane (~50 mol%)	MW = 16.0	Pure components: Exp. Vapor pressure & liquid density
N_2 (~9.9 mol%)	MW = 28.0	
CO_2 (~0.1 mol%)	MW = 44.0	
Light alkanes (~40 mol%)	MW_n= 46.6 → n-alkane correlation	

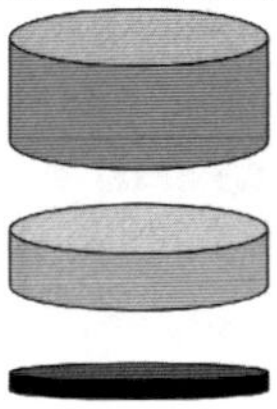

Saturates (~72.3 wt%)	MW_n= 255
Aromatics + Resins (~25.2 wt%)	MW_n= 222
Asphaltene (~2.5 wt%)	MW_n= 1700

RICE

21

Introduction to the PC-SAFT Eqn. of State

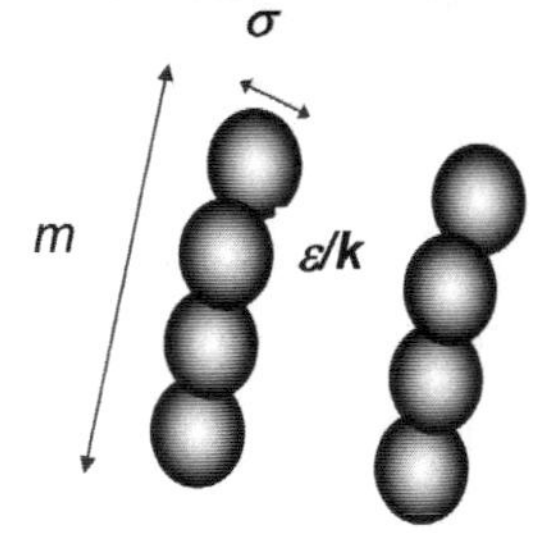

$$\frac{A^{res}}{RT} = \frac{A^{seg}}{RT} + \frac{\Delta A^{chain}}{RT} + \frac{\Delta A^{assoc}}{RT}$$

Chapman, Jackson, and Gubbins, Mol. Phys. 65, 1057 (1988)
Chapman, Jackson, Gubbins, and Radosz, I&EC Res, (1990)
Gross & Sadowski, Ind. Eng. Chem. Res., 40, 1244-1260 (2001)
Gonzalez, et al. Energy & Fuels 19, 1230-1234 (2005)

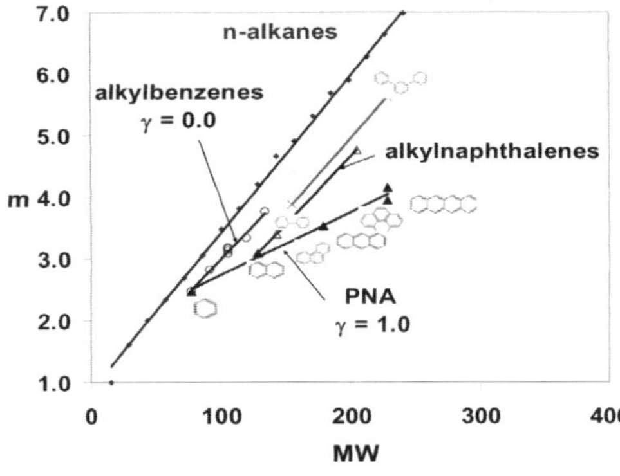

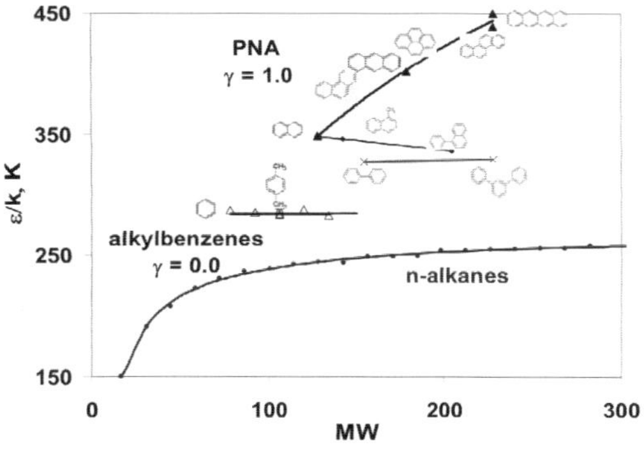

SAFT EoS available in InfoChem UK, VLXE and Calsep
SAFT parameters are well correlated to molecular weight

RICE

PC-SAFT Differential Liberation Predictions

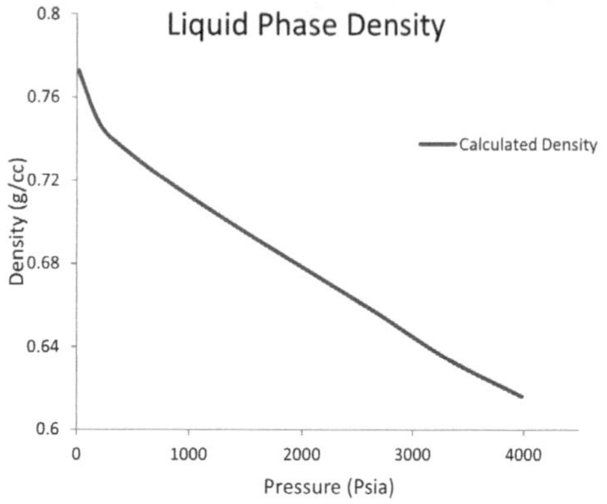

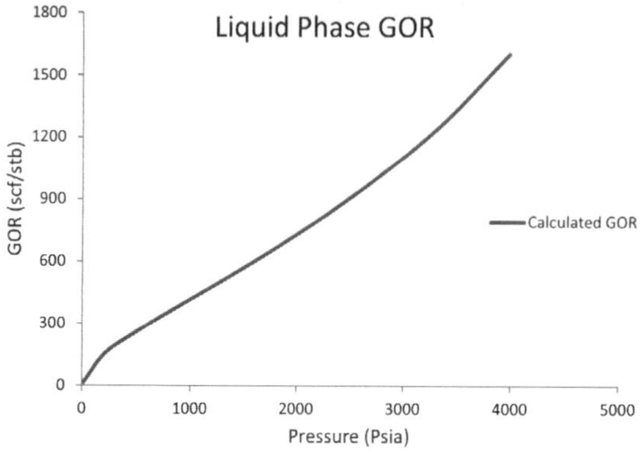

The continuous lines are the PC-SAFT predictions of differential liberation experiment at 250 F

Without volume translation

23

PC-SAFT Differential Liberation Predictions

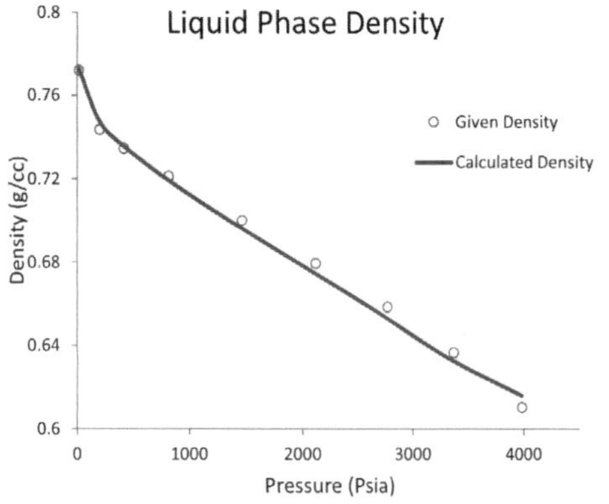

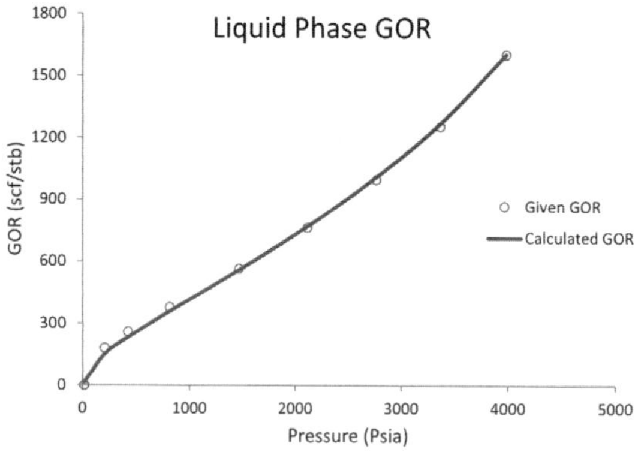

The continuous lines are the PC-SAFT predictions of differential liberation experiment at 250 F

No volume translation required

24

Comparison of PC-SAFT and Cubic EoS

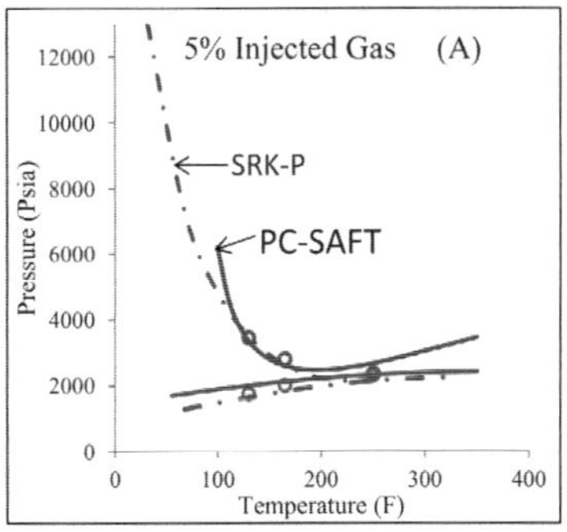

(Crude A)

Characterized using PC-SAFT and SRK-P EoS

- Will PC-SAFT work better than Cubic EOS?
- Will a specific set of PC-SAFT parameters be sufficient to capture the phase behavior of the system at a different condition?

Predictions of PC-SAFT and Cubic EoS

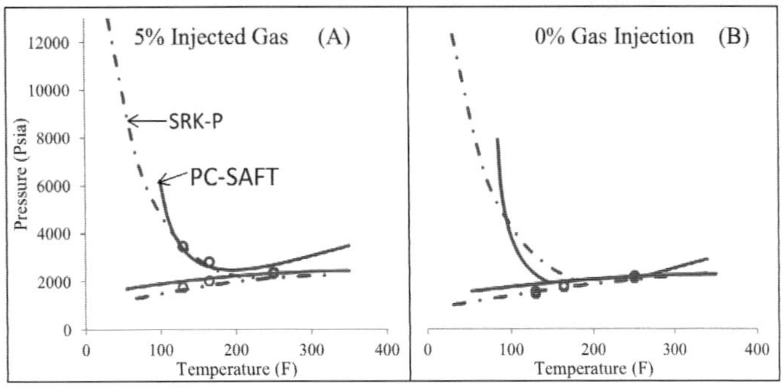

Crude A

Better performance of PC-SAFT is visible

- Will PC-SAFT with proposed characterization procedure be able to predict phase behavior for higher amounts of gas injected?

PC-SAFT vs Cubic EoS Prediction on Effect of Gas Injection

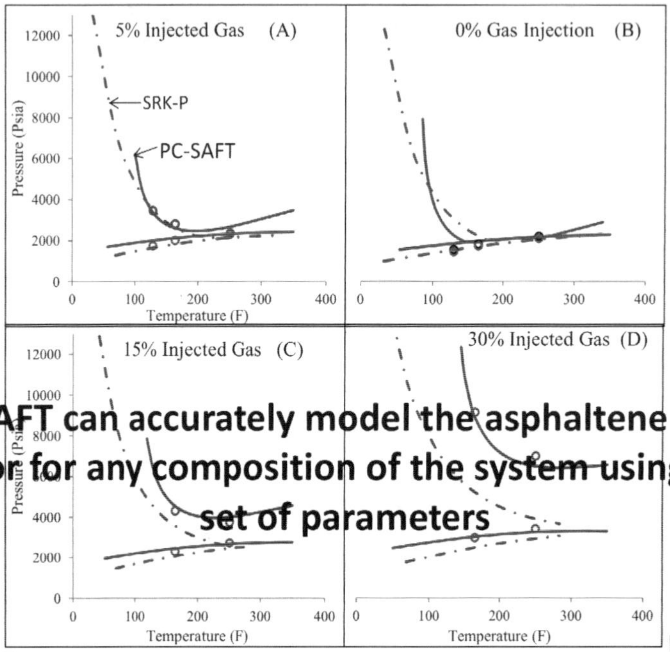

Crude A

PC-SAFT can accurately model the asphaltene phase behavior for any composition of the system using a single set of parameters

RICE

27

Oil Based Mud Contamination

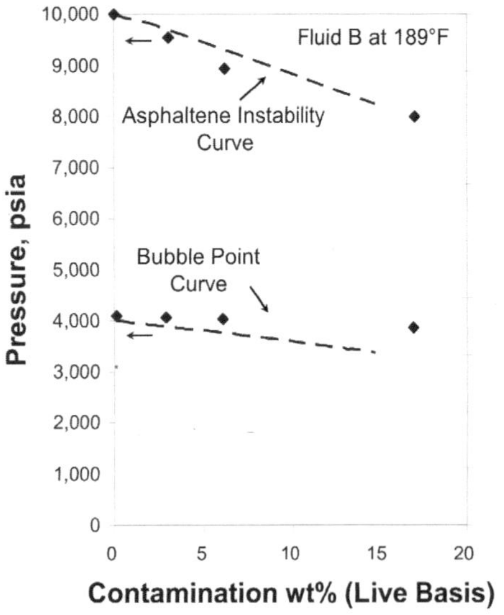

Doris Gonzalez - Schlumberger

28

Pressure Depletion Predicted by SAFT

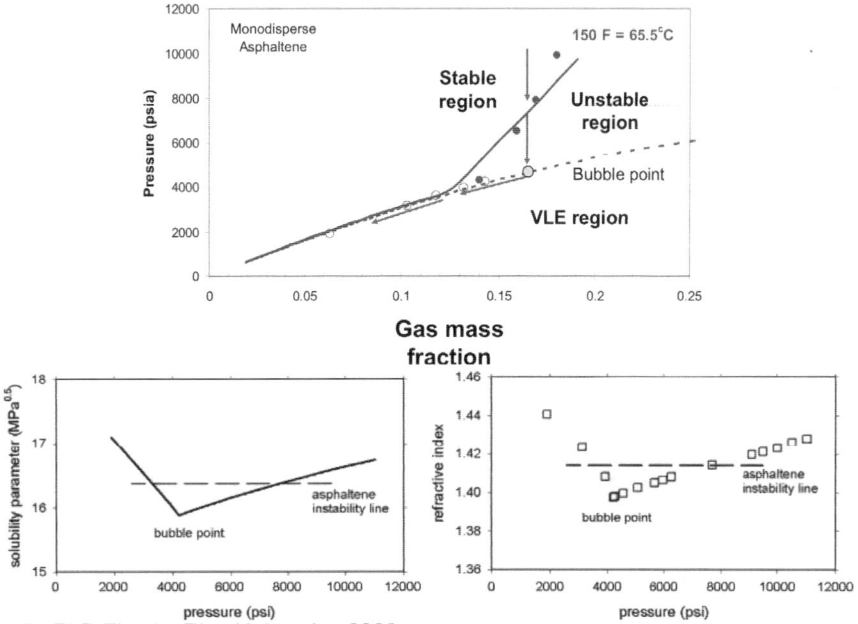

Ting, D. PhD Thesis. Rice University; 2003.

Reservoir Connectivity

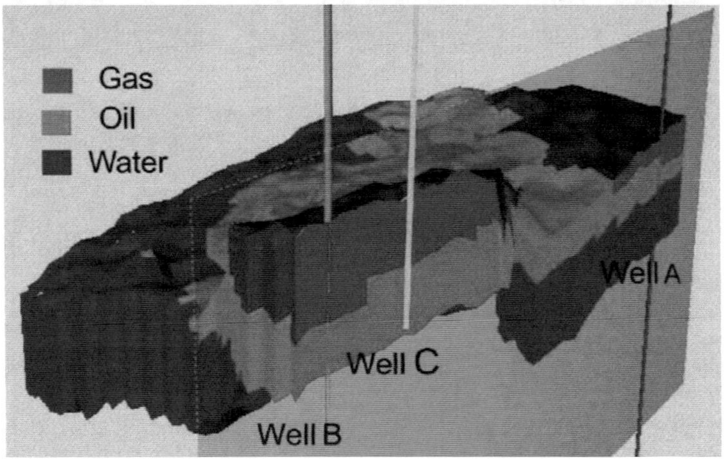

- Understanding reservoir connectivity helps in effective sweep of oil for a given number of wells
- Pressure communication can be used only to understand compartmentalization

Zao J Y, et al., Journal of Chemical & Engineering Data 2011. 56(4):1047-1058

RICE

30

Asphaltene Grading

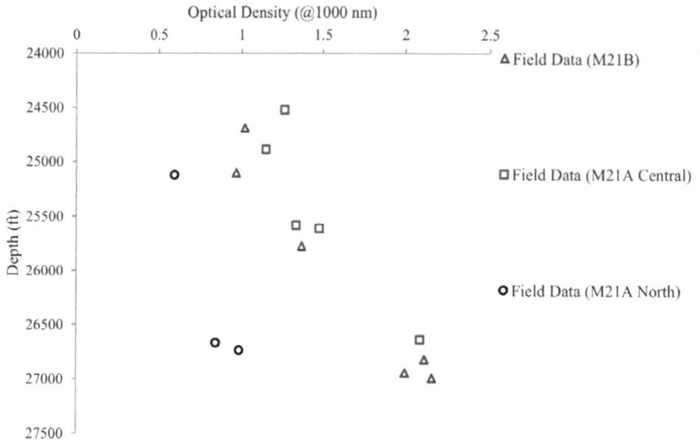

Tahiti field, Offshore in Gulf of Mexico

Heavy oil, isothermal reservoir at equilibrium

Optical density measured using infra red wavelength during down-hole fluid analysis

31

Freed DE et al. Energy and Fuels. 24:3942-3949

Predicting Asphaltene Compositional Grading

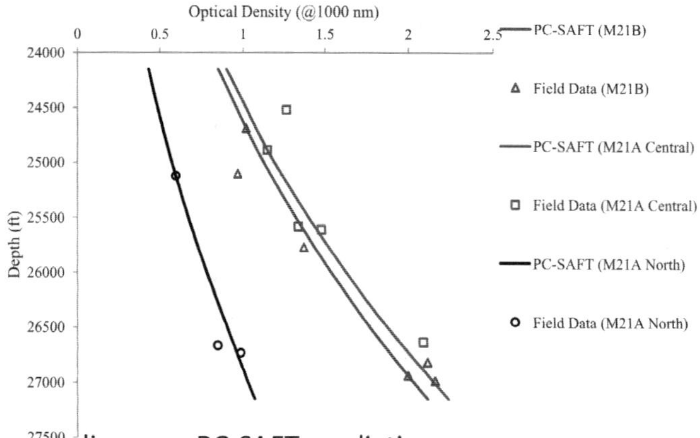

Tahiti field

- All continuous lines are PC-SAFT predictions
- All zones belong to the same reservoir as the gradient slopes are nearly the same
- The curves do not overlap implying each zone belongs to different compartment
- Partial molar volume of asphaltene = 1934 cm^3/mol. It corresponds to a spherical particle size of 1.83 nm

32

Freed DE et al. Energy and Fuels. 24:3942-3949

Predicting Asphaltene Compositional Grading

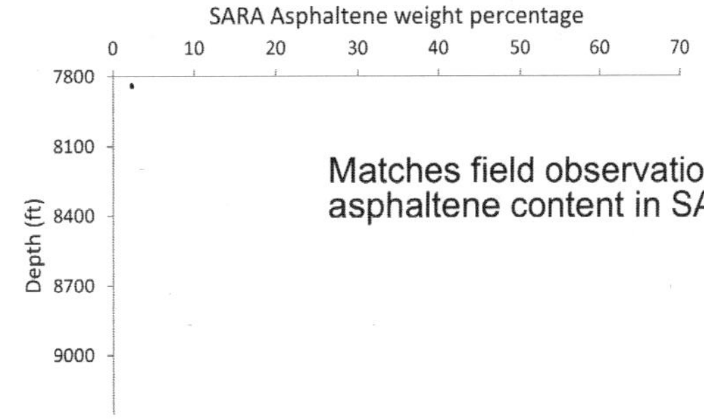

Matches field observations and tar-mat's asphaltene content in SARA

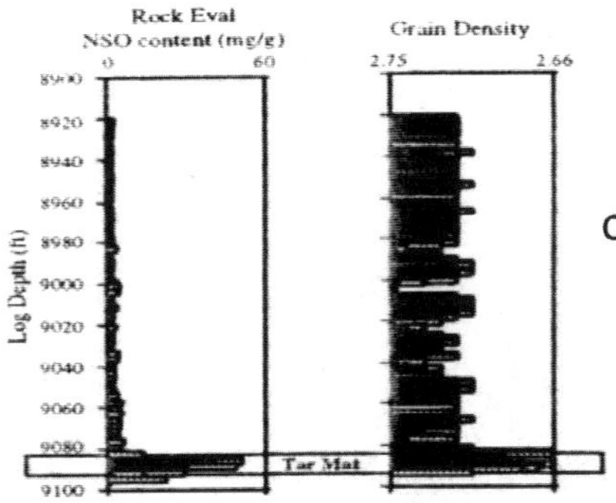

S field, onshore, light oil reservoir

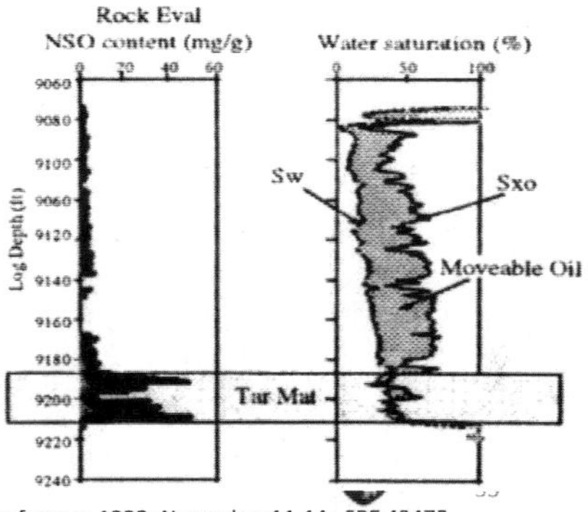

33

Carpentier B et al. Abu Dhabi International Petroleum Exhibition and Conference 1998; November 11-14; SPE 49472

Summary

- Understand model approximations and limitations
- Choose pseudo-components, but retain important components
- Check laboratory data and validate model
- Use model in a predictive mode (e.g. oil based mud contamination and tar mat formation)

- SAFT2011 Workshop in Pau, France, Oct. 24-25, 2011 – Sponsored by Total

34

RICE

Acknowledgments

GLOBAL DEEPWATER TECHNOLOGY DEVELOPMENT · DEEPSTAR PROJECT

RESEARCH & DEVELOPMENT
OIL SUB-COMMITTEE

Schlumberger

New Mexico Tech

Chevron

THE PETROLEUM INSTITUTE

From Molecules to Systems
Rice Chemical and Biomolecular Engineering

VLXE
POLYMER & COMPLEX FLUIDS

INFOCHEM

RICE

Observation about Molar Refractivity

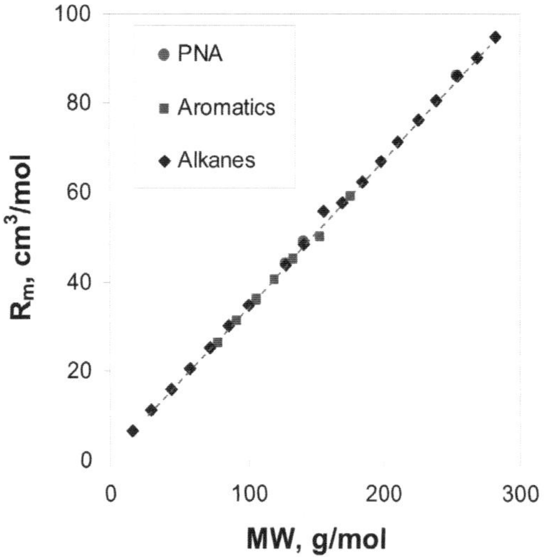

$$R_m = \left(\frac{n^2-1}{n^2+2}\right)\frac{1}{\rho}MW$$

$$\left(\frac{n^2-1}{n^2+2}\right)\frac{1}{\rho} \sim \text{constant}$$

$$\boxed{\left(\frac{n^2-1}{n^2+2}\right)\frac{1}{\rho} \sim \frac{1}{3}}$$

Vargas, F; Gonzalez, D. et al. *Energy & Fuels*. 23, 1147-1154. (2009)
Data from: CRC Handbook of Chemistry and Physics.

36

1/3 Rule for Crude Oils

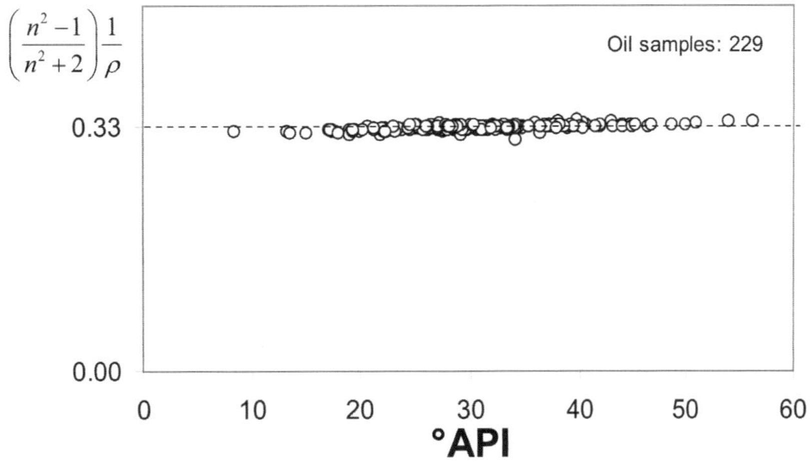

Vargas, F; Gonzalez, D. et al. *Energy & Fuels.* 23, 1147-1154. (2009)
Exp. data from: Wang, J.X.. Personal Communication. (2008).

37

Hydrocarbons/H_2S

RICE UNIVERSITY

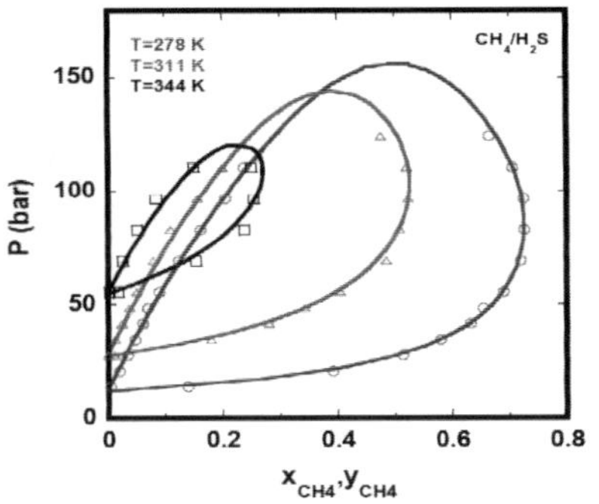

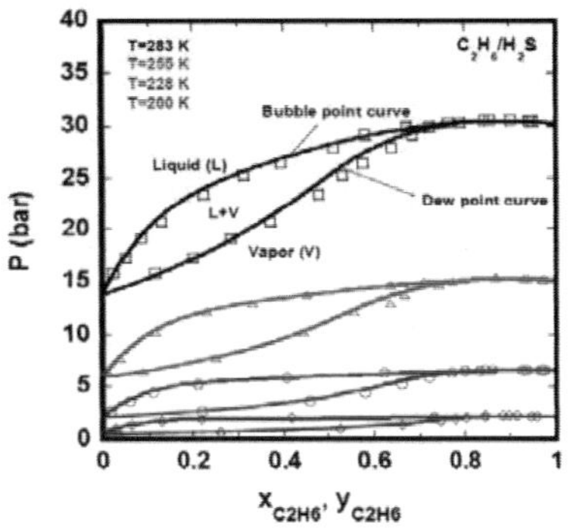

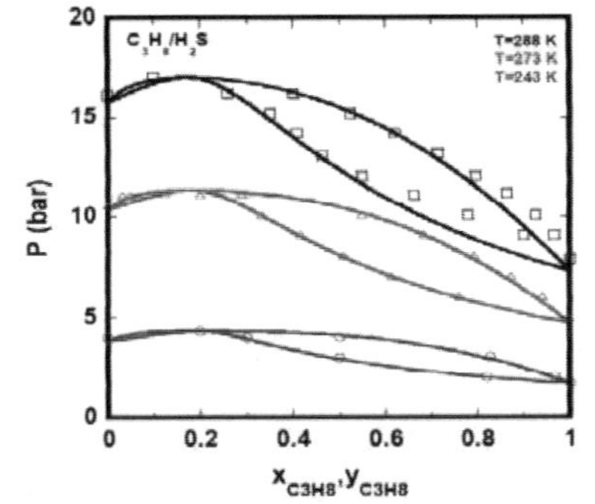

- **PC-SAFT results agree well with experimental data**
- **Model predicts the location of the azeotrope well for C_2H_6 and C_3H_8 at all temperatures**

38

Exp. Data from: Kohn, J.P., and F. Kurata, *AIChE J.* **4**, 211 (1958). Kalra et al. *J. Chem. Eng. Data.* **22**, 85 (1977). Carroll et al. *Fluid Phase Equilibria*, **81**, 187 (1992)

Different PVT Experiments

- Constant Composition Expansion

- For bubble pressure

- Constant Volume Depletion

- Time dependent PVT properties of condensates and volatile oils

- Constant Mass Expansion

- Saturation pressure at reservoir temperature and GOR during lifetime of reservoir

- Differential Liberation

- Volumetric changes in well stream

- Separation Test

- First idea about relative volumetric amounts of gas and oil

- Swelling Test

- How reservoir fluid reacts to gas injection

- Viscosity Test

- To measure oil viscosity at constant temperature and decreasing pressure

- Slim tube experiment

- Minimum miscibility pressure

- Multiple contact

- Understand phase equilibria near a well with gas injection

RICE

39

PC-SAFT Differential Liberation Predictions

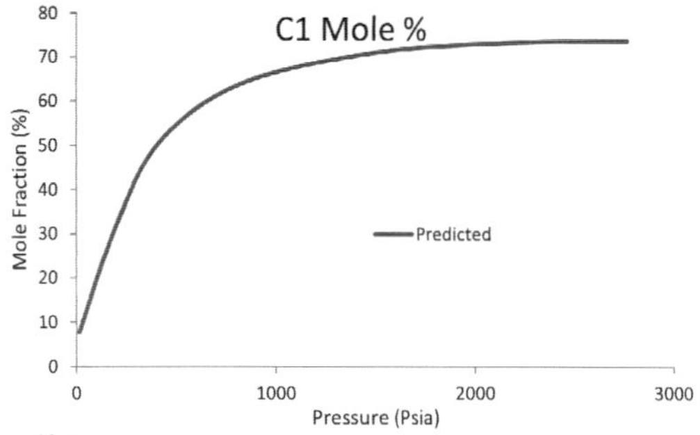

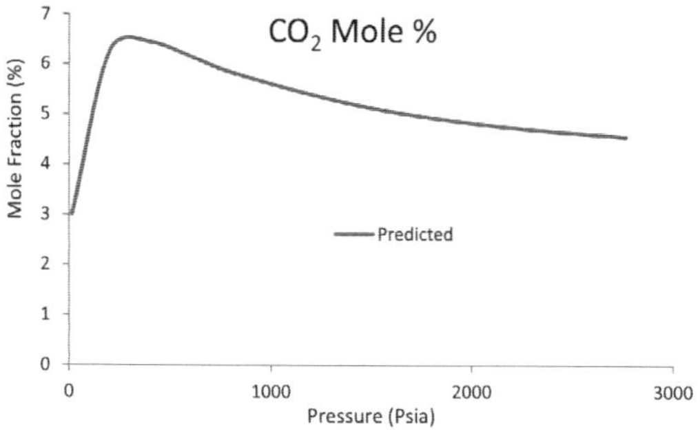

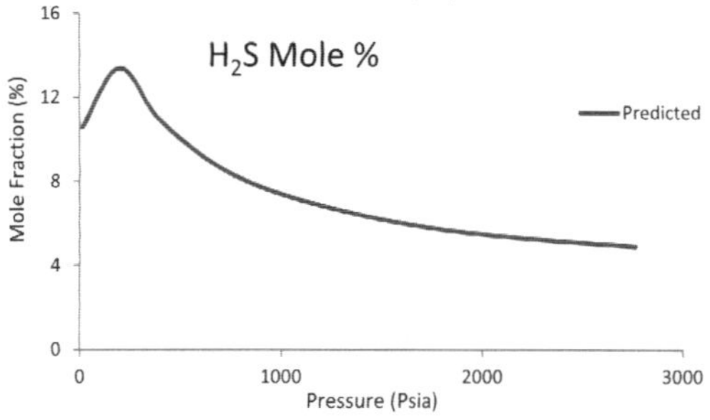

The continuous lines are the PC-SAFT prediction of liberated gas composition during differential liberation experiment at 250 F

RICE

40

PC-SAFT Differential Liberation Predictions

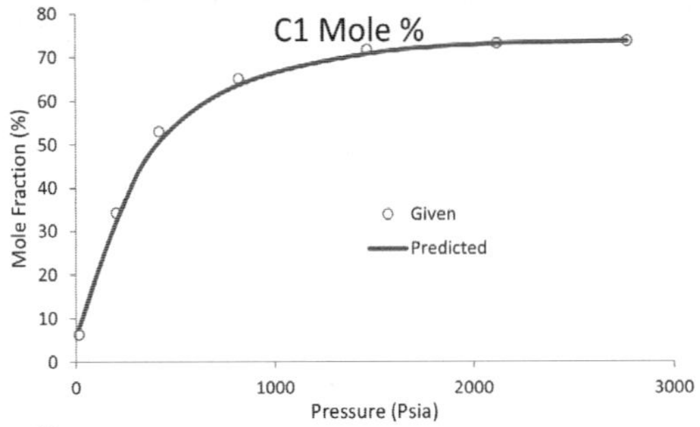

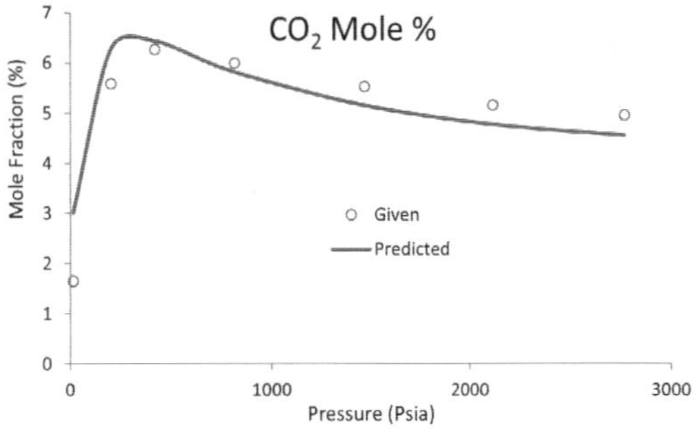

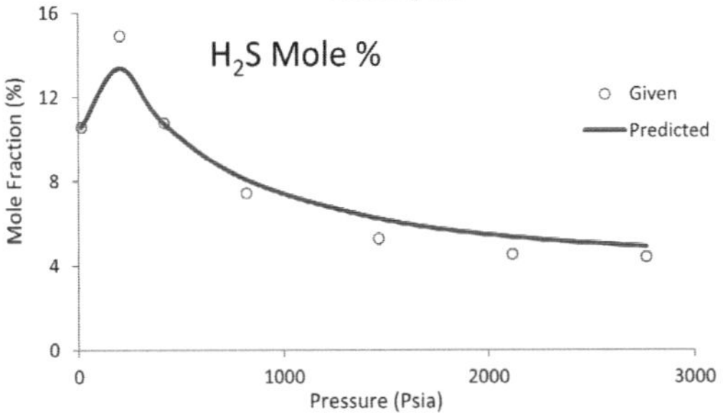

The continuous lines are the PC-SAFT prediction of liberated gas composition during differential liberation experiment at 250 F

RICE

41

Water Content Comparison Data vs Models

Methane + Water

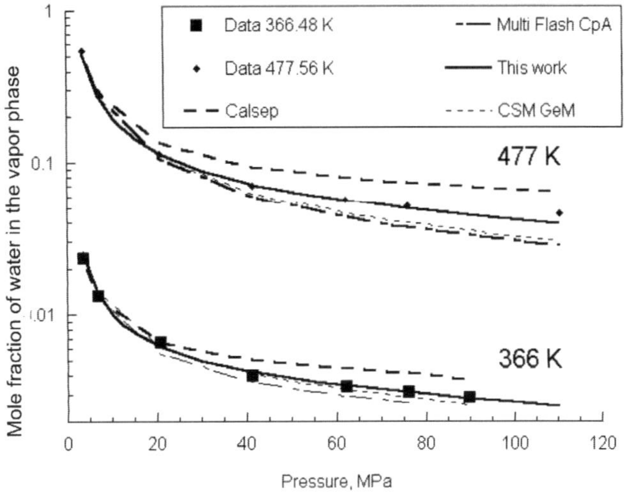

CSM GeM version 1.0 March 2002 – Colorado School of Mines
Multi Flash CpA – Multiflash Version 3.5 (February 2006) Infochem
Calsep – PVTsim Version 15

42

Yarrison, Cox, and Chapman, Ind. Eng. Chem. Res., 45, 6770-6777 (2006); GPA RR-200.

Multiphase Flow and Flow Assurance

Wicks, M., Shell

Flow Assurance is the use of fluid and solid mechanics, heat transfer, mass transfer, fluid properties, chemical additives, and phase behavior applied to production and transportation of oil, gas, and produced water, and to associated problems caused by wax, hydrates, asphaltenes, foam, corrosion, erosion, and sand. Practitioners use single- and multi-phase models for flow in one dimension and computational fluid mechanics models for flow in two or three dimensions, to help design the flow systems, measure and analyze their performance. A survey is given of the physical problems encountered and limitations of the tools used.

AIChE/SPE Joint Flow Assurance Workshop

State of the State of the Art of Multiphase Flow Prediction

Cem Sarica
The University of Tulsa

September 26-28, 2011

Outline

- Relevance
- Historical Review
- State of the Art
- Current Efforts
- Future

Relevance

- Steady State Multiphase Prediction Correlations and Models are Widely Used in the Design and Operation of Many Wells, Flow Lines and Facilities

Fluid Flow Projects AIChE/SPE Flow Assurance Workshop, Sept. 26-28, 2011

Relevance ...

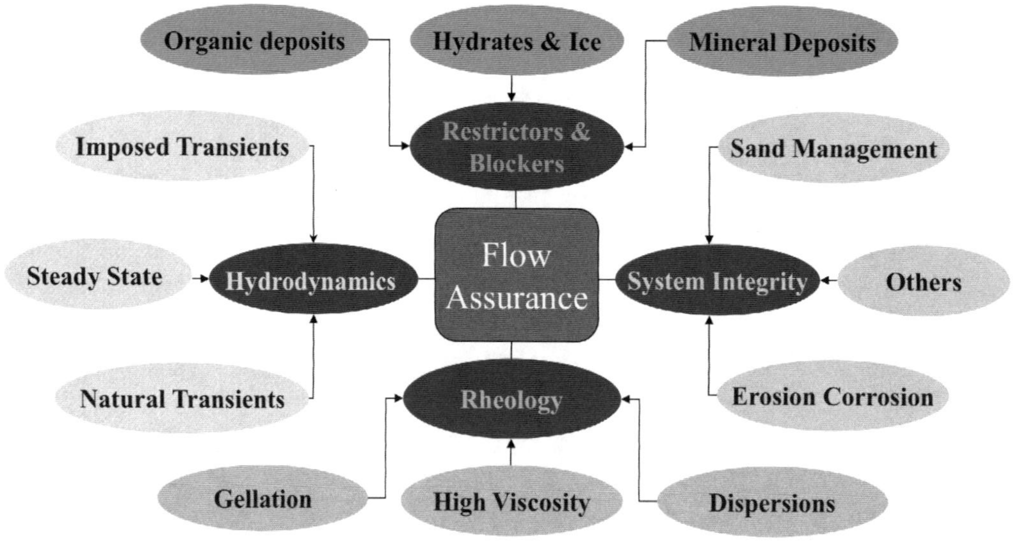

Fluid Flow Projects AIChE/SPE Flow Assurance Workshop, Sept. 26-28, 2011

Historical Review

- Correlation Approach
  - Homogeneous Flow
  - Pseudo Homogeneous Flow
  - Flow Pattern Dependent Flow
- Mechanistic Approach

Pseudo Homogeneous Approach

- Correlations Developed in (60's and 70's)
  - No Flow Pattern Consideration
  - Pressure Gradient and Holdup Prediction through Correlations

$$\frac{dp}{dL} = \frac{f_m \rho_m v_m^2}{2dg_c} + \rho_s \frac{g}{g_c} \sin\theta$$

$$\rho_s = H_L \rho_L + (1 - H_L)\rho_g$$

Pseudo Homogeneous Approach ...

- Examples
  - Hagedorn & Brown, Gray and Dukler/Eaton/Flanigan
- Hagedorn & Brown and Gray
  - Widely Used With and Without Various Modifications
  - Comparative Studies Show That They can be as Good as Flow Pattern Dependent Correlations

Flow Pattern Dependent Approach

- Correlations Mostly Developed in 70's
  - Flow Pattern and Flow Pattern Dependent Slip or Holdup, Friction Factor Correlations

$$\frac{dp}{dL} = \frac{f_m \rho_m v_m^2}{2dg_c} + \rho_s \frac{g}{g_c} \sin\theta$$

$$\rho_s = H_L \rho_L + (1 - H_L)\rho_g$$

Flow Patterns Dependent Approach ...

- Examples
  - Beggs and Brill, Duns and Ros, and Others
  - Literature is Awash with Correlations
- Correlations are Widely Used With and Without Various Modifications

Mechanistic Approach

- Emerged in 80's
- One Dimensional Flow Hydrodynamics Considered Based on Flow Structures
  - Models to Predict Flow Structures (Patterns)
  - Flow Pattern Dependent Holdup and Pressure Gradient Predictions Models
  - Usually Simplifications and Convenient Hypotheses are Generously Employed
  - Models are Closed with Empirical Closure Relationships

Mechanistic Approach ...

- Examples
  - Hasan and Kabir, Ansari, Xiao, Kaya, Gomez, and More
- Generally Better Performance Than Correlations
- Incorporated in Most Commercial Software and Widely Used

Fluid Flow Projects

AIChE/SPE Flow Assurance Workshop, Sept. 26-28, 2011

Mechanistic Approach ...

- Shortcomings
  - No Single Model Applicable for All Inclination Angles
  - Separate Flow Pattern Prediction and Pressure Gradient and Holdup Models
    - May Lead to Abrupt Changes and Discontinuities Near Flow Pattern Transitions

Fluid Flow Projects

AIChE/SPE Flow Assurance Workshop, Sept. 26-28, 2011

State of the Art

- Unified Models Emerged in 00's
  - Mechanistic Models Applicable for Entire Range of Inclination Angles
  - Idea from Barnea (1987) – Unified Flow Pattern Prediction
- Current Examples
  - TUFFP Unified Model
  - OLGA Steady State (More in Next Presentation)
- Incorporated in Widely Used Software

Unified Models

- Typically Honor Basic Mass and Momentum Conservations
- Needs Closure Relationships
- TUFFP Unified Model
  - Single Model for All Flow Patterns
    - Smoother Transition Between Flow Patterns
  - Based on Slug Flow Hydrodynamics

Unified Models ...

- Slug Flow Combines all Flow Patterns
  - Film Zone ~ Stratified or Annular Flow
  - Slug Body ~ Dispersed or Bubbly Flow

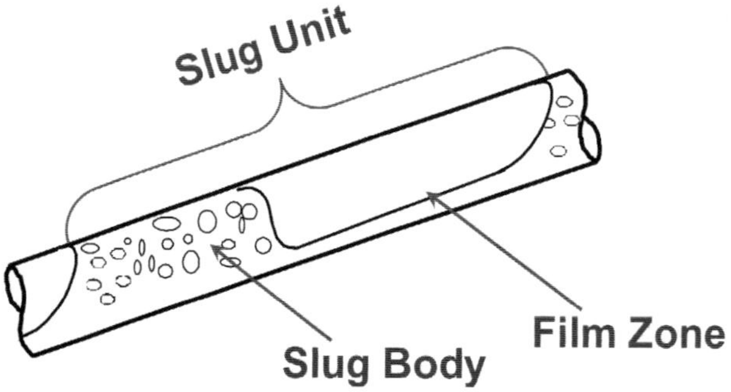

Fluid Flow Projects

AIChE/SPE Flow Assurance Workshop, Sept. 26-28, 2011

Unified Models ...

- Slug Flow Surrounded by Other Flows

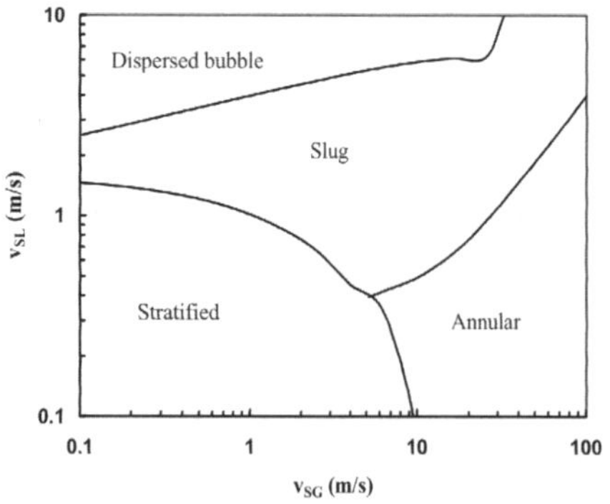

Flow Pattern Map (Air-Mineral Oil, 2-in. ID, -5° Inclination)

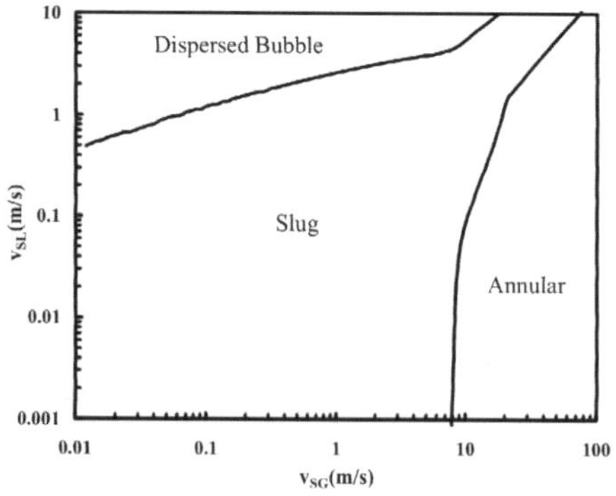

Flow Pattern Map (Air-Water, p = 3 atm, T = 20 °C, d = 25.4 mm, ◊ = 90°)

Fluid Flow Projects

AIChE/SPE Flow Assurance Workshop, Sept. 26-28, 2011

Unified Models …

- Closure Relationships
  - Examples: *Wall and Interfacial Friction Factors, Translational Velocity, Liquid Slug Holdup, Bubble Slip Velocity, Wetted Wall Fraction, Entrainment Fraction, and Slug Length and Frequency*
  - Empirical in Nature
  - Inherently Depends on Experimental Data
    - Mostly Developed for Typical Conditions
  - Can Still Fail When Used Out of Developmental Range: Similar to General Multiphase Correlations

Current Efforts

- Improvement of Existing Models
  - Development of Better Closure Relationships
  - Rigorous Incorporation of Energy Conservation
    - Flow Pattern Dependency
  - Better Modeling of Three-phase Flow
  - Specific Flow Conditions (High Viscosity Oil, Impact of MEG, etc.)
  - Up-scaling
    - Size and Pressure

Closure Relationships

- **Slug Length**
  - Liquid Viscosity Effect (Al-safran *et al.*, 2011)

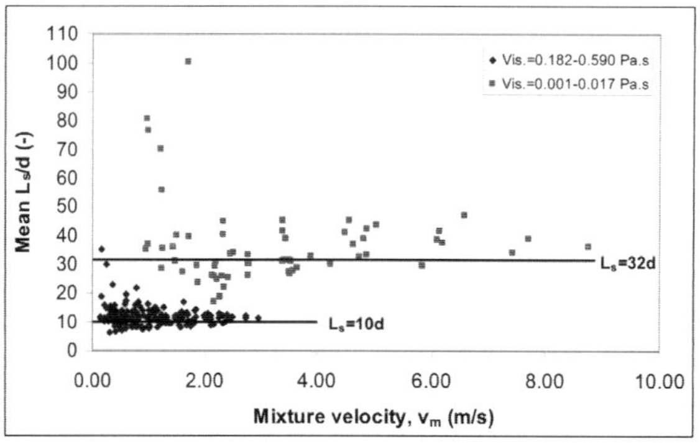

Fluid Flow Projects

AIChE/SPE Flow Assurance Workshop, Sept. 26-28, 2011

Closure Relationships ...

- **Slug Length Distribution**
  - Viscosity Effect (Al-safran *et al.*, 2011)

v_{SL}=0.3 m/s
v_{Sg}=1.5 m/s

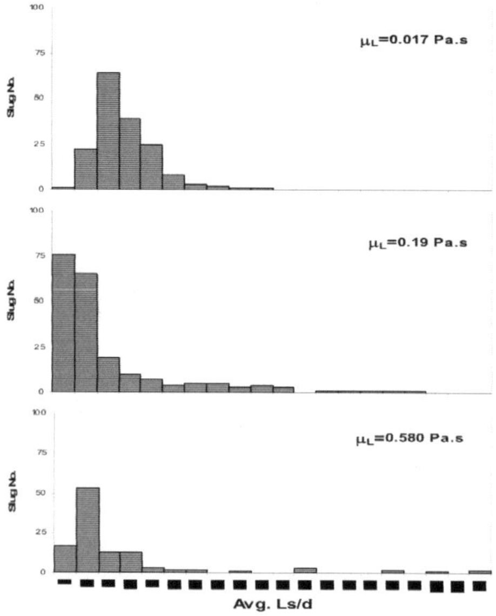

Fluid Flow Projects

AIChE/SPE Flow Assurance Workshop, Sept. 26-28, 2011

Closure Relationships ...

- Drift Velocity (Gokcal *et al.*, 2009)

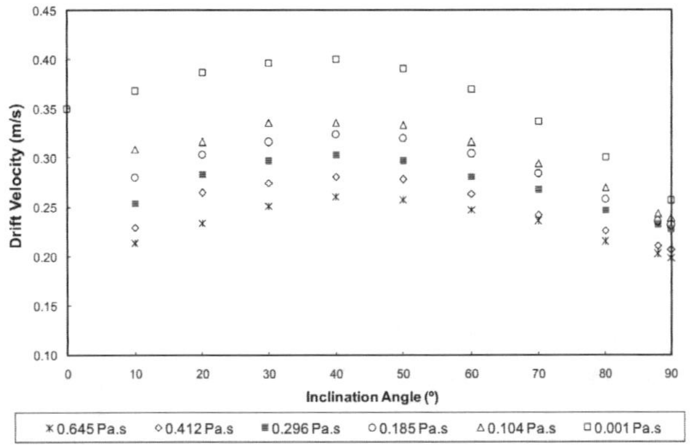

Fluid Flow Projects

AIChE/SPE Flow Assurance Workshop, Sept. 26-28, 2011

Closure Relationships ...

- Slug Liquid Holdup (μ_o=0.587 Pa·s), Kora (2010)

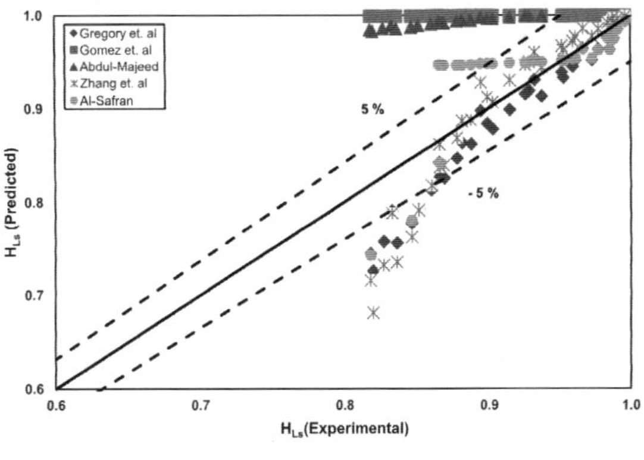

Fluid Flow Projects

AIChE/SPE Flow Assurance Workshop, Sept. 26-28, 2011

Size Up-scaling

- **Flow Behavior**
  - Bubbly Flow Does Not Exist
    - Vertical Flow d<2 in. (Taitel *et al.*, 1980)
  - Slug Flow May Not Exist
    - Vertical Flow d>6 in. (Omebere-Iyari & Azzopardi, 2007)
    - Depends on Upstream Conditions

Size Up-scaling

- **Flow Behavior**
  - Can Observe New Flow Patterns
    - 6 in. ID Low Liquid Loading Flow
    - Stratified with Channel Water and Water in Oil Dispersion

Side View

Bottom View

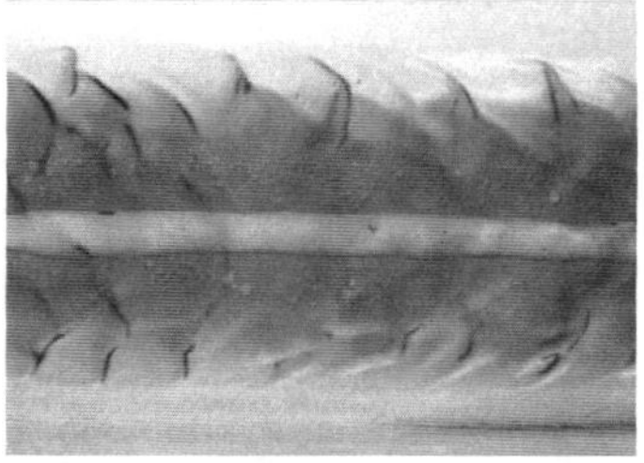

v_{SG} = 10 m/s, LL = 300 $m^3/MMsm^3$, WC = 0.2

Pressure Up-scaling

- Flow Behavior Changes
  - High Pressure – High Gas Density – Lower In-situ Gas Velocities
  - Annular Flow
    - Existence of Droplet Field in Question

Fluid Flow Projects

AIChE/SPE Flow Assurance Workshop, Sept. 26-28, 2011

Future

- Near Future
  - New and Improved Closure Relationships through Targeted Experimentation
  - Incremental Improvement on Modeling
- Long Term
  - Comprehensive Models with Ability to Simulate Anything and Everything and Do it Fast and Accurately
    - Requires Dynamic Understanding of Interaction Between Chemistry of Fluids and Flow Dynamics
    - Allows Fast Forwarding to See Many Realizations of Production and Transport Scenarios

Fluid Flow Projects

AIChE/SPE Flow Assurance Workshop, Sept. 26-28, 2011

Concluding Remarks

- Steady State Prediction Tools are Mostly Reliable with Room to Improve
- Modular Model Structures are in Place
- Further Improvements Mostly Expected from Closure Relationship Developments
- Experimental Studies may still Reveal more to Change the Models beyond Closure Relationship Development

Fluid Flow Projects

AIChE/SPE Flow Assurance Workshop, Sept. 26-28, 2011

SPT

OLGA – Past, Present, and Future

Dick Shea

AIChE/SPE Meeting, Houston, Sept 26, 2011

OLGA

be dynamic

OLGA and I (OLGA and me??)

- **Background – Process Engineer for Mobil R&D Corp**
- **Process Designer for Statfjord B and C**
- **During "C" studies (1978), asked to look at multiphase flowline from C to A**
- **Found out state-of-the-art stunk**
- **Checked available data vs predictions – pressure drops +/- 50%, holdups +/- 600%, slug length – predicted = 2 ft long, measured = 1000 ft long**

SPT

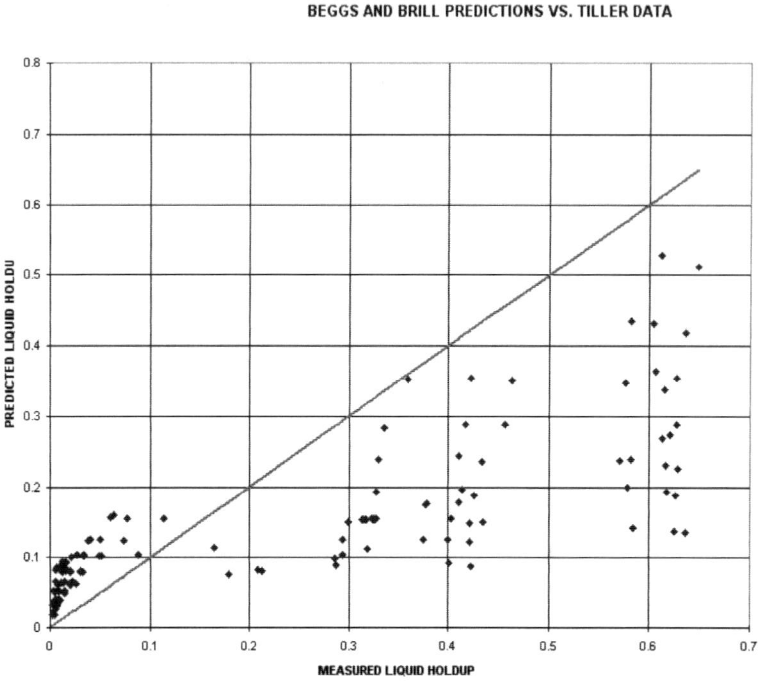

OLGA be dynamic SPT

Statfjord Report

- Invented my own methods – better than existing, but not near what we needed
- Checked pipeline (including doing dynamic analysis of slugging on Stat A topsides)
- Wrote report and sent to partners
- Discussed with Statfjord partners (including Exxon, Shell, Statoil, Norsk Hydro, Saga, Conoco)
- All agreed that there needed to be a lot of work done on flow assurance

OLGA

be dynamic

Mobil R&D Plan

- Dick named head of R&D project on multiphase flow
- For 15 years, gave same priority list to management and affiliates
- For 15 years, they agreed with list - #1 R&D priority in Mobil affiliates
- Main reasons – improved economics with subsea tiebacks, elimination of platforms, deep water installations, hilly terrain pipelines, etc.

SPT

OLGA

be dynamic

Mobil R&D Plan

- Get high quality lab data in large diameter, high pressure test loops
- Get good field data
- Compare data against best available methods
- Develop improved prediction methods for pressure drop, liquid holdup, and flow regimes
- Develop a transient multiphase flow simulator

SPT

OLGA

be dynamic

SPT

Mobil R&D Plan

- **Get high quality lab data in large diameter, high pressure test loops –** 8" Tiller Test Loop, 8 & 16" BHRA Test Loops
- **Get good field data –** Field Tests in North Sea and Nigeria, UK National Data Base
- **Compare data against best available methods –** Developed large multiphase flow data base, compared against all available methods
- **Develop improved prediction methods for pressure drop, liquid holdup, and flow regimes –** Developed proprietary models, sponsored development of OLGAS
- **Develop a transient multiphase flow simulator –** Sponsored OLGA

OLGA

be dynamic

SPT

OLGA Transient Multiphase Flow Program

- **Designed to investigate flow assurance issues such as:**
  - Line sizing
  - Slugging
  - Pigging
  - Startup and Shutdown
  - Rate Changes
  - Effects on Process Equipment
  - Development of Operating Procedures
- **Originally primarily pipeline design program, but now used almost as much for well simulations**

OLGA Original Development

- OLGA developed at IFE, based on earlier nuclear code Ramona
- Statoil sponsored original development of OLGA beginning in 1980
- OLGA Development merged with Tiller Test Loop work in 1984 as SINTEF/IFE Multiphase Flow Project
- Project ran from 1984 to 1993 with 5-10 sponsors
- Used data from Tiller test loop to initially tune OLGA

SPT

- SPT (then Scandpower) given marketing rights to OLGA in 1992
- Dick brought in to start SPT Office in Houston in 1993
- OLGA version at that time – limited features and very difficult to use
- Goals were to make OLGA:
  - Easier to use
  - More powerful
  - More accurate

OLGA

be dynamic

SPT

THE OLGA TEAM

The Scandpower personnel working with OLGA software are the following:

Jon Rasmussen has the technical responsibility for OLGA within Scandpower. Jon has a degree in Nuclear Engineering from RWTH, Aachen, Germany, 1962 and a long background with two-phase modelling. He took part in the first phase of the OLGA development at IFE. The commercialization of the code at Scandpower and the consulting work for clients has given him an extensive experience in OLGA applications. He was a member of the steering committee during the main OLGA development period and has also used the code quite extensively. His capacity is now used both for performing OLGA simulations for clients and for marketing and client support in North and Latin America.
Email: jr@scandpower.no

Zheng Gang Xu has a PhD in Petroleum Engineering from NTH in Trondheim 1988. Besides activities on OLGA code development and applications, Zheng has the responsibility for the PVT material data generation code PVTOL.
Email: xu@scandpower.no

Richard H. Shea is our principal OLGA consultant on the American Continent. Dick has worked for many years as multiphase expert for Mobil before he joined Scandpower in 1993. Dick has a MSc in Chemical Engineering from Illinois Institute of Technology in 1969.

Bjarte Håtveit is our OLGA system engineer and is also responsible for pre- and post-processor software development and support. Bjarte has a degree in Mechanical Engineering from NTH in 1992.
Email: bjh@scandpower.no

Kjell Ø. Petersen has participated in the development of the interactive user interface Interact-OLGA. Like most of the team members he is engaged in OLGA maintenance and support as well as applications. Kjell has a PhD in Fluid Mechanics from Ecole Centrale de Lyon, France 1992.
Email: kp@scandpower.no

Norbert Hoyer was earlier engaged in OLGA development at IFE. Since 1986 he has been working at Scandpower on general methods development in multiphase flow simulation. Norbert is presently putting the final touch on his PhD thesis in Fluid Mechanics.
Email: nh@scandpower.no

Lars Hovden is mainly engaged in OLGA maintenance and client support activities. Lars has a degree in Mechanical Engineering from NTH in 1990.
Email: lh@scandpower.no

Xin Chen Billdal is working with OLGA development and applications. Xin has a PhD in Fluid Mechanics from NTH in 1993.
Email: xin@scandpower.no

Odd Vesterhaug is coordinating the global marketing and sales of OLGA software as well as managing other customer related activities such as training courses, exhibitions, etc. Odd has a degree in Engineering Physics from NTH in 1963.
Email: ov@scandpower.no

Ramesh Ladwa is working as an associated consultant to Scandpower with OLGA marketing in UK and other oversea areas.

All OLGA activities are organized in the Section for Dynamics and Flow analyses under the supervision of **Lars Moberg.**
Email: lm@scandpower.no

SCANDPOWER

SCANDPOWER A/S
P.O. Box 3 / Gåsevikv. 2
N-2007 Kjeller, NORWAY
Tel.: + 47 63 81 49 20
Fax: + 47 63 81 68 22

SCANDPOWER Inc.
101 Lakeforest Blvd, Suite 340
Gaithersburg, MD 20877
USA
Tel.: + 1 301 330-1100

SCANDPOWER Inc.
11111 Wilcrest Green, Suite 250,
Houston, Texas 77042,
USA
Tel.: + 1 713 260-7078

SCANDPOWER Ltd.
11 Lambourne Close,
Thruxton, Nr Andover,
SP11 8LS, UK
Tel.: + 44 264 77 20 72

OLGA

be dynamic

SPT

OLGA '94-97

- **Many major improvements in OLGA**
  - **Keyword Input**
  - **3 Phase Flow**
  - **Slug tracking**
  - **Steady State Preprocessor**
  - **First GUI**
  - **Migrated from Workstations to PC**
  - **Linked OLGAS to many steady state programs**

OVIP

- **Started OLGA Verification and Improvement Project (OVIP) in 1996**
- **SPT and IFE**
- **Goals of OVIP**
  - Compare OLGA with lab and field data
  - If OLGA doesn't work well in a given area:
    - Take additional data
    - Improve the models
  - Develop tools for sponsors
    - Showing program accuracy
    - Documenting comparisons
- **OVIP still going strong with 10 sponsors**

OVIP

- **Excellent contributions from sponsors**
- **Extensive Data Base**
  - >15000 lab data points
  - Data on hundreds of wells and pipelines
- **Data from IFE flowloops in areas where OLGA was weak**
- **Considerable improvement in model predictions for many different systems**
- **Model improvements end up in commercial code**

OLGA Improved Power and Usability

- **Added many features**
  - Compositional tracking
  - Network capabilities
  - FEMTherm
  - CSM HyK (Hydrate Kinetics Module)
  - Wax
  - Corrosion
  - Wells
  - Single component module
  - New GUI's

OLGA Improved Accuracy

- **OVIP**
- **Horizon I (Improved Stratified Flow Model (HD))**
- **Horizon II (Improvements to Slug Flow Model)**
- **Sponsorship and Participation in TUFFP and TMF**

OLGA HD

- Major improvement in modeling of stratified flow
- Uses velocity profiles for each phase to estimate interfacial shear stresses
- Much more fundamentally based than previous model
- Improved predictions for OVIP field and lab data

The HD Model applies a 2D cross sectional velocity distribution to obtain the frictions

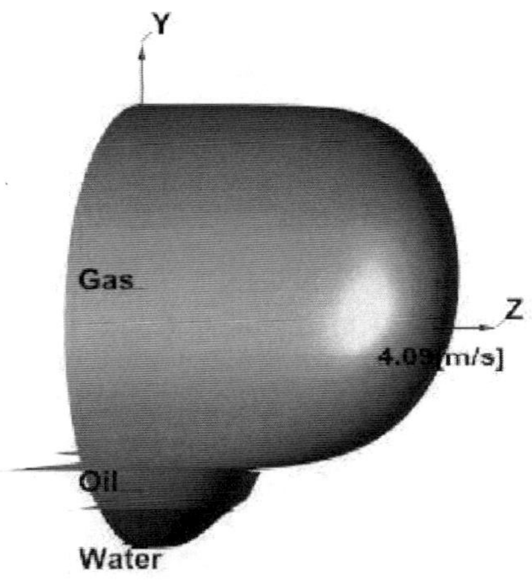

OLGA

be dynamic

OLGA-7

- **Released in June, 2011**
- **Includes:**
  - OLGA HD Stratified Flow Model
  - Plug In Modules
  - New GUI
  - Improved Multiphase Pump Module
  - Risk Management and Optimization Module (MEPO)
  - OLGA for Wells GUI

SPT

OLGA

be dynamic

OLGA Future Developments

- **Continue to improve code and make it easier to use**
- **Development efforts include:**
  - Implement new and improved models from Horizon II and OVIP
  - Improve execution speed
  - Continue to improve pump and valve models
  - Link to reservoir simulator (ROCX)
  - Improvements in Well Modeling (GLV, ESP)
  - Improved integration capabilities with other software
  - Improvements in Drilling Modeling
  - Add features as requested by users

SPT

LedaFlow Multidimensional Modeling of Two-Phase Slug Flow in Pipes

LedaFlow®

Alireza Ashrafian, Sjur Mo, Wouter Dijkhuizen, Stein Tore Johansen
SINTEF

TOTAL KONGSBERG SINTEF ConocoPhillips

LedaFlow®

Overview

- LedaFlow models
- Model concepts
- 1D slug modeling
- Leda Q3D model
- Leda Q3D verification and applications
- Surface chemistry and flow evolution
- Summary

LedaFlow models

1D models developed during the last 10 years

LedaFlow®

- 3-phase point model for fast steady state simulations
- 3-phase transient flow model
  - UnitCell based concept
  - Slug capturing for improved performance of slugging flows
  - Non-equilibrium thermal and compositional models
    - Phase (zone) specific enthalpies and temperatures'
    - Phase specific compositional model
  - Thermal models (soil, half buried pipes)
  - Pigging module
  - Single component tracking (PVT tables)
- Steady state / initialization
- Network solver
- Well inflow model and control valves/controllers

SINTEF

The 6th AIChE/SPE Joint Workshop on Challenges in Flow Assurance
September 26-28, 2011, Houston, TX USA

Technology for a better society 3

Animation: Dynamic evolution of gas fraction

LedaFlow®

- Oscillations in downcomer
- Development of hydrodynamic waves
- Partially filling of riser

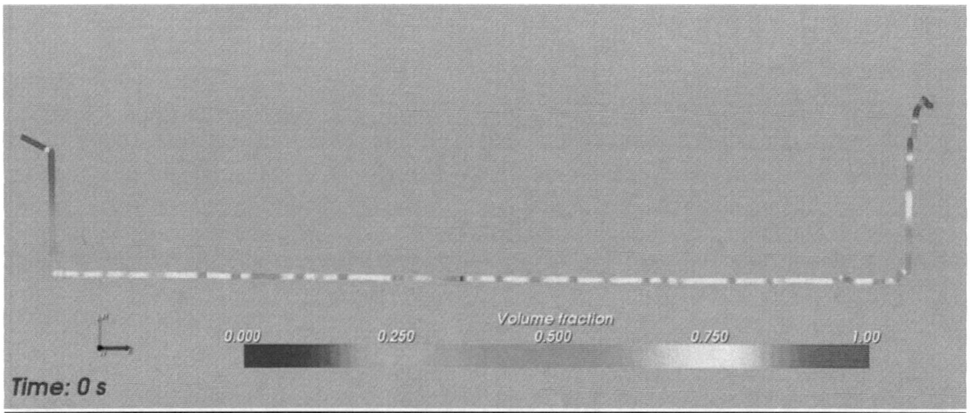

SINTEF

The 6th AIChE/SPE Joint Workshop on Challenges in Flow Assurance
September 26-28, 2011, Houston, TX USA

Technology for a better society

LedaFlow®

Case with slug capturing and higher order schemes

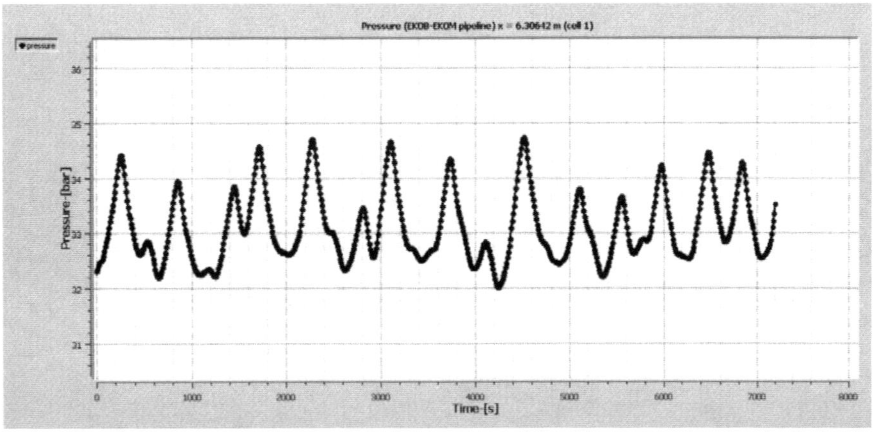

P amplitude = 2.2bar
Period = 9min

Pressure amplitude close to observations

SINTEF | The 6th AIChE/SPE Joint Workshop on Challenges in Flow Assurance, September 26-28, 2011, Houston, TX USA | Technology for a better society

LedaFlow®

Field data: only temperature available

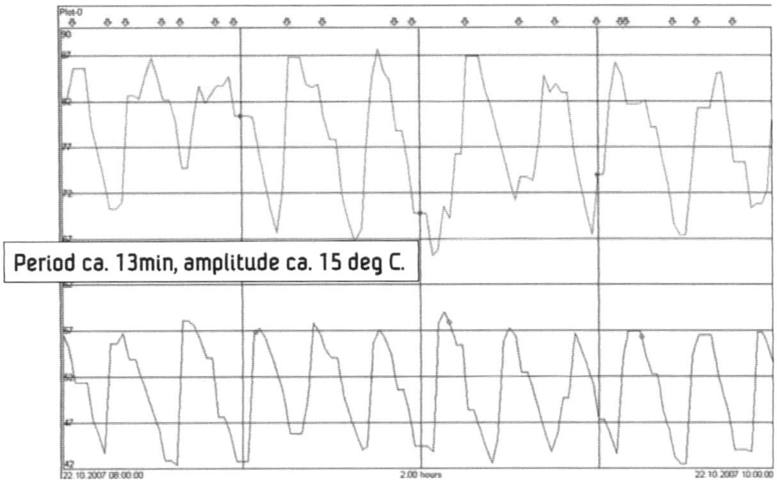

SINTEF | The 6th AIChE/SPE Joint Workshop on Challenges in Flow Assurance, September 26-28, 2011, Houston, TX USA | Technology for a better society

Temperature at seabed, before riser

LedaFlow®

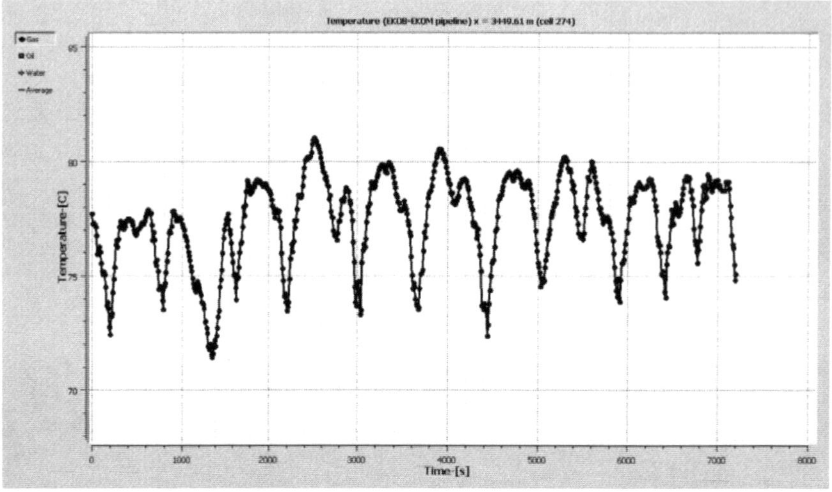

Temperature amplitude ca. 8 deg C vs. 15 deg C in field case.

LedaFlow Quasi 3D model

Verification and applications

LedaFlow®

What is the Q3D model ?

LedaFlow®

- Based on a full 3D and 3-phase model
- 9 fields of gas, oil and water (continuous fields, bubbles and droplets)
- 3-phase turbulence
- Dynamic evolution of droplet and bubble sizes
- Reduced to 2D by integration over transversal cross section (speed)

Grid (cell configuration)

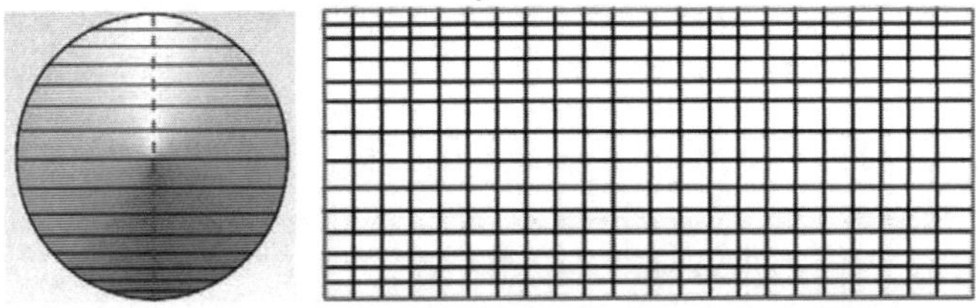

SINTEF

The 6th AIChE/SPE Joint Workshop on Challenges in Flow Assurance
September 26-28, 2011, Houston, TX USA

Technology for a better society 9

Overview – Q3D modeling

LedaFlow®

- New developments
  - Generic mass source/sink
  - Compressible flow and PVT tables
- Validation
- Applications
- Surface chemistry – a new dimension !

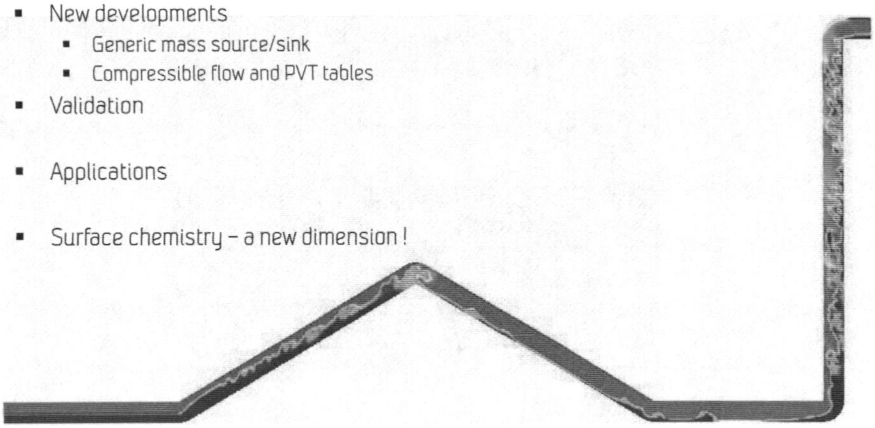

SINTEF

The 6th AIChE/SPE Joint Workshop on Challenges in Flow Assurance
September 26-28, 2011, Houston, TX USA

Technology for a better society

LedaFlow®

Q3D Point mass source

- Mass source is a generic tool.
- Example: Gas-lift:

Gas injection point

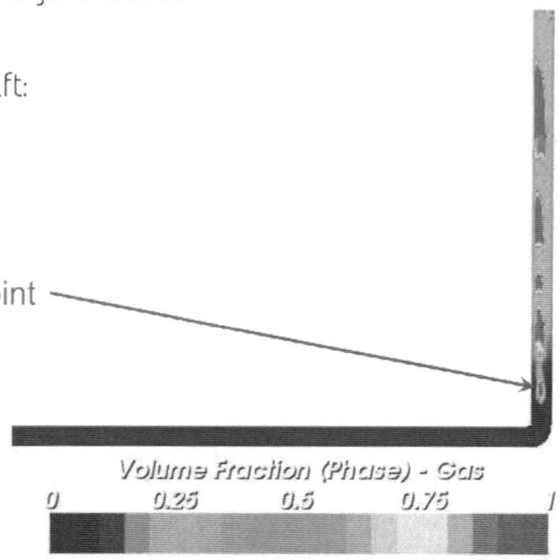

LedaFlow®

Q3D mass source – 3-phase

Oil injection point

Gas inlet

Pipe initially filled with water (50%)

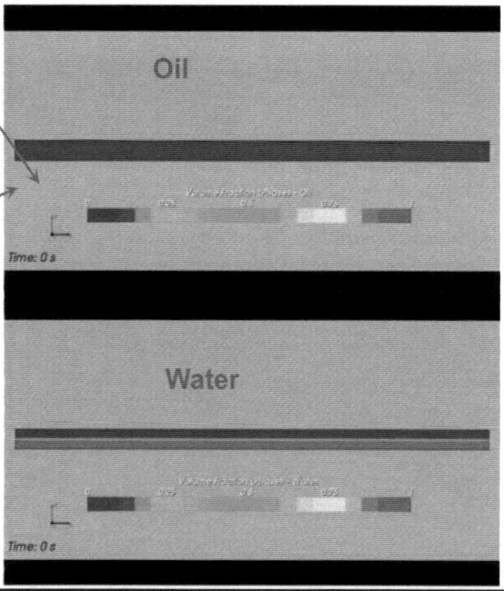

Gas Compressibility Improvements

LedaFlow®

- Improvements in
  - Volume inlet BC
  - Wall-Q3D BC
- Test case:
  - Pressure waves in a closed pipe

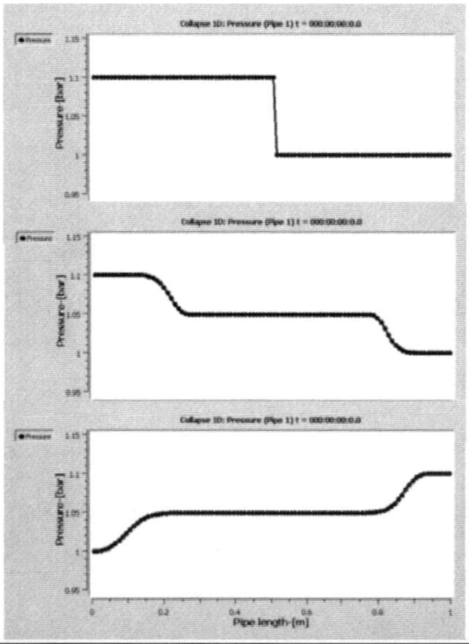

SINTEF | The 6th AIChE/SPE Joint Workshop on Challenges in Flow Assurance September 26-28, 2011, Houston, TX USA | Technology for a better society

Turbulent Single-Phase Pipe Flow

- Velocity profiles
  - Different Re
  - Princeton superpipe data[1]
- Turbulent kinetic energy
  - DNS data
  - Slice-averaged

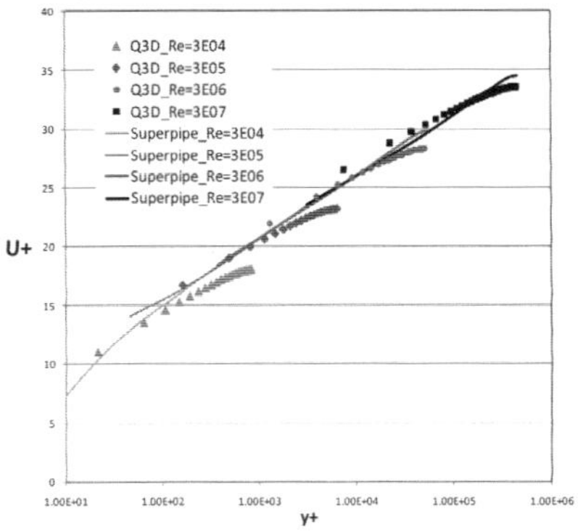

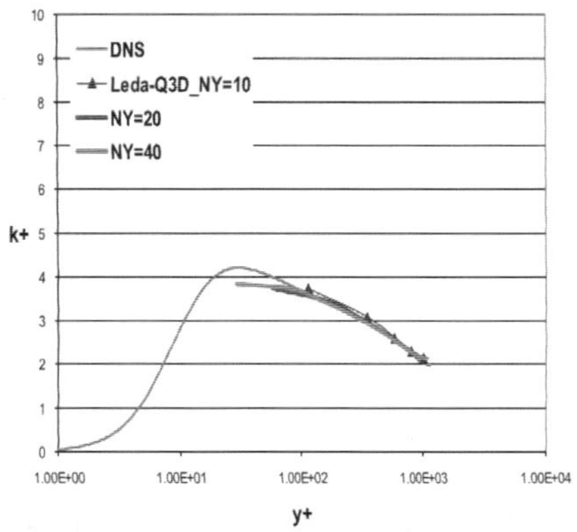

SINTEF | The 6th AIChE/SPE Joint Workshop on Challenges in Flow Assurance September 26-28, 2011, Houston, TX USA | Technology for a better society

Validation: Friction factor

LedaFlow®

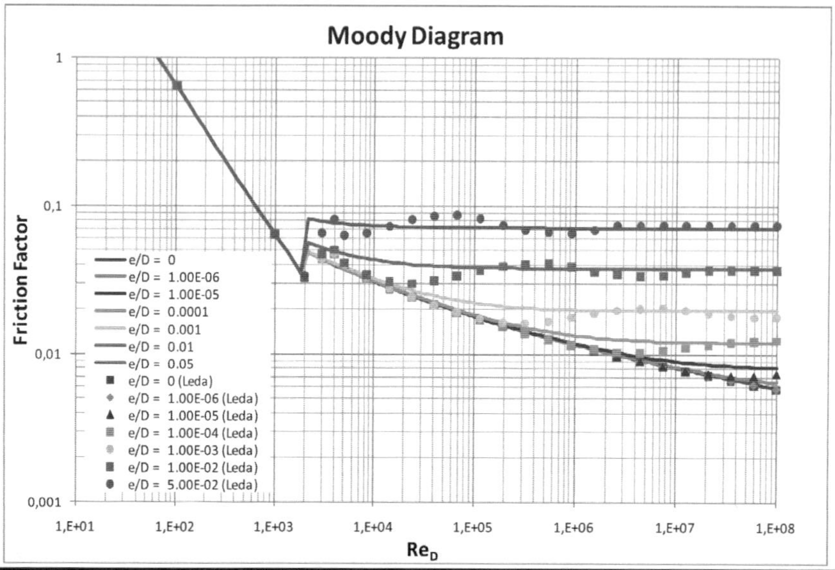

LedaFlow®

Rayleigh-Taylor instability (60x40 mesh) in Q3D

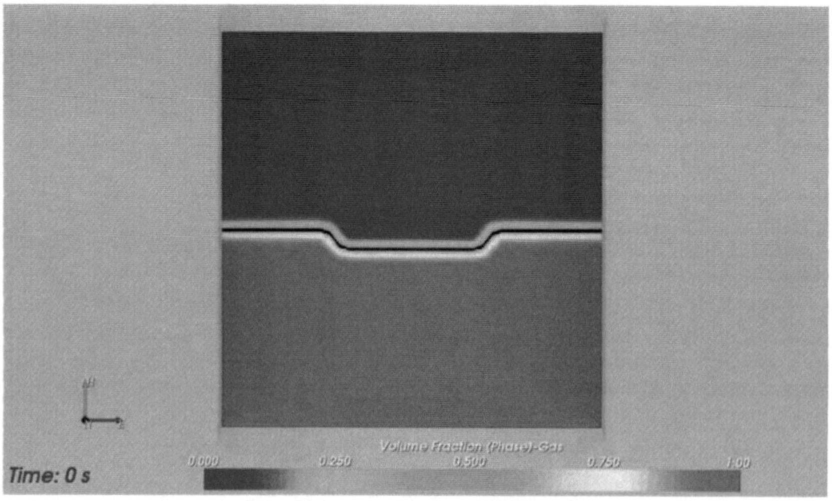

Inviscid flow example

Symmetry tests – 3 phase

LedaFlow®

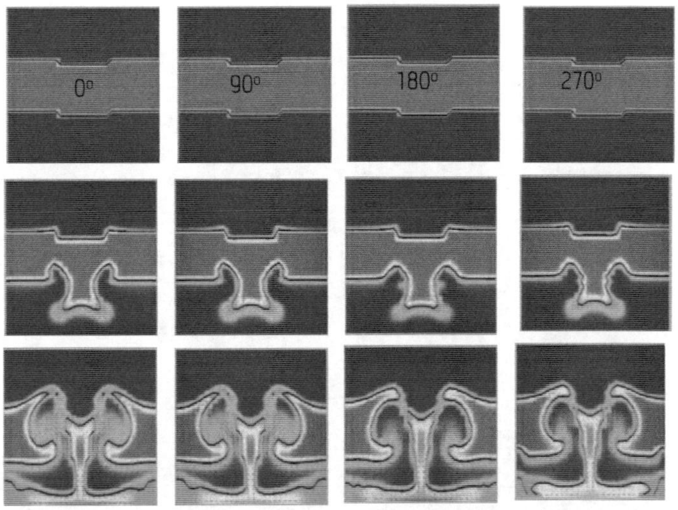

SINTEF

The 6th AIChE/SPE Joint Workshop on Challenges in Flow Assurance
September 26-28, 2011, Houston, TX USA

Technology for a better society

Kelvin-Helmholtz instability

LedaFlow®

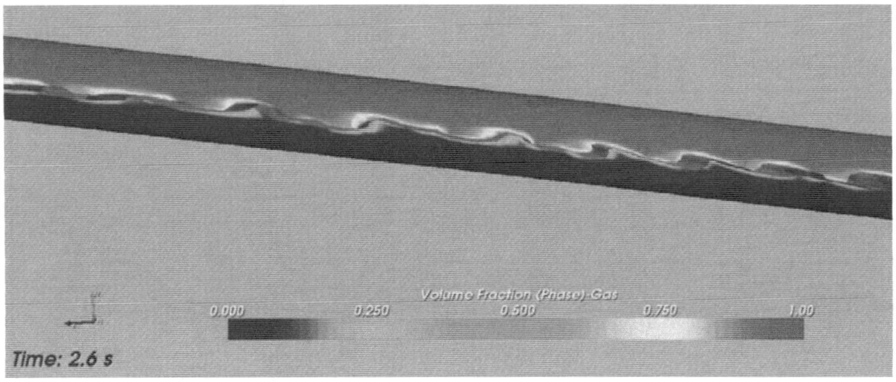

Oil-water in 2m long and 5.6 cm diameter pipe

SINTEF

The 6th AIChE/SPE Joint Workshop on Challenges in Flow Assurance
September 26-28, 2011, Houston, TX USA

Technology for a better society

Kelvin-Helmholtz instability (o/w)

LedaFlow®

Q3D model:
56 mm diameter pipe

Experiment:
50x50mm channel

Taylor bubble propagation in viscous liquid

LedaFlow®

- Large-diameter pipe, various inclinations

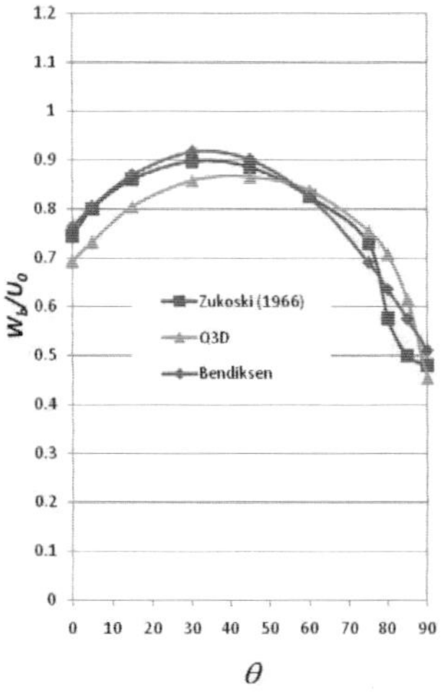

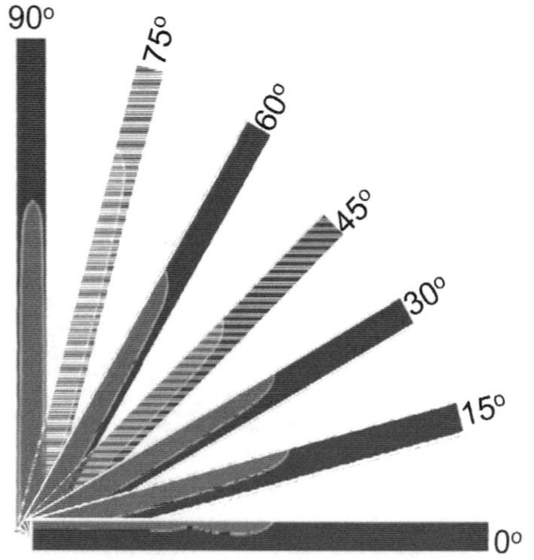

LedaFlow®

Taylor bubble diameter scaling at 45° inclination

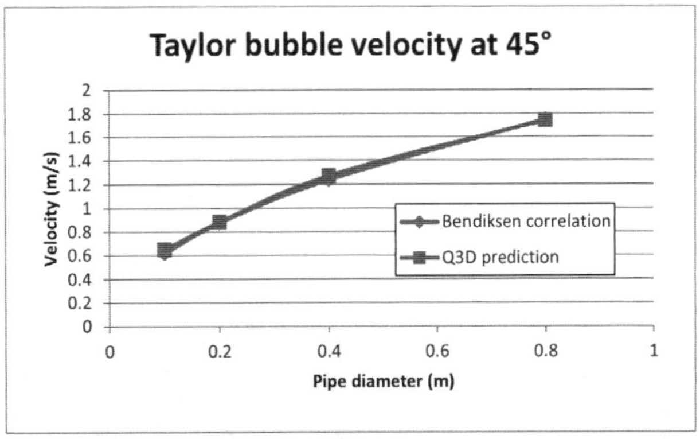

LedaFlow®

Applications

- Slug flows (2-phase)
- Two phase flow in a terrain
- 3-phase flow in risers
- Slug flows; 3-phase inclined flows
- 3-phase flows at higher inclinations
- Water accumulation in inclines
- Slug-flow in riser with bend
- Fluid displacement

SINTEF
The 6th AIChE/SPE Joint Workshop on Challenges in Flow Assurance
September 26-28, 2011, Houston, TX USA
Technology for a better society

TMF slug flow case

LedaFlow®

- TMF slug flow exercise

  Ujang, et al., (2006), "Slug initiation and evolution in two-phase horizontal flow", *Int. J. Multiphase Flow*, **32**, 527-552.

- Air-Water Slug flow in horizontal pipe
  - D=77.92 mm, L= 37m
  - U_{sG} = 4.64 m/s; U_{sL} = 0.611 m/s

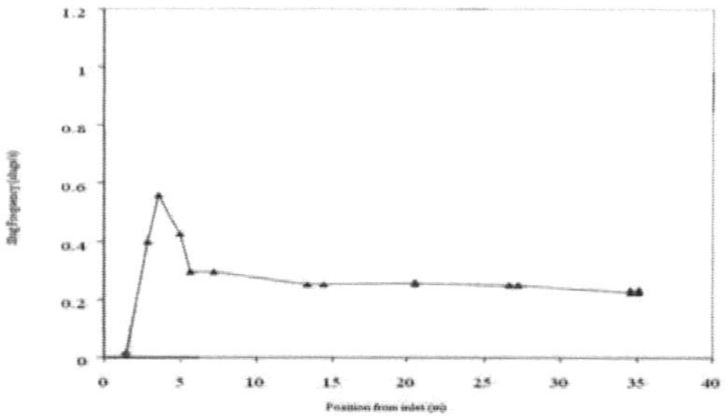

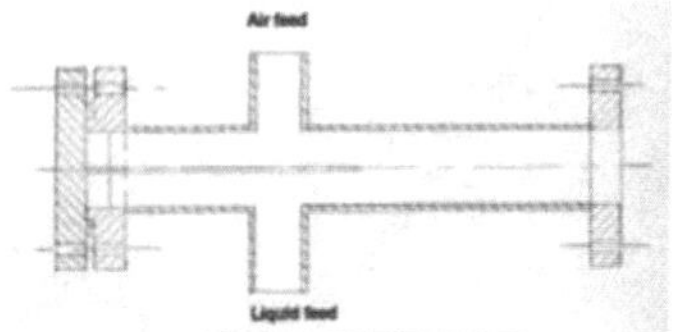

Inlet geometry

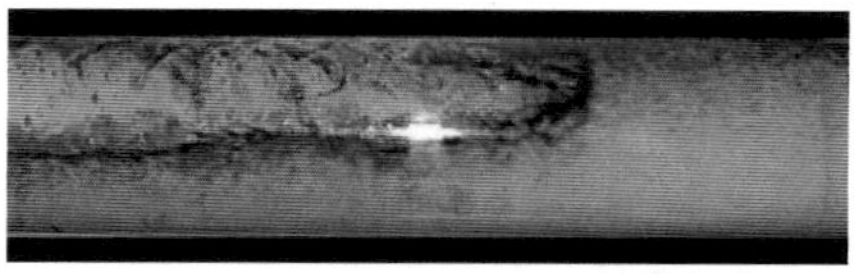

Slug in an air-water flow

Q3D on TMF slug flow case

LedaFlow®

- 30m long; NY=10, NX=2440 → **Grid Aspect Ratio~1.5**
- Compressibel gas
- No perturbations imposed at the inlet! (fully stratified flow)
- Simulation time 58 sec; running on 4 processors

- Snapshot from the flow:
  - Pipe diameter in the figures is magnified 10 times

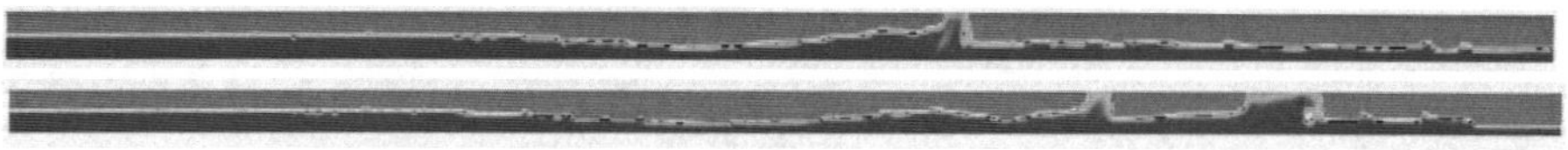

Q3D on TMF slug flow case

LedaFlow®

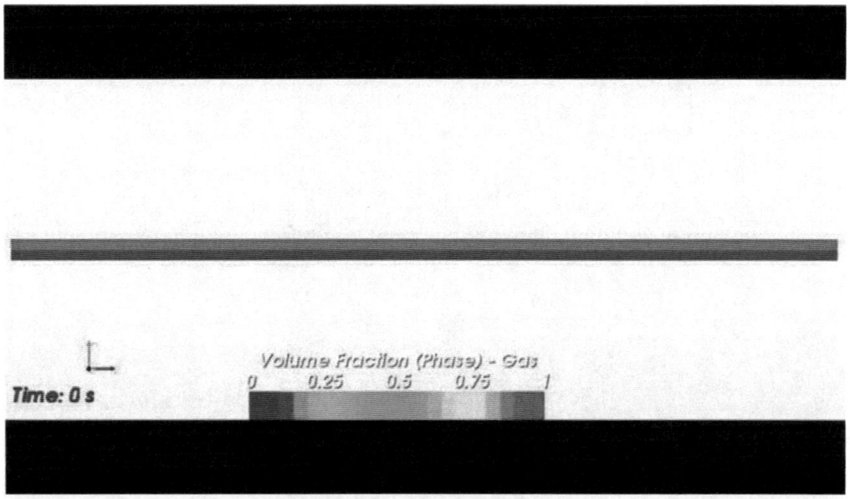

SINTEF

The 6th AIChE/SPE Joint Workshop on Challenges in Flow Assurance
September 26-28, 2011, Houston, TX USA

Technology for a better society

Time-series of volume fraction

LedaFlow®

- Experiments

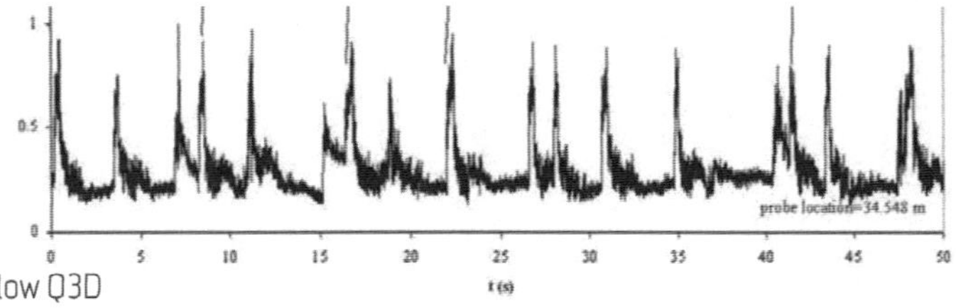

- Leda-Flow Q3D

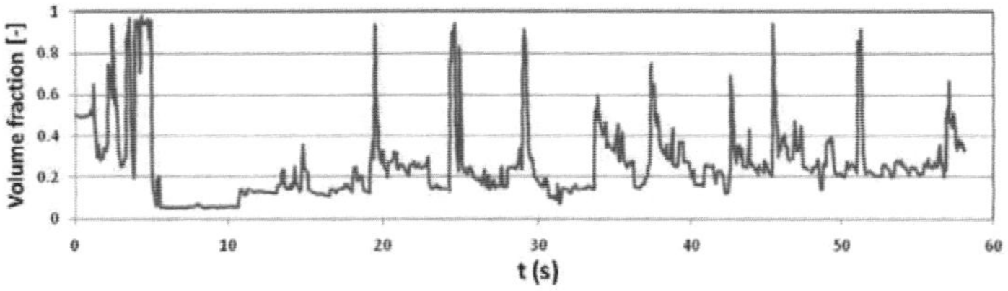

SINTEF

The 6th AIChE/SPE Joint Workshop on Challenges in Flow Assurance
September 26-28, 2011, Houston, TX USA

Technology for a better society

Slug frequency

LedaFlow®

- Liquid hold up threshold for counting a wave as a slug = 0.8
- Waves passing this criteria are simply counted in a period of 30 s

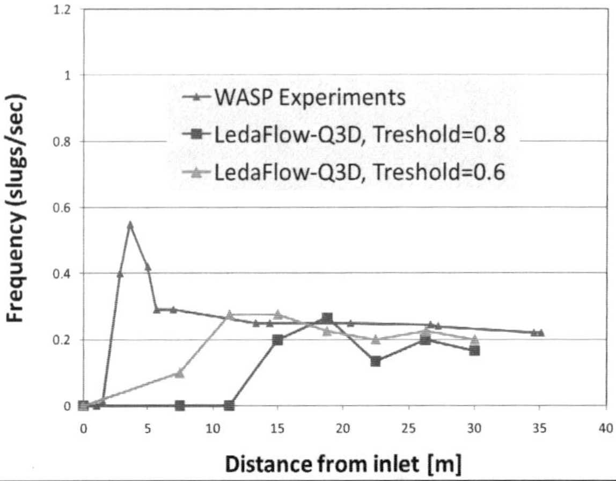

SINTEF

The 6th AIChE/SPE Joint Workshop on Challenges in Flow Assurance
September 26-28, 2011, Houston, TX USA

Technology for a better society

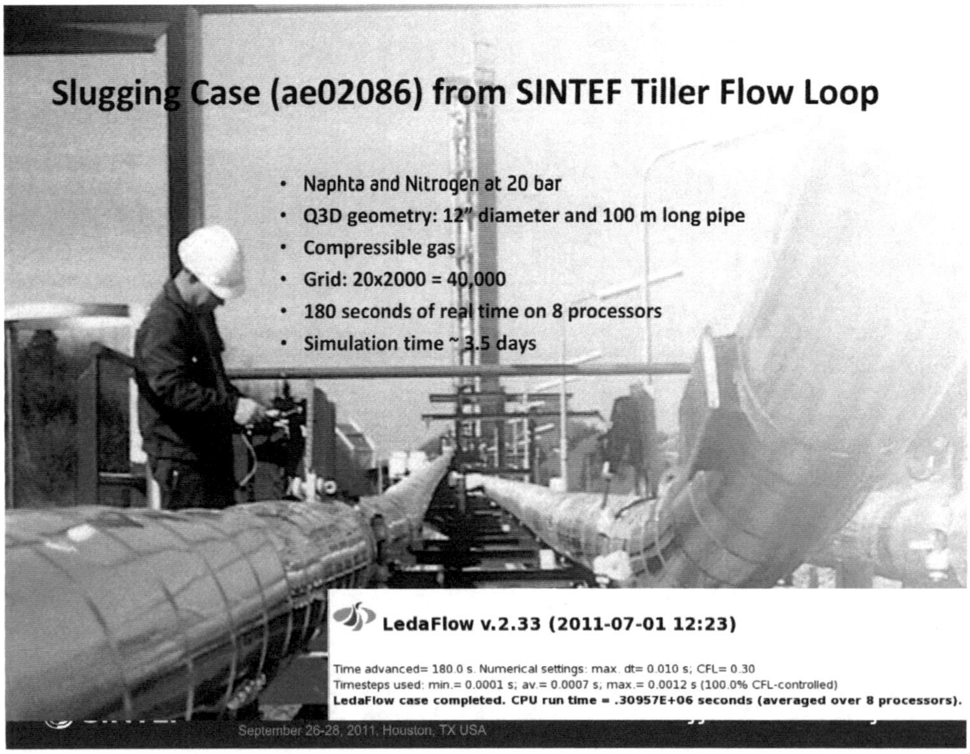

Snapshot from the flow

LedaFlow®

- Diameter magnified x5

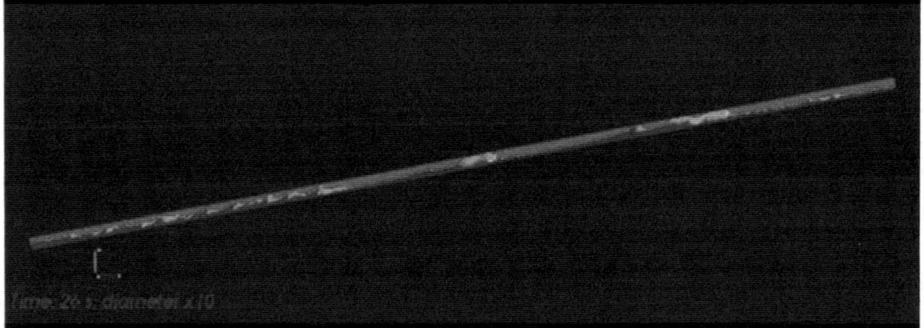

- Slug flow regime is created ca. 50 m from the inlet!

LedaFlow

Tiller ae02086

Oil Holdup Signal @ x~90m from pipe inlet

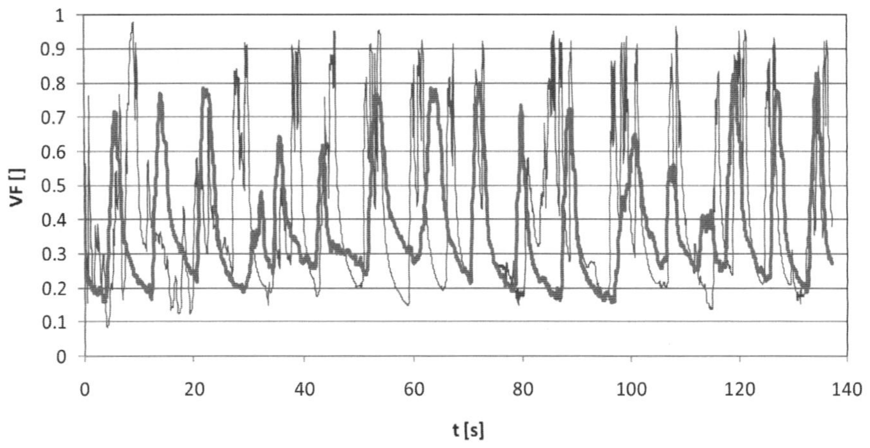

Flow regime indication

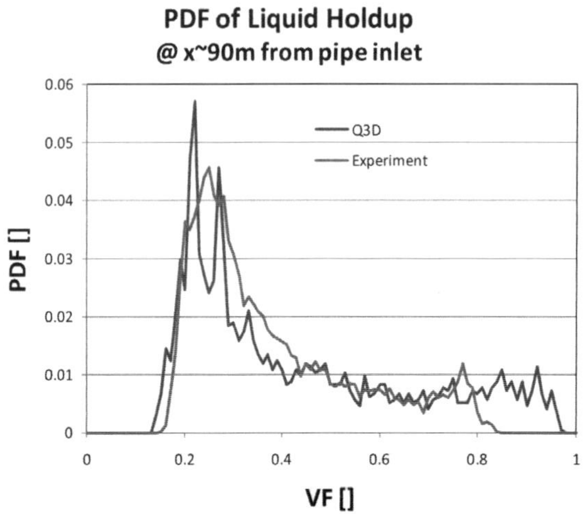

SINTEF The 6th AIChE/SPE Joint Workshop on Challenges in Flow Assurance September 26-28, 2011, Houston, TX USA Technology for a better society

Two phase flow in a terrain

(L=66 m and riser height 20 m)

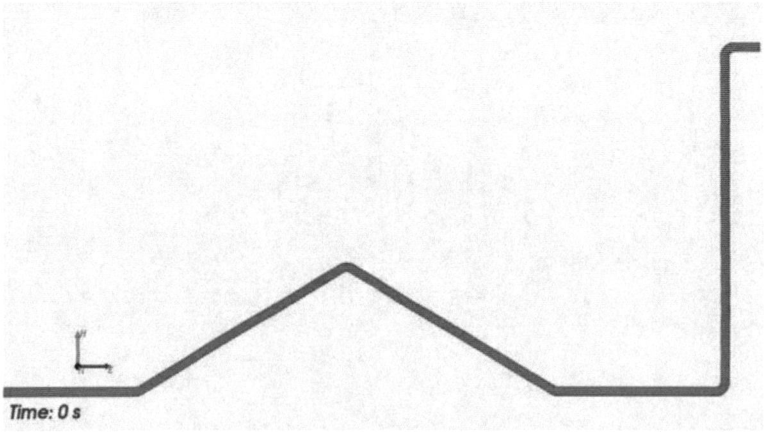

SINTEF The 6th AIChE/SPE Joint Workshop on Challenges in Flow Assurance September 26-28, 2011, Houston, TX USA Technology for a better society

3-phase flow in risers

LedaFlow®

Gas Water

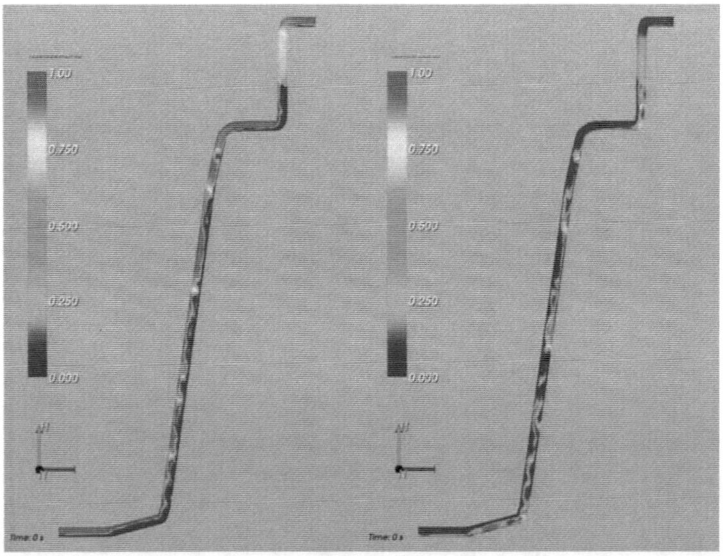

Flow domain:
- L= 264 m
- D=0.35 m

Grid (coarse):
- 300x10

Q3D on 3-phase flows at higher inclinations

LedaFlow®

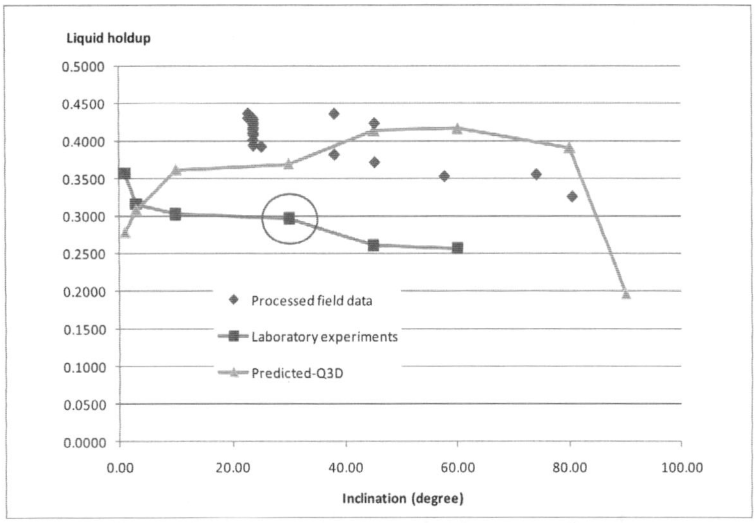

Q3D on 3-phase flows at higher inclinations

LedaFlow®

30 degree flow case from previous slide: Gas

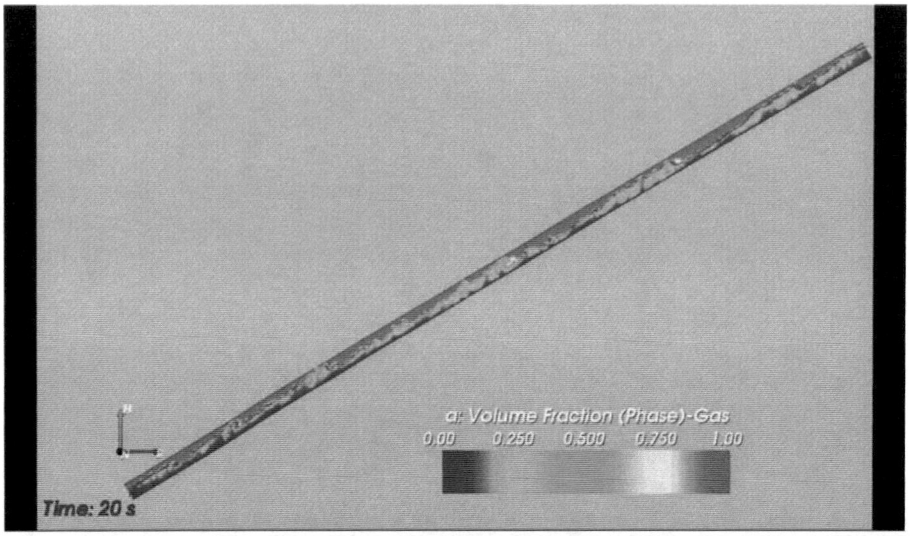

SINTEF

The 6th AIChE/SPE Joint Workshop on Challenges in Flow Assurance
September 26-28, 2011, Houston, TX USA

Technology for a better society

Q3D on 3-phase flows at higher inclinations

LedaFlow®

30 degree flow case from previous slide: Water

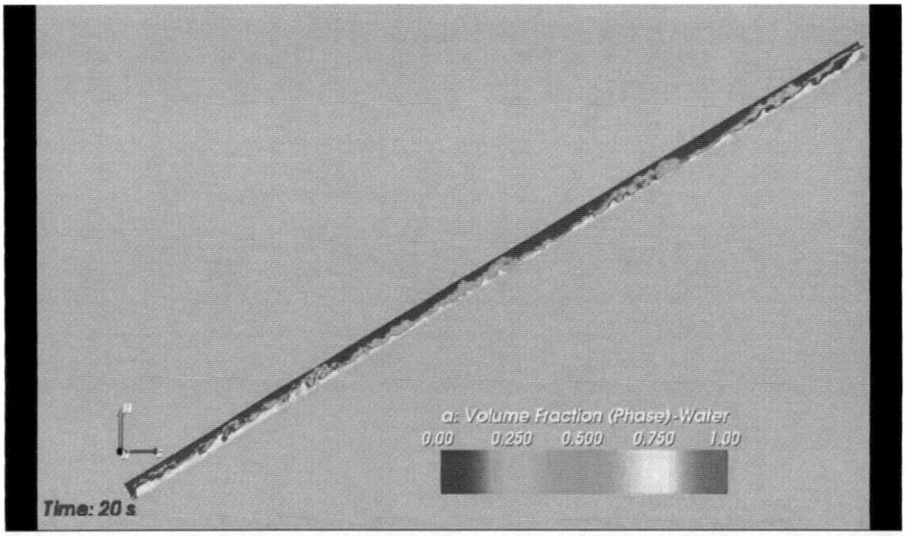

SINTEF

The 6th AIChE/SPE Joint Workshop on Challenges in Flow Assurance
September 26-28, 2011, Houston, TX USA

Technology for a better society

Water accumulation in inclines

LedaFlow®

L=100 m,D=0.5 m, Uso=1.0 m/s, Usw=0.05 m/s
Water droplet size = 0.5 mm

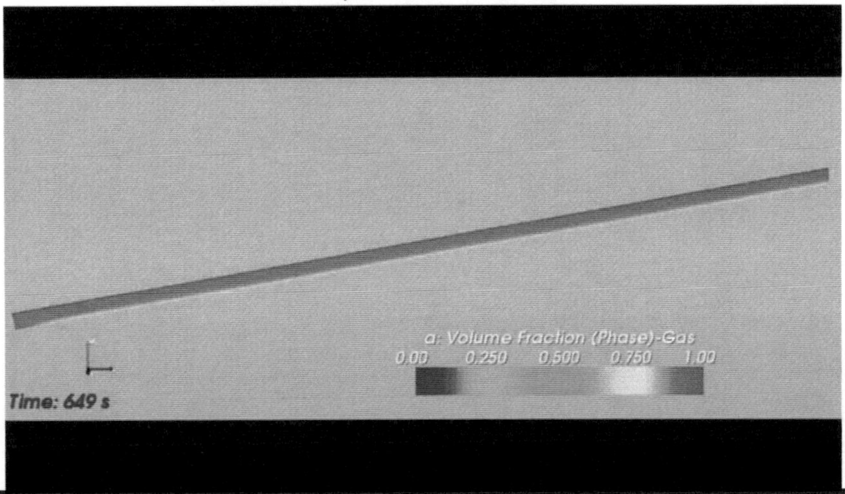

3-phase separation

LedaFlow®

L=10 m,D=0.5 m, Uso=0.2 m/s, Usg=0.1 m/s, Usw=0.1 m/s
Inlet water droplet size = 1.0 mm, bubble size = 2.0 mm

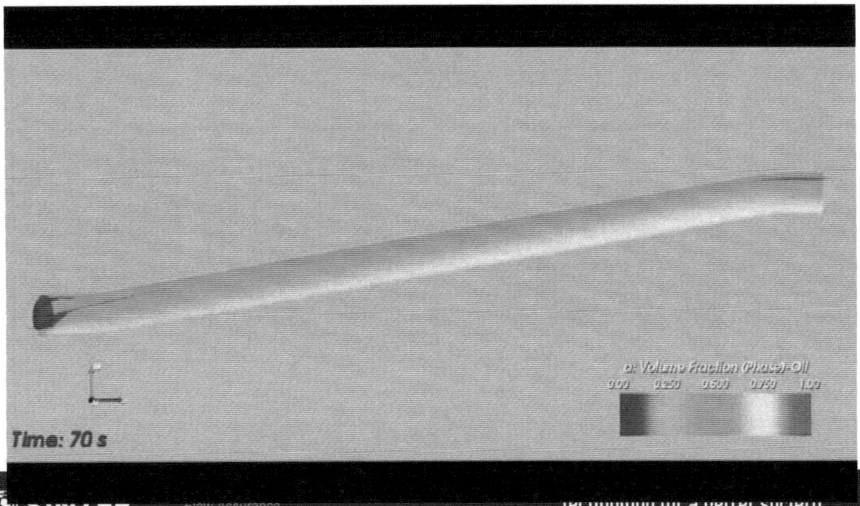

Slug flow in riser with bend

LedaFlow®

- Same geometry (D=67mm, $L_{vertical}$=6m
- 2m horizontal section
- 4500 Grid cells (coarse grid, aspect ratio~8):
  NX=300, NY=15 with refinement towards walls
- Dynamic particle size
- Real time: 60 sec
- CPUs=1
- Wall time ~ 3h

Time advanced= 60.0 s. Numerical settings: max. dt= 0.010 s; CFL= 0.50
Timesteps used: min.= 0.0000 s; av.= 0.0016 s; max.= 0.0100 s (100.0% CFL-controlled)
LedaFlow case completed. CPU run time = 11502. seconds.

Flow Regime Indication from Q3D data

LedaFlow®

- Statistical analysis can be performed just as in experiments
  - PDF, auto- and cross-correlations, PSD (power spectral density)
  - Example: Bubbly flow in vertical pipe:

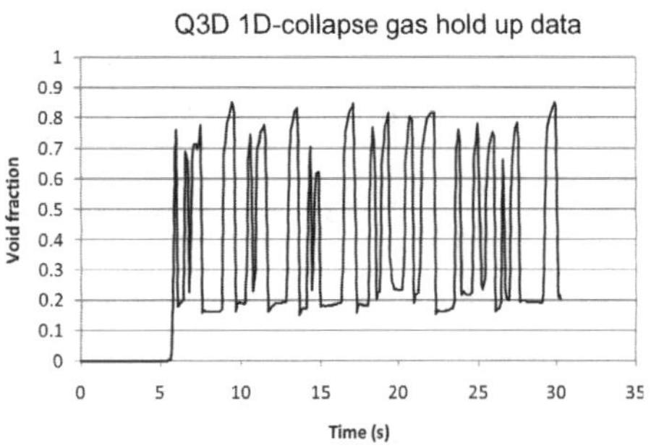

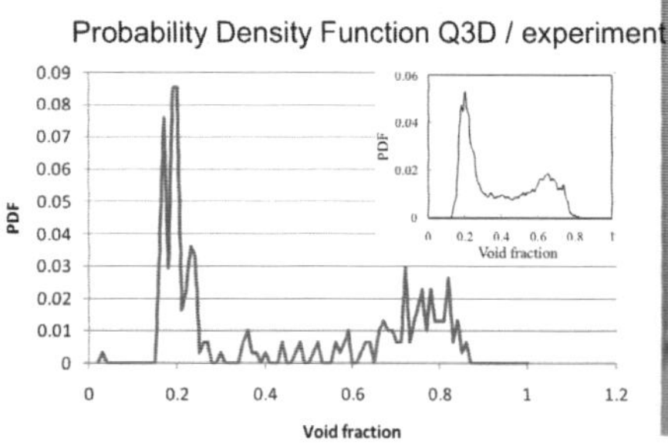

Air/Oil
Usl=0.25
Usg=0.06

Slug/bubbly Flow - Q3D Simulation

LedaFlow®

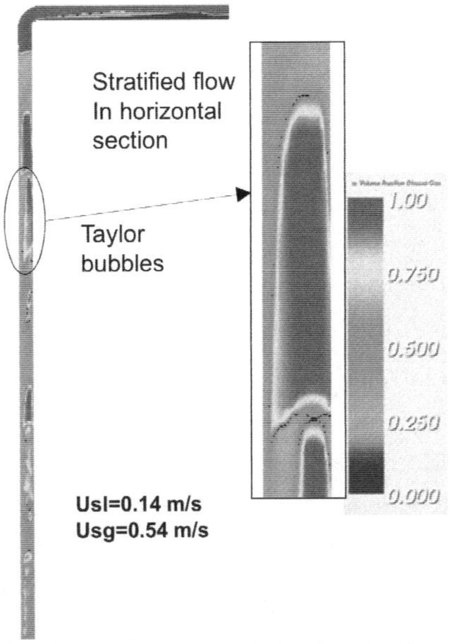

Experiments

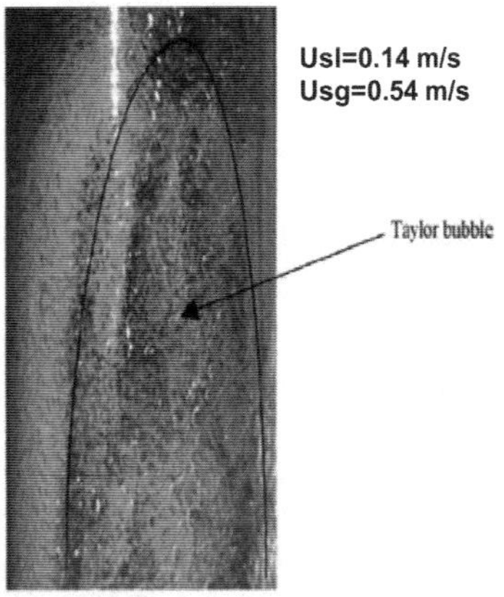

Fig. 8. Video image of slug flow for a riser at liquid and gas superficial velocities of 0.14 and 0.54 m/s, respectively.

LedaFlow®

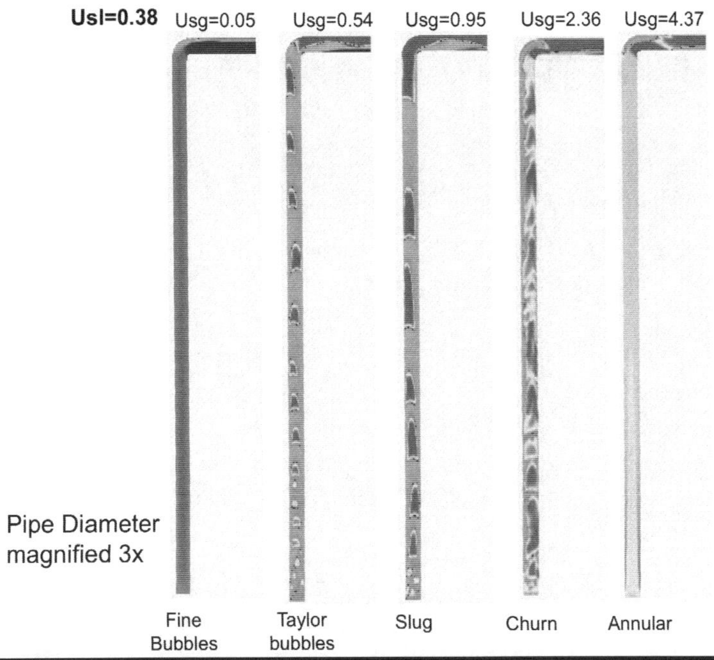

Transition to Churn Flow

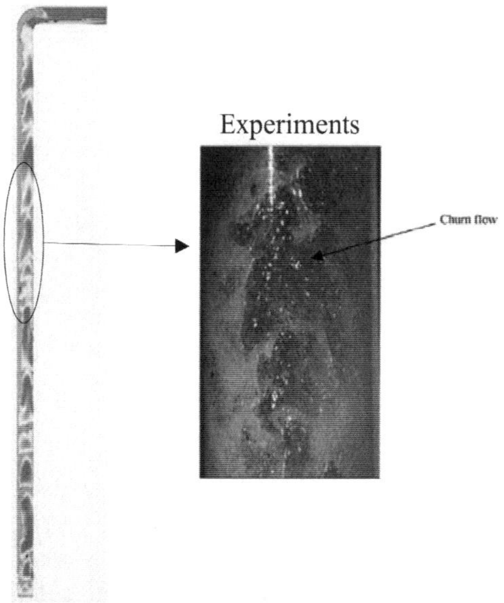

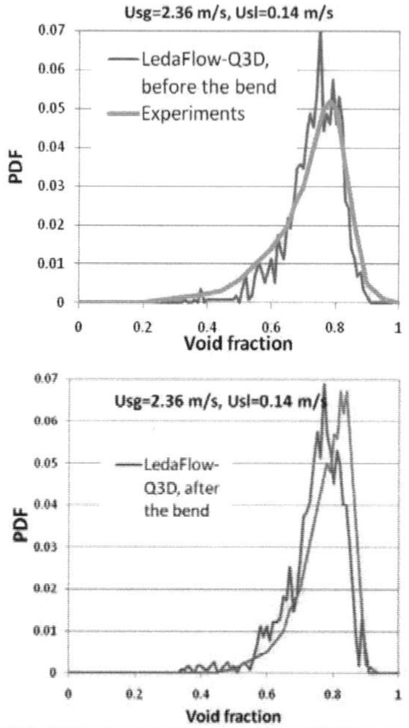

Average 1D void fraction Usl=0.25m/s, air/silicone, D=3inch, 6m riser, Patm

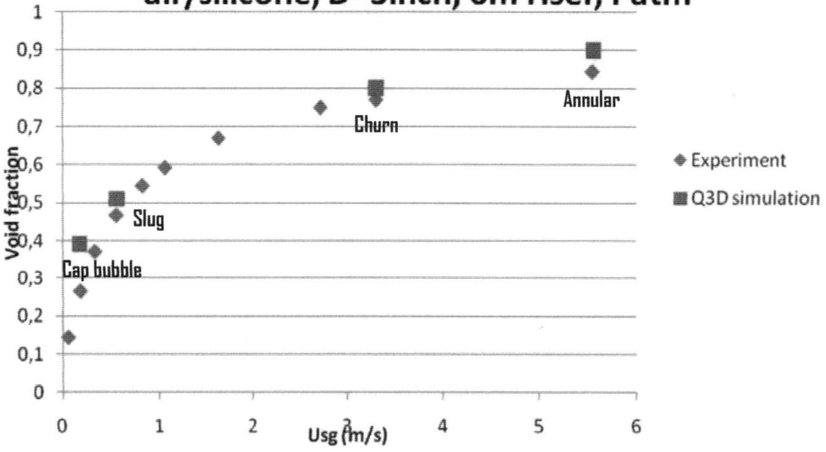

L. Szalinski, L.A. Abdulkareem, M.J. Da Silva, S. Thiele, M. Beyer, D. Lucas, V. H. Perez, U. Hampel , B.J. Azzopardi
Comparative study of gas–oil and gas–water two-phase flow in a vertical pipe,
Chem Eng Sci **65 (2010) 3836–3848**

Two-Phase Flow in Bends

LedaFlow®

- Vertical to horizontal bend

LedaFlow-Q3D

Taylor bubble moving outside the bend

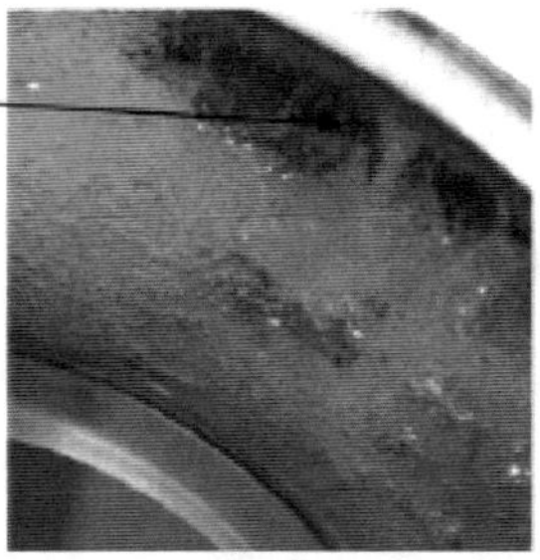

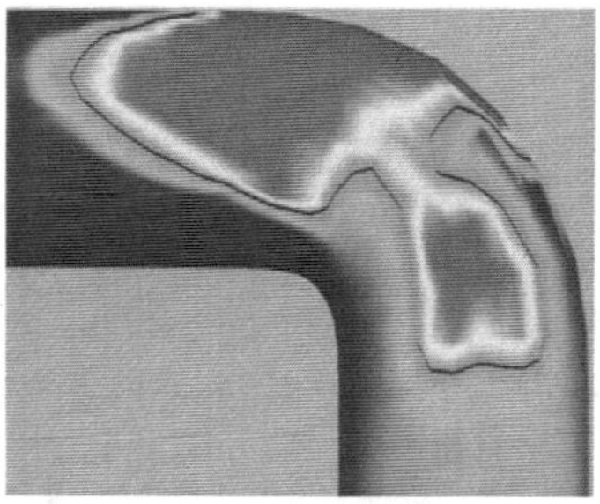

Displacement Operations in Pipes and Jumpers

Liquid displacement in pipes

- **Water flushing with oil**
- 90 mm 17m pipeline, horizontal and slightly inclined
- Oil injection velocity = 0.01 – 0.4 m/s

Validation

- Displacement curve

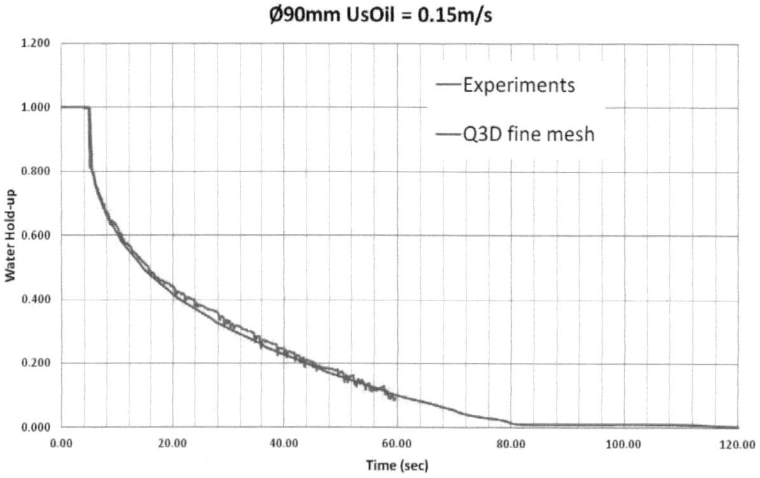

SINTEF
The 6th AIChE/SPE Joint Workshop on Challenges in Flow Assurance
September 26-28, 2011, Houston, TX USA
Technology for a better society

Various injection speeds

UsF: Front velocity
Usi: Oil injection velocity
Fr = Usi / SQRT(g.D)

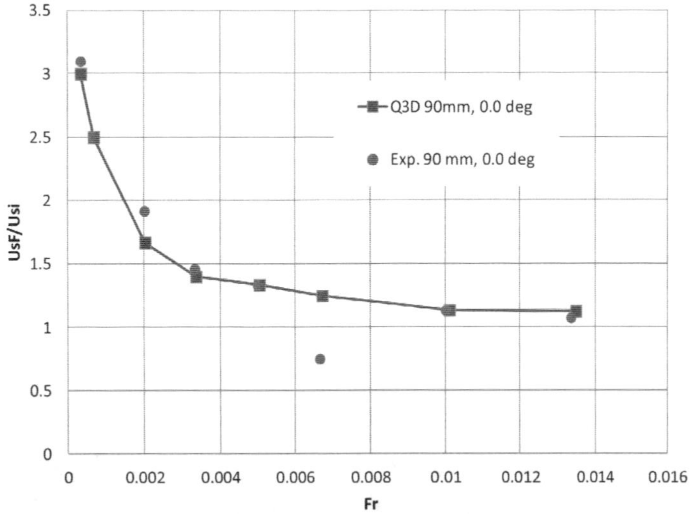

Liquid displacement in jumpers

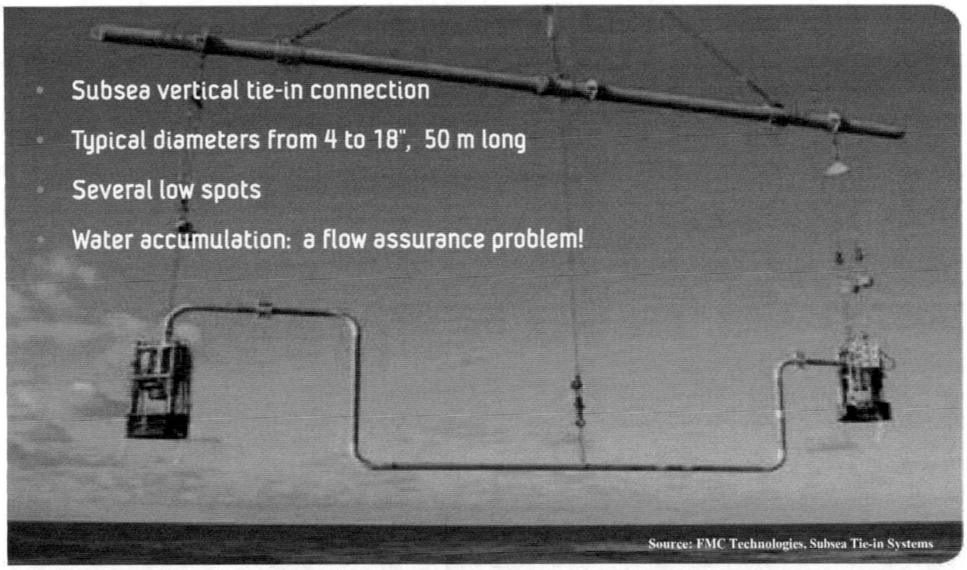

SINTEF
The 6th AIChE/SPE Joint Workshop on Challenges in Flow Assurance
September 26-28, 2011, Houston, TX USA
Technology for a better society

Example: Tie-in configuration

LedaFlow®

- 10" pipe (25.4 cm)
- Horizontal

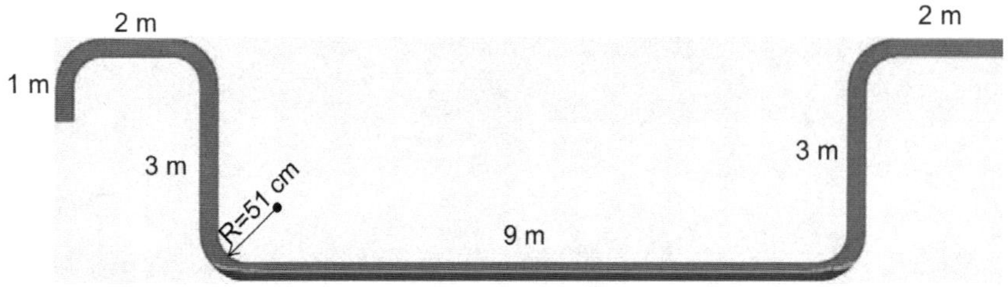

- Fluids and operating pressure: Water & Air @ 1 bara

Gas purge operation

LedaFlow®

- Initial water loading: 50% in the base
- Gas superficial velocity Usg=1.2 - 9 m/s

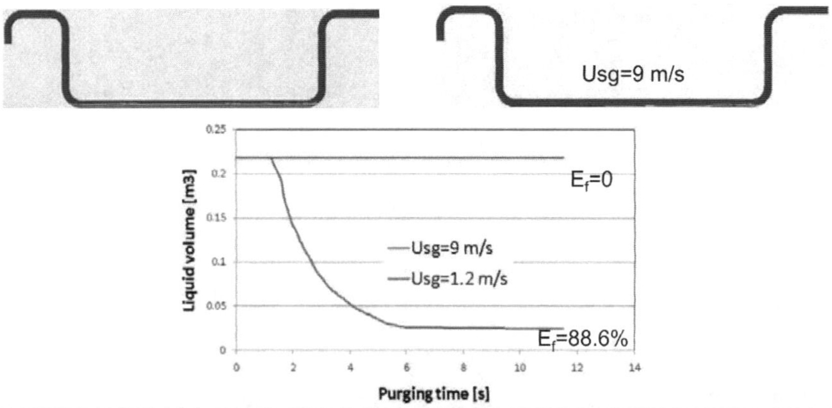

MEG flushing simulations

LedaFlow®

- Water in the jumper is displaced by MEG
- Different jumper inclination (-5, 0, +5 degrees)
- Us_{MEG}= 0.14 m/s (Q=25 m^3/hr)
- Water cut at restart =100% (water filled jumper)

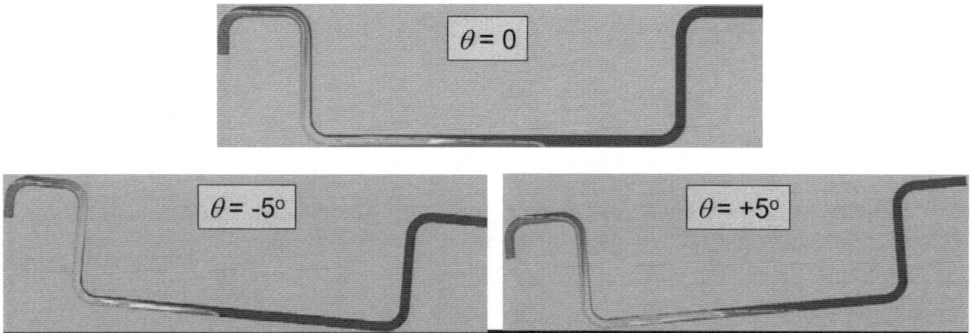

MEG distribution in the jumper after 180 s

LedaFlow®

- 0
- -5
- +5

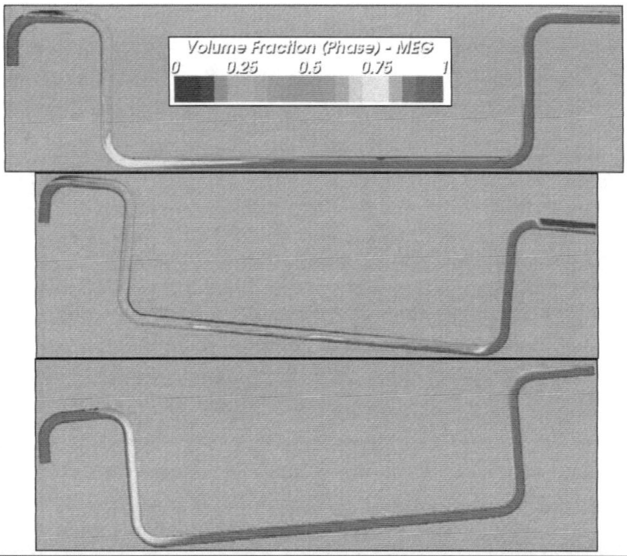

Water volume in the jumper

LedaFlow®

- Jumper volume= 0.95 m^3
- Jumper inclination has significant effect on flushing efficiency.

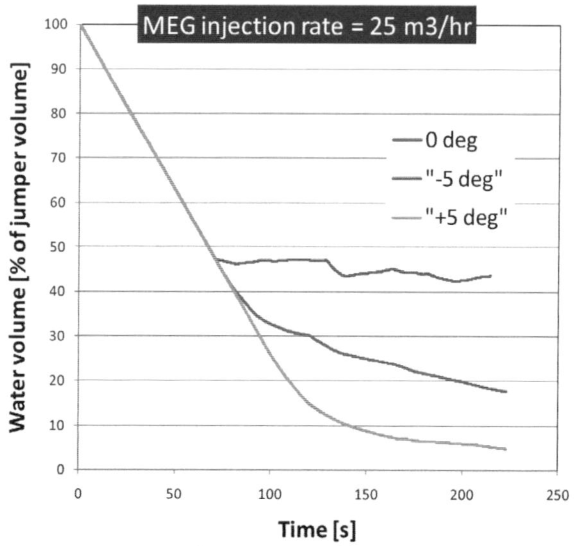

SINTEF

The 6th AIChE/SPE Joint Workshop on Challenges in Flow Assurance
September 26-28, 2011, Houston, TX USA

Technology for a better society

LedaFlow Q3D and effects of surface chemistry

Reproduces closely the slug frequency, pressure gradient and liquid volume fraction in a transitional slug flow with **unstable** coalescing bubbles

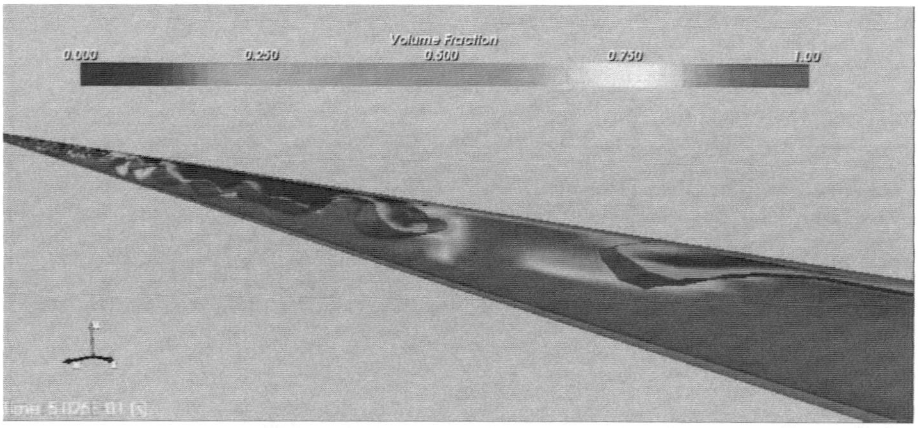

SINTEF

The 6th AIChE/SPE Joint Workshop on Challenges in Flow Assurance
September 26-28, 2011, Houston, TX USA

Technology for a better society

56

Complex fluids (stabilized bubbles)

as predicted by LedaFlow Q3D

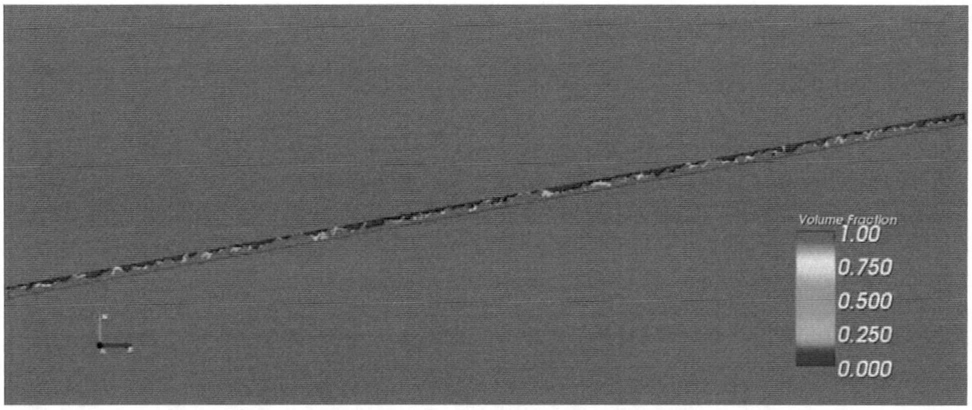

Flow regime, holdup and pressure drop is dramatically effected by bubble stability !

No "normal" physical properties are changed!

Summary on1D transient model (slug modeling)

LedaFlow®

- Slug Capturing can reproduce industrially observed slugging which has not been possible to simulate with other methods
- A new method for representing sharp fronts (slugs) is under development
- Both the standard transient model (Unit Cell based) and Slug Capturing is under constant improvement in order to handle the hardest slugging cases

Summary: The prospects of multidimensional modeling

LedaFlow®

- Q3D model can reproduce critical experiments
- Generic mass source is available (gas lift, fluid injection)
- Various validation cases have been performed for some fundamental problems (Taylor bubble, slug flow in horizontal and inclined pipes)
- Validations also performed for geometries which are relevant for jumper geometries (flow in vertical risers with bends)
- LedaFlow-Q3D shows very promising performance

SINTEF The 6th AIChE/SPE Joint Workshop on Challenges in Flow Assurance September 26-28, 2011, Houston, TX USA Technology for a better society

Application to jumper flushing

LedaFlow®

- Different scenarios of jumper flushing can be simulated by Q3D
- Detailed time dependent data about the dynamics of the fluids motion in the jumper can be obtained from the Q3D simulations
- Results from such simulations are now being validated by comparing to experimental data
- Gas flushing of liquids from risers (direct applications or improvement of 1D closures)

SINTEF The 6th AIChE/SPE Joint Workshop on Challenges in Flow Assurance September 26-28, 2011, Houston, TX USA Technology for a better society

Multi dimensional modeling: Final remarks

LedaFlow®

- Q3D results shows that multi-D models will play an important role in predicting pipe flow phenomena in the years to come
- Acts as a virtual laboratory. Can be used to interpret experiments or improve on 1D model closures
- Can explain phenomena that have not been understood earlier
- Can compute industrially relevant flows much faster (40 x) than full 3D models, with equal or better accuracy
- Potential direct applications:
  - Slug physics
  - 3-phase separation
  - Flow assurance (emulsions, jumper flushing, water accumulation, hydrates, wax)
- The model allows for representation of surface chemistry effects that cannot be represented by static surface tension

Merging Flow Assurance and Compact Separation Performance in Remote Applications

Gene Kouba
Research Consultant
Chevron

The 6th AIChE/SPE Joint Workshop - Challenges in Flow Assurance
Compact/Modular Processing for Remote Applications
Houston, September 26-28, 2011

Flow Line to Separation

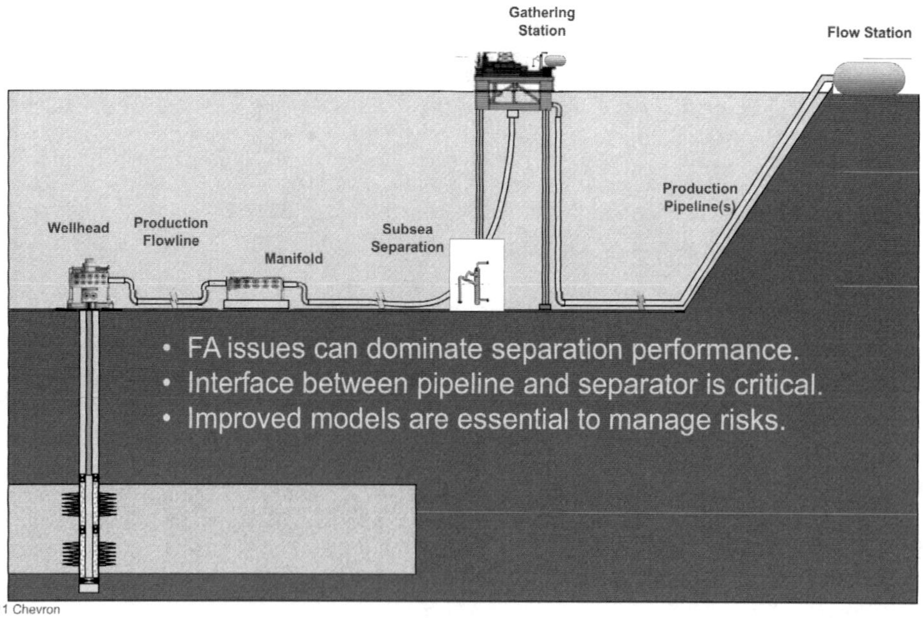

Separation/Flow Assurance Technology Qualification

How do we qualify integrated systems for new, remote, and expensive applications?

Reliability Concerns

- Robustness
- Maintainability
- Separation performance
  - Contamination of separated phase
    - Liquid carry-over
    - Gas carry-under
    - Oil in water

Operational Envelope

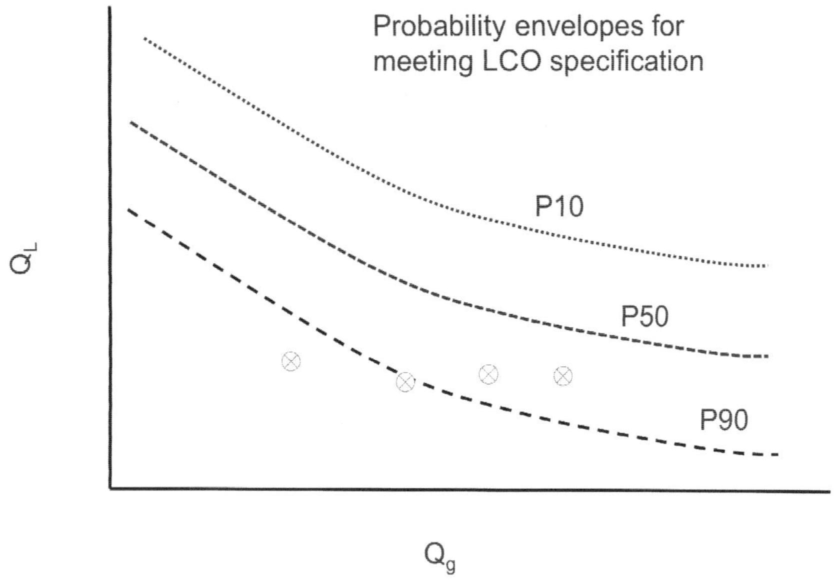

Operational Envelope

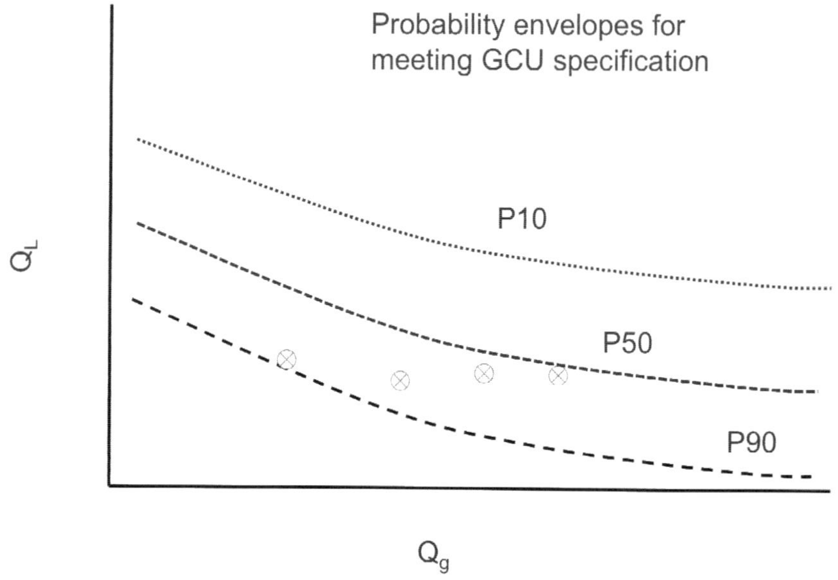

Value of Performance Probability Predictions

- System Performance Probability Model
  - Design & Optimization
  - Sensitivity study
  - Evaluation of value of information
  - Decision analysis among alternatives

Flow Assurance Issues

- Characterization of fluids entering separator
  - Slugs
    - Length, holdup, frequency, entrained bubble size
  - Foam
    - Quality, stability, bubble size
  - Mist
    - Entrainment, droplet size
  - Emulsions
    - Holdups, drop size, stability, viscosity
  - Solids
    - Size, concentration

Integration

- Mechanical
  - CFC, SD, pipe size & inclination
  - Low shear chokes & valves
  - Internals – vanes, mesh pads, cyclones
- Chemical
  - Foam, mist, drag
  - Wax, paraffin, hydrates, scale, corrosion
- Thermal
  - External heat exchange
  - Internal reactions

Typical Flow Patterns

- Slugs
  - Length, holdups, frequency
  - Entrained bubble/droplet size

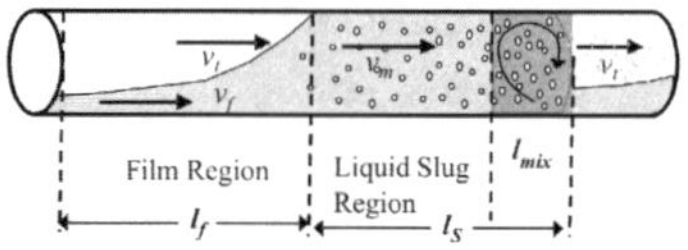

Slug Flow

- Pseudo-Slugs
  - Entrained droplet holdup & size
  - Length, holdup, frequency

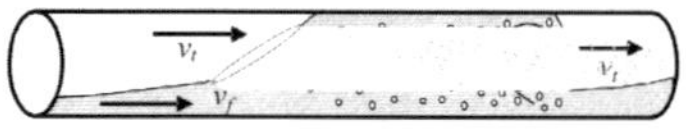

Pseudo-Slug Flow

- Annular Mist
  - Droplet holdup, droplet size

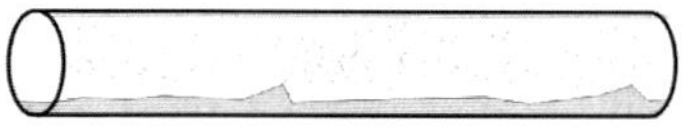

Annular Mist Flow

Inlet Conditioning to GLCC©

- **Inclined inlet** promotes stratification & downward annular momentum to liquid.
- **Dual inlets** allow pre-separation and expand slug capacity.
- **Churn Flow Coalescer** is effective at knocking out mist.
- All components are pipe.

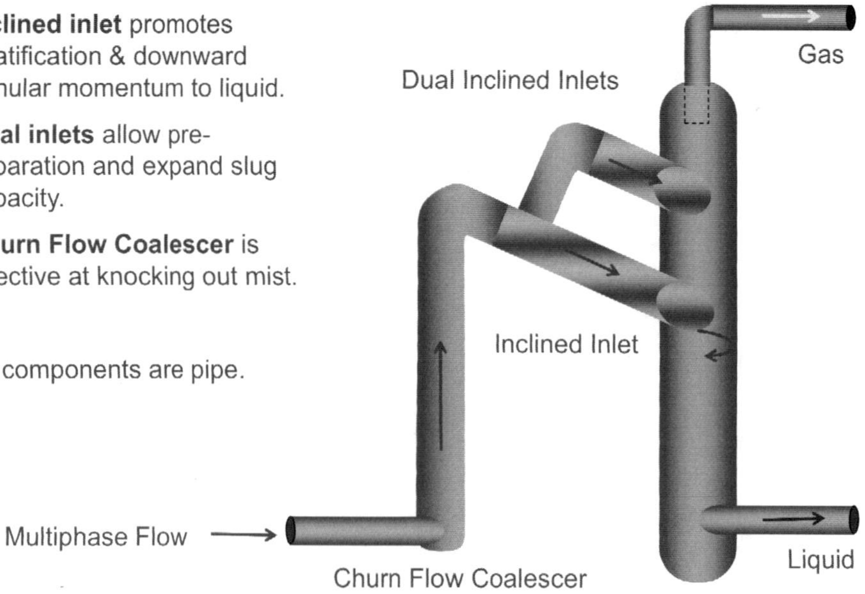

What's the problem?

Lack of adequate characterization of developing region.

- The churn flow coalescer and inlets operate are in developing flow.
- Models for developing flow regions are needed.
- Current steady-state & transient models need improvement.

How well do we characterize the multiphase flow entering the CFC?

Liquid Holdup in Slugs

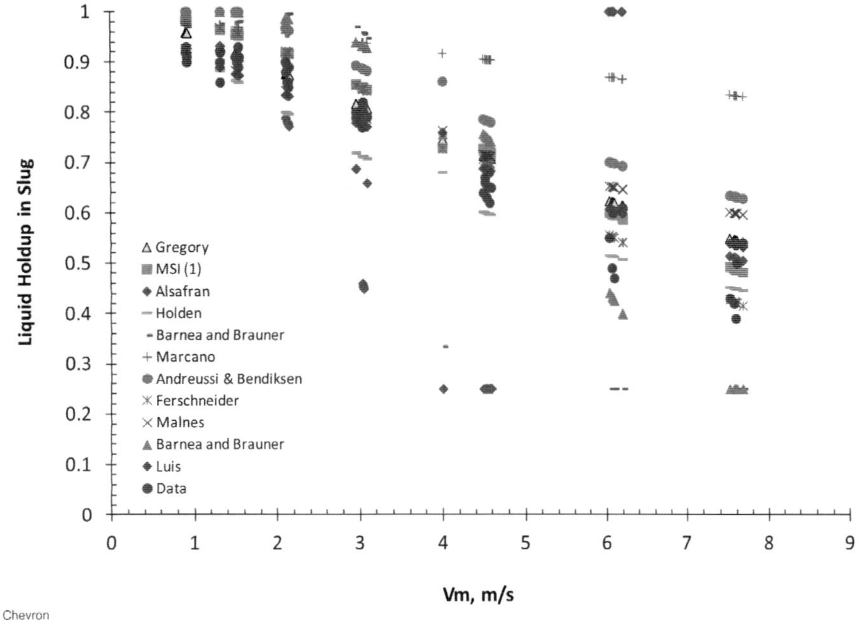

Droplet Entrainment Predictions

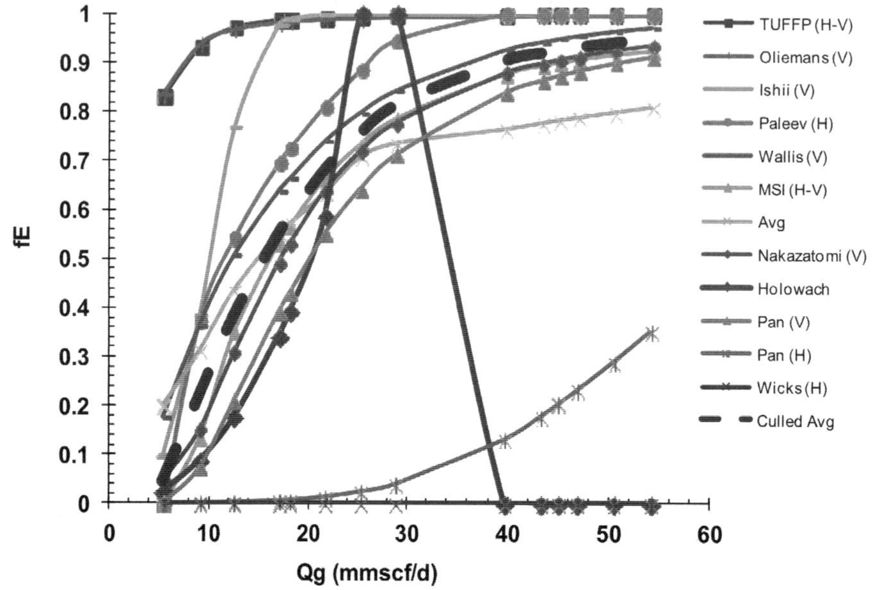

 13

Droplet Event Map

What happens upon droplet impact with wall or film surface?

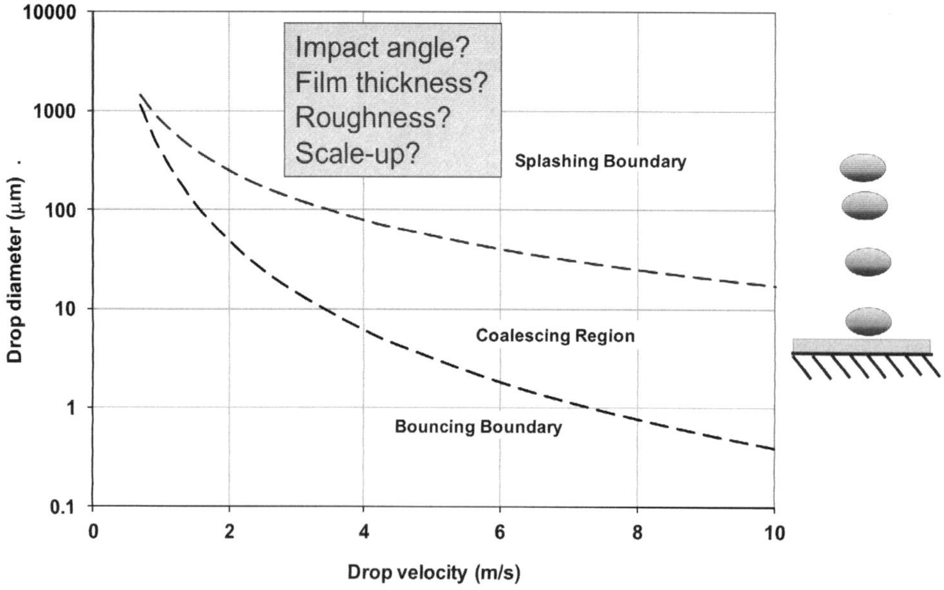

 14

Droplet Size Distribution

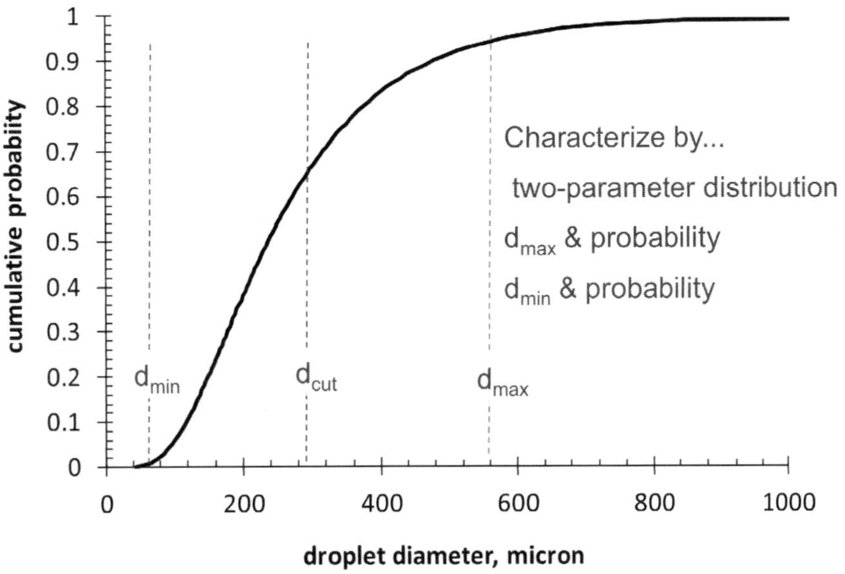

 15

Range of LCO Predictions

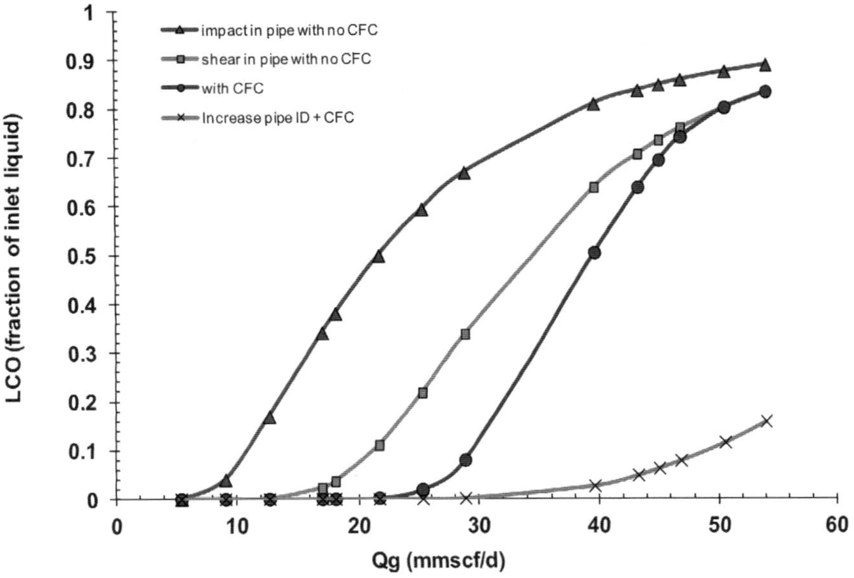

 16

LCO Probability Envelope

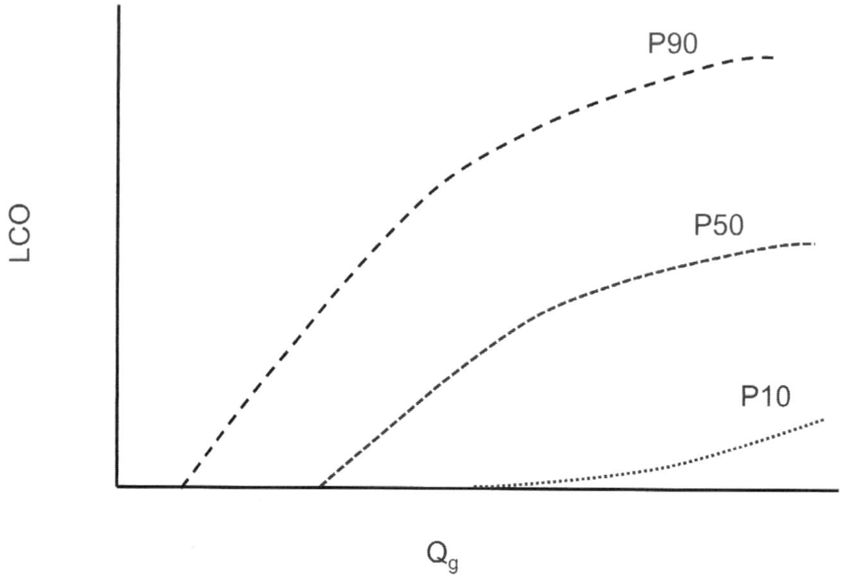

 17

Gaps and Needed Improvements for Remote Separation Applications

Gaps

- Characterization of ...
  - Churn flow, pseudo-slugs and waves
  - Foam
- Analysis of multiphase developing regions in short pipe sections.
- Equipment
  - Coalescing components
  - Eductors for commingling wells

Improvements

- Slug Characterization
  - Size and holdup
  - Mapping
- Bubbles & Droplets
  - Entrainment & size distribution
  - Impact event mapping
- Equipment reliability
  - Water separation
  - Defoaming
  - Mist extractors
  - Sand handling
  - Low shear chokes, valves, etc.

 18

Vision of Future

- Improvements in predictions of closure relationships and dynamic processes.
  - A much better understanding of uncertainties in these predictions.
- Use of simulation benchmarked against "accessible" data will enable confidence in deployment of subsea separation equipment.
- Confidence envelope for achieving a performance spec and reliable operation.

 19

How Will We Make Our Decisions About Subsea Systems?

Take a chance?

Follow current thinking of others?

Use skill & knowledge with a clear vision?

 20

The Impact of Subsea Processing on Integrated Flow Assurance Solutions

Shaw, C., FMC Technology

It has been well documented how an integrated approach to flow assurance can reduce the impact on the surface production facilities resulting in increased availability (uptime) of the satellite production. It is also well known how a well designed chemical program properly executed can assure production availability from subsea tie-backs. What is perhaps less well documented is the effect of subsea processing specifically on flow assurance issues. Much has been said about the reduced back-pressure and increased flow but not so much emphasis has been placed on its effect on flow management and surface facilities. There are a number of benefits and some minor issues that must be addressed when adding pumps and separators for seabed processing, this presentation will address the benefits and issues using examples taken from real applications and study work.

Tulsa University Separation Technology Projects (TUSTP)

Compact Separation Technology – From the Lab to the Field

by

Drs. Ram S. Mohan and Ovadia Shoham

The 6th AIChE/SPE Joint Workshop
Houston, TX
September 26-27, 2011

1-1

Topics

- Introduction – TUSTP
- Compact Multiphase Cyclonic Separators
- Research Program
- Field Applications
- Conclusions

1-2

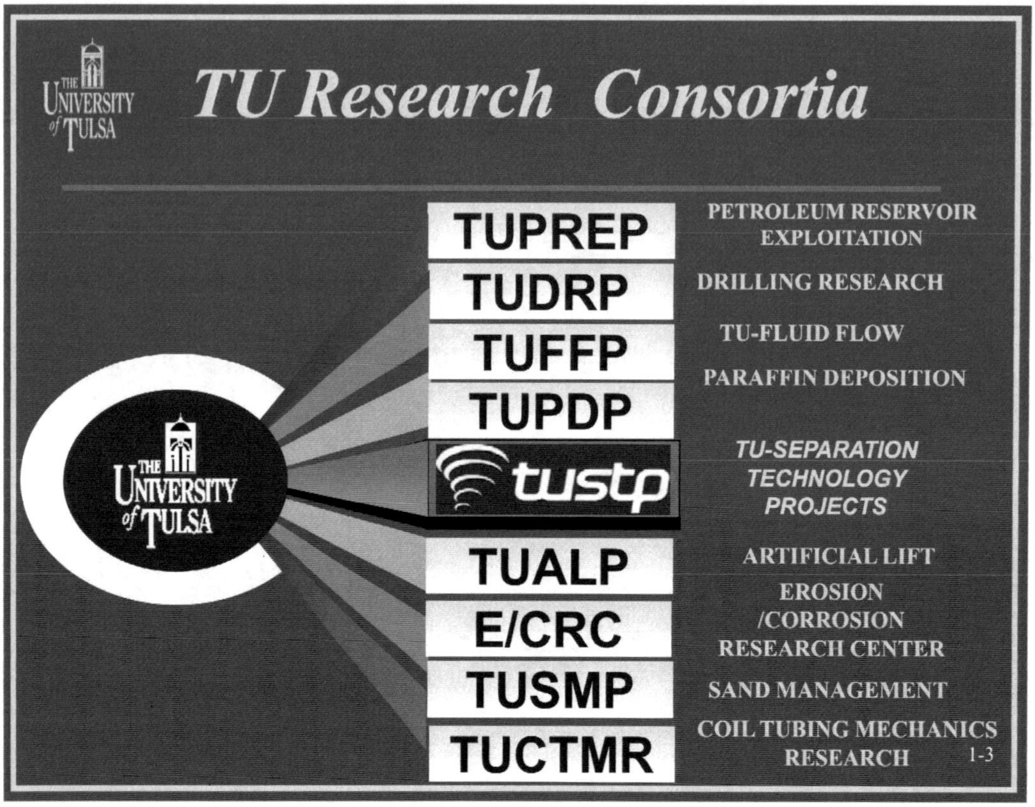

TUSTP Mission and Vision

- **Advance the State-of-the-Art of Compact Multiphase Cyclonic Separation Technology for Gas/Oil/Water/Sand Flow**
- **Long-term Cooperation between University, Industry and Government is Envisioned to Better Understand, Analyze, and Design Compact Separators and Separation Systems**

1-4

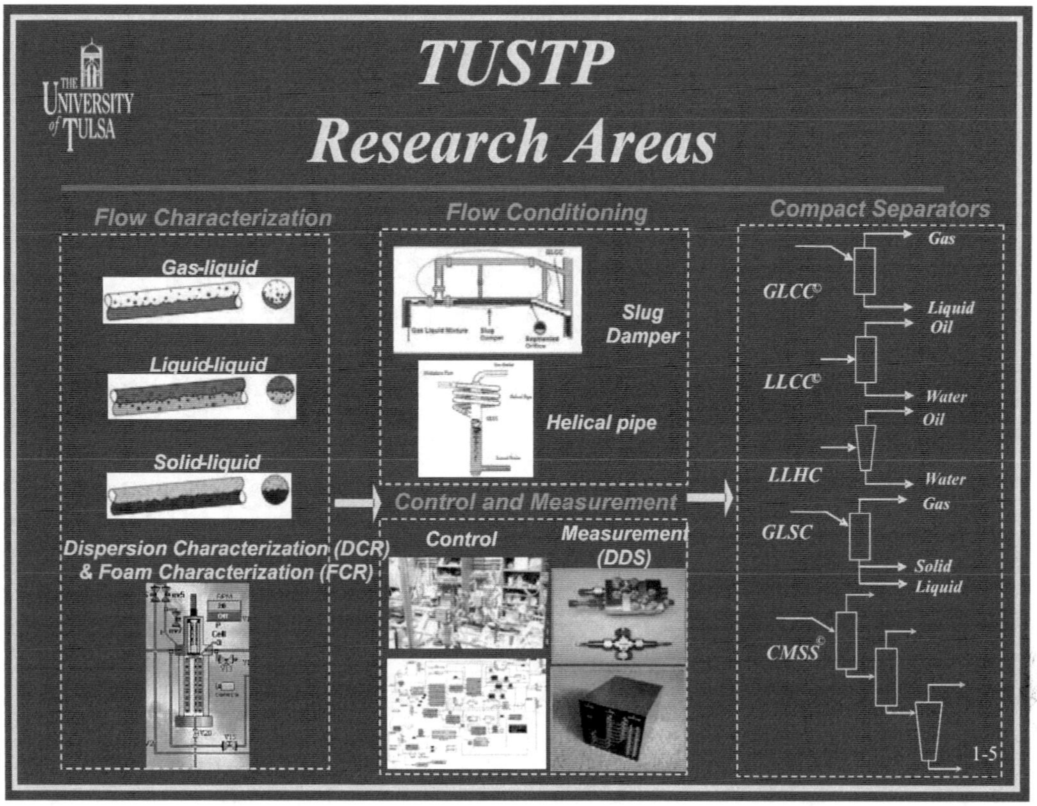
THE UNIVERSITY of TULSA
TUSTP
Research Areas
Flow Characterization
Gas-liquid
Liquid-liquid
Solid-liquid
Dispersion Characterization (DCR) & Foam Characterization (FCR)
Flow Conditioning
Slug Damper
Helical pipe
Control and Measurement
Control
Measurement (DDS)
Compact Separators
GLCC©
Gas
Liquid
Oil
LLCC©
Water
Oil
LLHC
Water
Gas
GLSC
Solid
Liquid
CMSS©
1-5

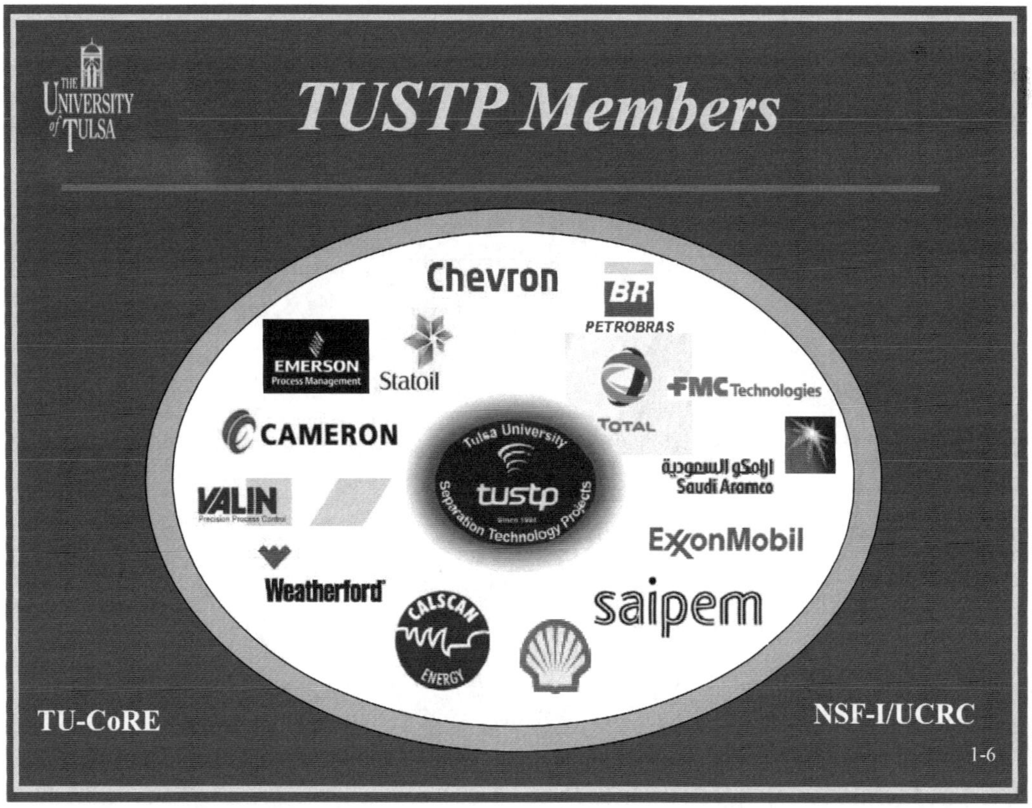
THE UNIVERSITY of TULSA
TUSTP Members
Chevron
BR PETROBRAS
EMERSON Process Management
Statoil
FMC Technologies
TOTAL
CAMERON
Tulsa University Separation Technology Projects
tustp
أرامكو السعودية
Saudi Aramco
VALIN Precision Process Control
ExxonMobil
Weatherford
CALSCAN ENERGY
saipem
TU-CoRE
NSF-I/UCRC
1-6

TUSTP Home Page
http://www.tustp.org

THE UNIVERSITY of TULSA

1-7

Topics

THE UNIVERSITY of TULSA

- Introduction – TUSTP
- Compact Multiphase Cyclonic Separators
- Research Program
- Field Applications
- Conclusions

1-8

Compact Multiphase Cyclonic Separators

- Industry has relied in the past mainly on conventional vessel-type separators
  - bulky
  - heavy
  - expensive
- Alternatives are sought, such as the Compact Cyclonic Separators, which are simple, possess low weight, low cost and efficient

1-9

How the GLCC© Works

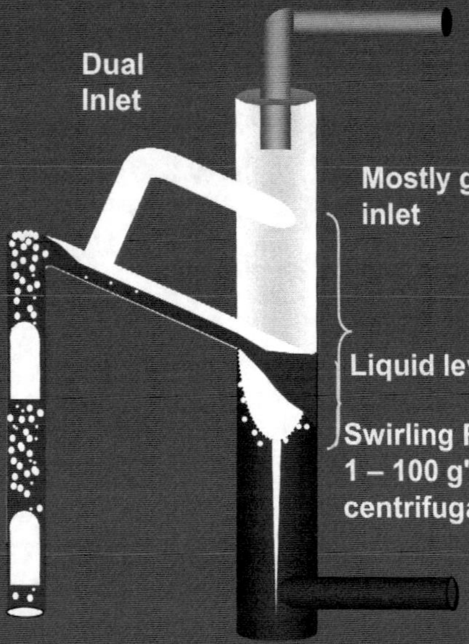

BHRG 7th Int. Conf. On Multiphase Flow, June, 1995.

1-10

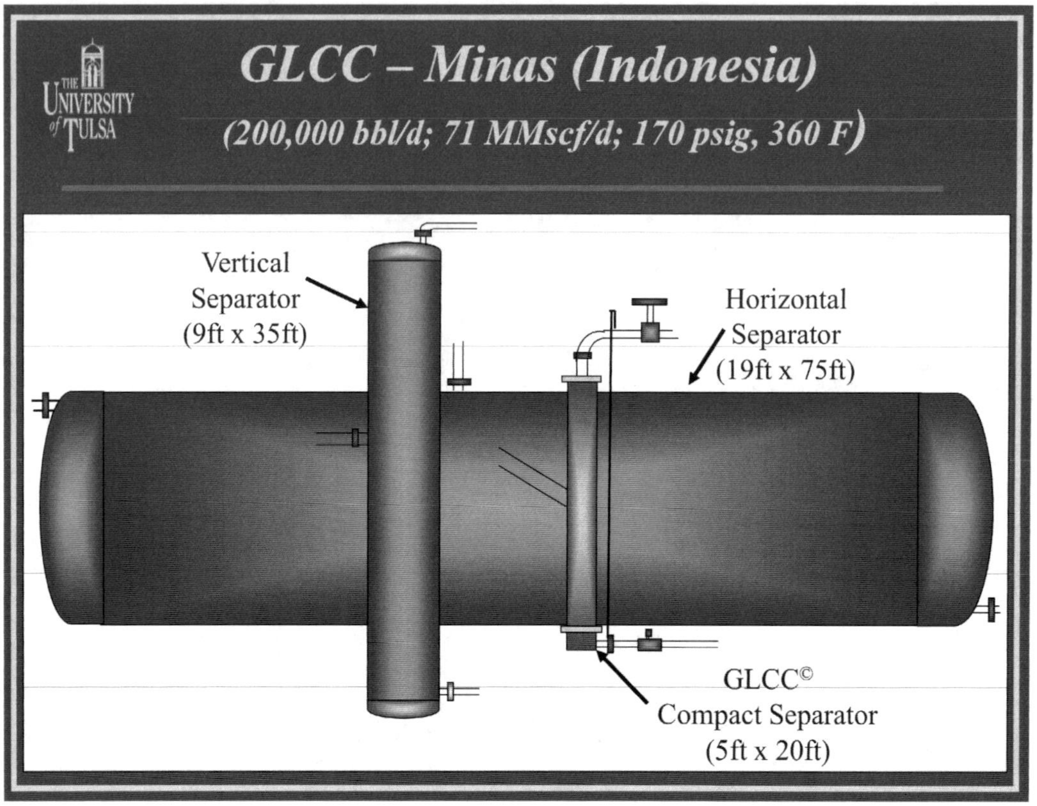
THE UNIVERSITY of TULSA
GLCC – Minas (Indonesia)
(200,000 bbl/d; 71 MMscf/d; 170 psig, 360 F)
Vertical Separator (9ft x 35ft)
Horizontal Separator (19ft x 75ft)
GLCC© Compact Separator (5ft x 20ft)

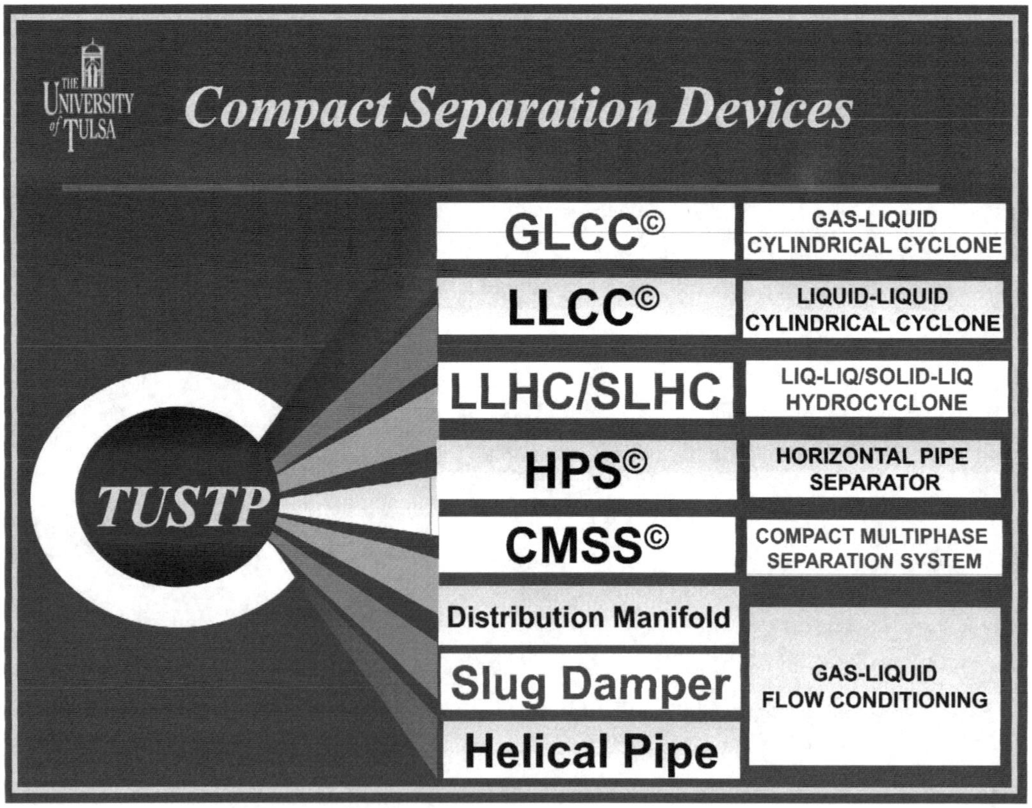
THE UNIVERSITY of TULSA
Compact Separation Devices
TUSTP
GLCC©
GAS-LIQUID CYLINDRICAL CYCLONE
LLCC©
LIQUID-LIQUID CYLINDRICAL CYCLONE
LLHC/SLHC
LIQ-LIQ/SOLID-LIQ HYDROCYCLONE
HPS©
HORIZONTAL PIPE SEPARATOR
CMSS©
COMPACT MULTIPHASE SEPARATION SYSTEM
Distribution Manifold
Slug Damper
Helical Pipe
GAS-LIQUID FLOW CONDITIONING

THE UNIVERSITY of TULSA

Topics

- Introduction – TUSTP
- Compact Multiphase Cyclonic Separators
- Research Program
- Field Applications
- Conclusions

1-13

THE UNIVERSITY of TULSA

Two-Phase Facility

1-14

THE UNIVERSITY of TULSA

3-Phase Flow Loop

1-15

THE UNIVERSITY of TULSA

Gas-Liquid Cylindrical Cyclone (GLCC©)

1-16

THE UNIVERSITY of TULSA

High Pressure GLCC© Test Facility at CEESI (up to 1500 psia)

1-17

THE UNIVERSITY of TULSA

GLCC© Operational Envelope for Liquid Carry-over

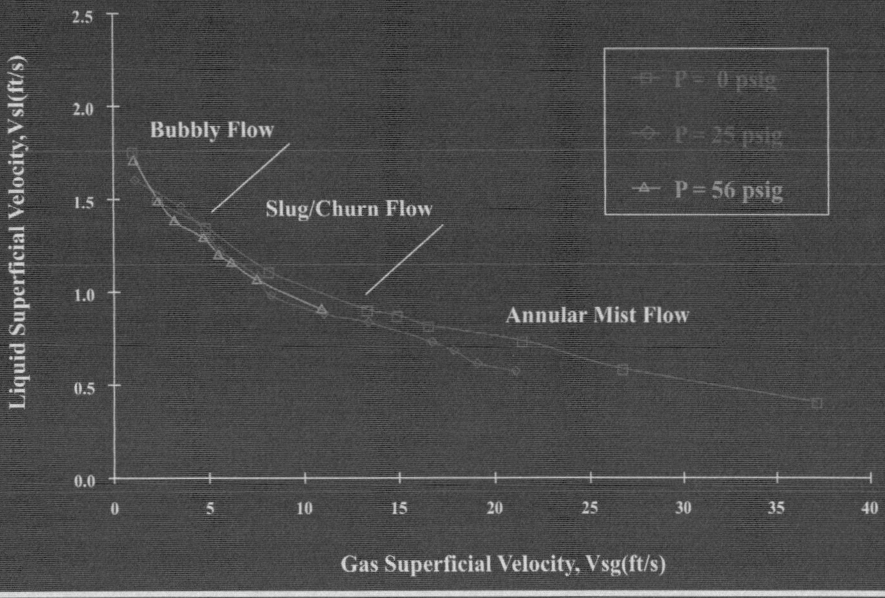

1-18

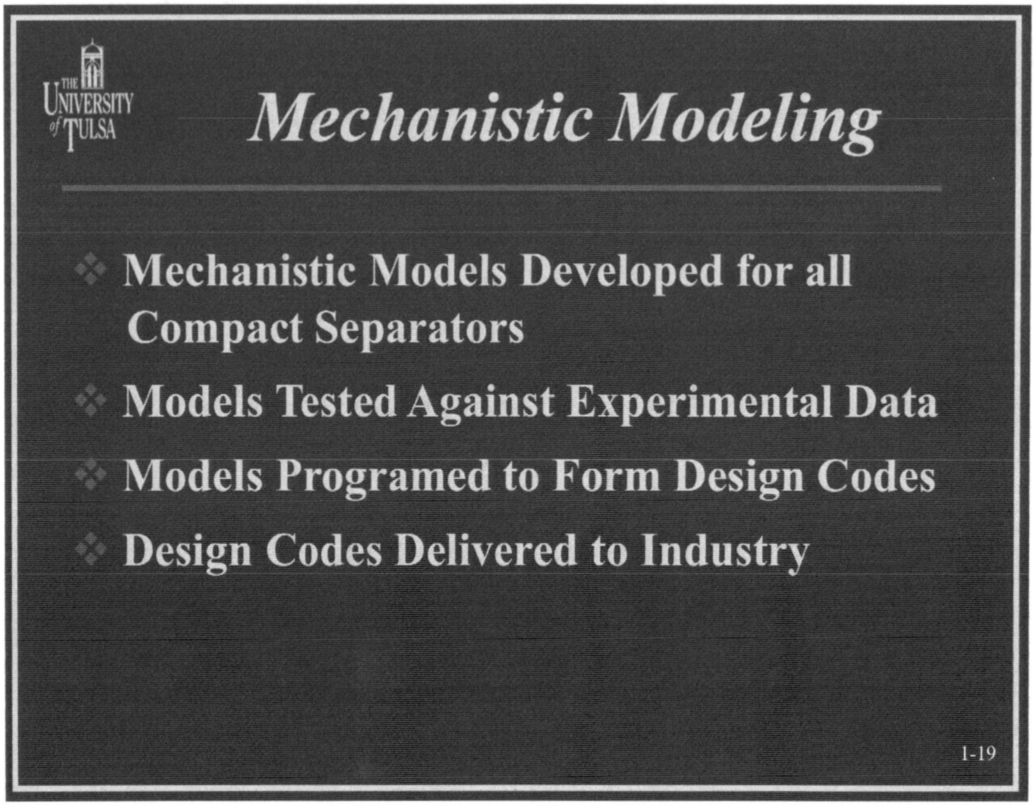
THE UNIVERSITY of TULSA
Mechanistic Modeling
Mechanistic Models Developed for all Compact Separators
Models Tested Against Experimental Data
Models Programed to Form Design Codes
Design Codes Delivered to Industry
1-19

THE UNIVERSITY of TULSA
TUSTP Deliverables
TUSTP
GLCC©vx8.7
GAS-LIQUID CYLINDRICAL CYCLONE
LLCC©vx1.1
LIQUID-LIQUID CYLINDRICAL CYCLONE
FlowpatOW-vx1.2
OIL-WATER FLOW PATTERN
DPDLOWDisp-vx1.0
Pressure Gradient OIL-WATER
Hydrocyclone-vx4.0
DEOILING &DESANDING HYDROCYCLONE
FLOPATNvx2.3
GAS-LIQUID FLOW PATTERN PREDICTION
Add-in Libraryvx1.31
Slug Damper vx2.1
GAS-LIQUID FLOW CONDITIONING

Topics

- Introduction – TUSTP
- Compact Multiphase Cyclonic Separators
- Research Program
- **Field Applications**
- Conclusions

1-21

Field Applications

- Metering Loop with Single-Phase Meters
- Metering Loop with Multiphase Meters (Gas Knockout)
- GLCC as a Pre-separator for De-bottlenecking
- GLCC as a Bulk Separator
- High Pressure GLCC for Wet Gas Application
- Portable GLCC
- Offshore Application
- Mining Application
- GLCC for Severe Slugging Applications
- Low API Gravity Application (12^0API)
- Other Applications
  - GLCC after First Stage Compressor
  - GLCC for Raw Gas Lift
  - GLCC as a Gas Scrubber
  - GLCC in a Multiphase Pumping Loop

1-22

History of GLCC© Installations

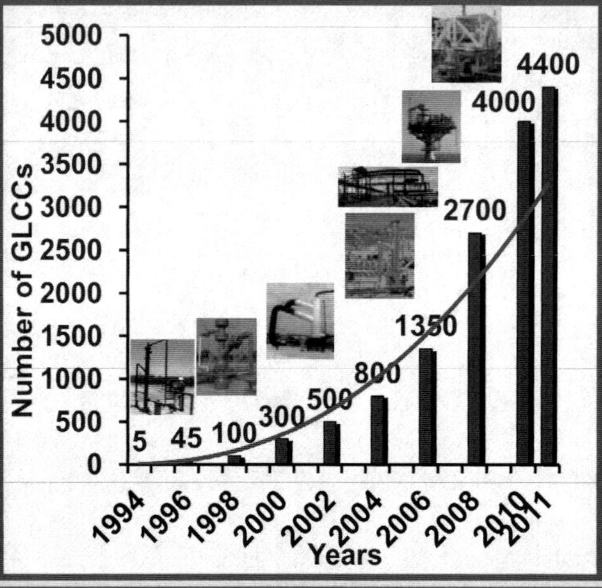

1-23

GLCC© MPFM Schematic

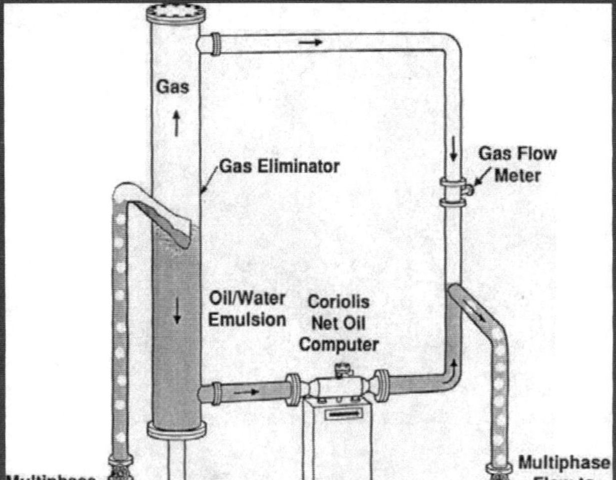

1-24

Chevron Multiphase Metering Loop

- Oklahoma
- 0.5 ft x 11 ft
- 1500 - 2000 bbl/d
- 500 - 800 Mscf/d
- 25 psia, 75 F

Automated Well Test Unit - Minas, Indonesia

- 3 ft ID, 11 ft tall
- 76 psia, 360 F
- 6,000 bbl/d, 10.0 MMscft/d
- Integrated liquid level control

Bulk Separation/Metering Loop Application - Minas, Indonesia

- 5 ft ID, 20ft tall
- 185 psia, 360 F
- 200,000 bbl/d, 71 MMscft/d
- Integrated liquid level control with Dead band filter

Bulk Separation Application Minas, Indonesia

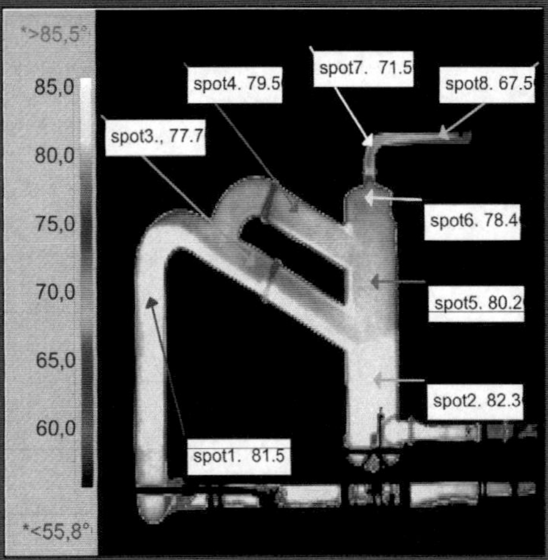

GLCC© (60-inch) Application in a Bulk Separation/Metering Loop

1-28

China Offshore Platform GLCC for Wet Gas Metering (CNOOC)

1-29

1st Subsea GLCC© - Petrobras

1-30

1st Compact Multiphase Separation System (CMSS©) – Saudi Aramco

THE UNIVERSITY of TULSA

31

Conclusions

THE UNIVERSITY of TULSA

- **Cooperative Research Program between University, Industry and Government**
- **Compact Separation Technology R&D- From Lab to Field**
- **Potential for Pronounced Impact on Oil and Gas Industry World-wide**

1-32

THE UNIVERSITY of TULSA
Before…..
…from this…
1-33

THE UNIVERSITY of TULSA
After…..
…to this !
1-34

Surveillance and Remediation of Wax and Asphaltene Impairment

6th AIChE/SPE Joint Workshop
Challenges in Flow Assurance
September 26-28, 2011
Houston, TX, USA

subsea
surveillance team

George Broze, Shell International Exploration and Production Inc.

Chris Holloway, Shell Exploration and Production Company

Impact of Organic Deposition

- Impact to topsides operations
  - Asphaltenes can fill up in inlet separator, level control bridle.
  - Wax can form an emulsion pad creating separation issues.
  - Both events lead to trips and downtime for cleanout.
  - Inject chemicals to mitigate (\$100K-\$1MM/yr).
- Impact to well production
  - Asphaltenes deposit in wellbore and tubing reduce inflow.
  - Well integrity and reserves at risk.
  - Inject chemicals to inhibit deposition during production.
  - Chemical soaks to remediate (\$100K-\$3MM).
- Impact to subsea flowlines
  - Asphaltene or wax deposition on pipe wall.
  - Due to flowline size (+6" ID), build up not detected immediately.
  - Chemical inhibition is effective, but soaks require huge volumes, downtime.
  - Major intervention to remediate plugs (\$3MM-\$30MM).

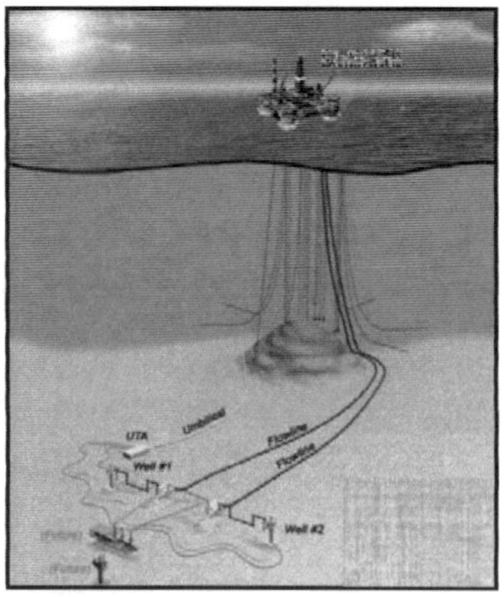

2

Conceptual Risk Mitigation

Prediction & Prevention

Remediation

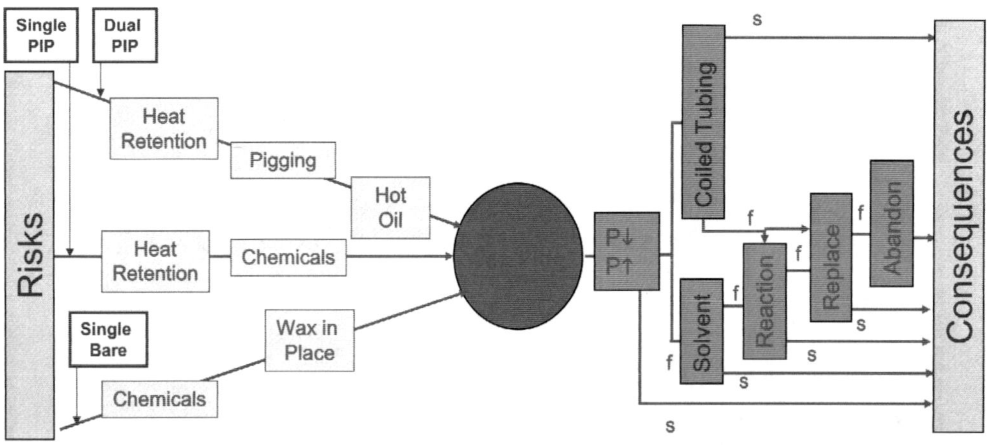

Design & Surveillance

Surveillance/Operations

3

Sampling and Fluid Properties

Fluid Sampling

The perfect sample would be acquired and transported in a single chamber and maintained at reservoir temperature and reservoir pressure.

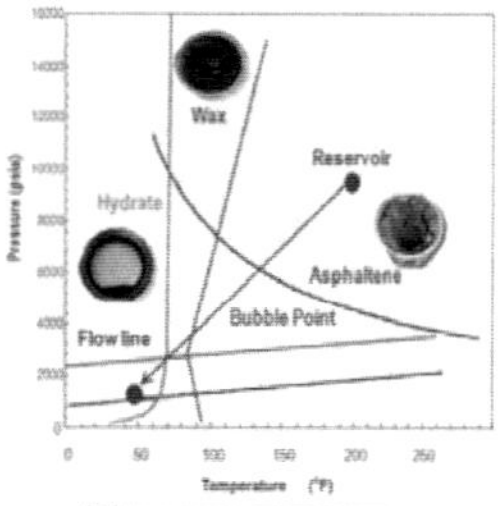

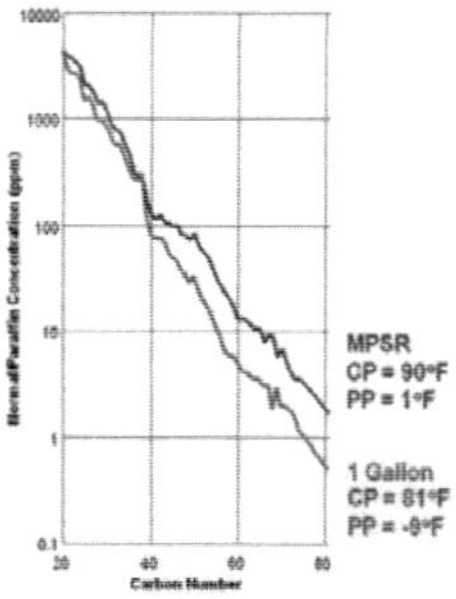

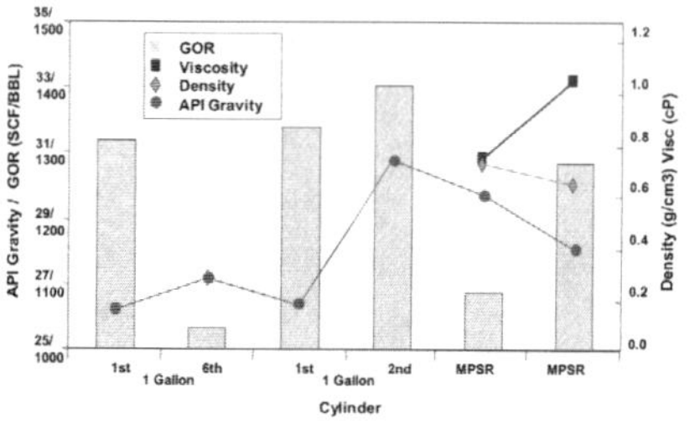

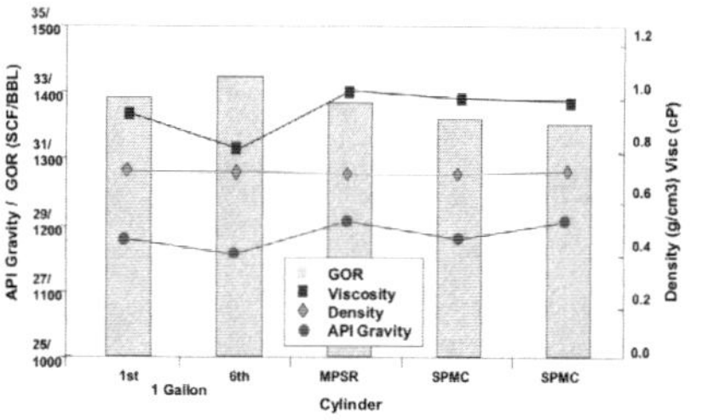

5

Fluid Surveillance

Produced fluid properties change.

Field C

- 25 mile single-flowline subsea tieback
- 2004 – startup – bad wax problems
- 2006 – burial mitigated wax
- 2008 – asphaltene problems began in flowline and topsides

Field P

- 23-mile dual subsea flowline, gas condensate since1995 with no wax problems
- 2011 – pigged before hydrotest – plugged line with wax!
- 2009 – condensate had changed color

Field S

- Operated above accepted WAT
- Blowdown for hydrates management generated plug in riser
- Coiled tubing job removed wax plug several hundred feet long

Track your fluid properties.

6

Prevention and Surveillance

Prevention & Surveillance

Strategy	Method	Wax	Asph	Comment
Prevention	Heat retention	✓	×	Effective for moderate-length subsea offsets; high CAPEX
	Inhibitors	✓	✓	Reduce but do not eliminate wax deposits; high OPEX
	Surfaces/ Coatings	✓	✓	May prevent deposits at low-to-medium τ_w; depends on surface and deposit type
	DEH Flowlines	?	×	High CAPEX; high power to keep wall above WAT
	Cold Flow	✓	×	Would prevent deposits in long tieback without CAPEX of dual PIP lines
	Pressure Maintenance	×	✓	Not always an option
	Control Commingling	?	✓	Production deferment
Surveillance	Fluid Sampling	✓	✓	Topsides has uncertainty; subsea requires design and vessel
	T, ΔP	✓	✓	The basics
	Tracers	✓	✓	Shown to be more accurate than flow-ΔP methods.
	Data Pig	✓	✓	Indicates location and qualitative amount; needs interpretation

8

Prevention Approach

- Design approach for prevention:
  - Fluid sampling and analysis in exploration phase.
  - Thermal insulation to keep subsea temperatures above critical wax deposition temperature.
  - Chemical inhibitor injection at wellbore and tree.
  - Pressure and temperature indication across flow path.
- Operations approach for prevention/indication:
  - Surveillance of ΔP and temperatures.
  - Sustain chemical injection where prescribed.
  - Proactive solvent treatments when deposition is suspected.
  - Routine fluid sampling to measure chemical effectiveness.
  - Properly addresses deposition if identified in the assessment phase.

9

Field S Flowline Plug

- Field S is a 2-well, 5.5-mile, single-flowline tieback designed with electrical heating for hydrate management during shut-ins.
- Fluid properties in project phase showed no expected deposition of wax or asphaltenes.
- EH used successfully several times. However, after a blowdown, a plug was detected while re-pressuring the flowline for restart.
- Uncertainty regarding plug (hydrates/wax/ sand/ mechanical). Hydrate plug was leading theory based on surveillance data available.
- Coiled tubing was deployed for flowline cleanout.

10

Remediation

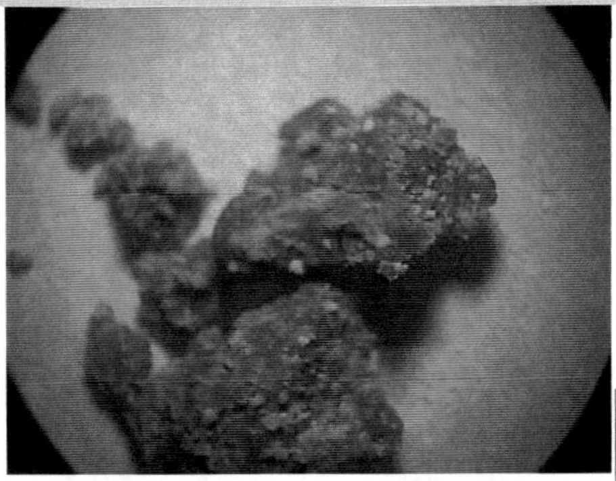

- 20 bbls of solids were removed from the flowline; > 900' in the riser
- HTSD & HTGC analysis showed that the deposit was ~60% wt wax and < 5% inorganic.

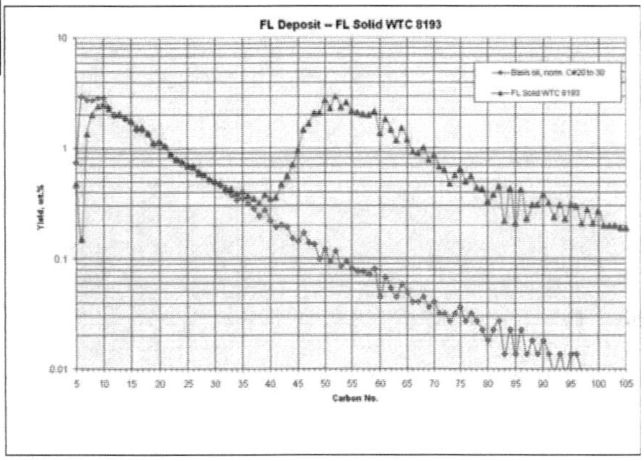

- This event led to more proactive surveillance and equal focus for deposition risk, including assets with no prior indicators.
- Developed methods to analyze data over longer time periods (years) to seek out deposition indicators.

Surveillance Methods – ΔP in Subsea Flowline

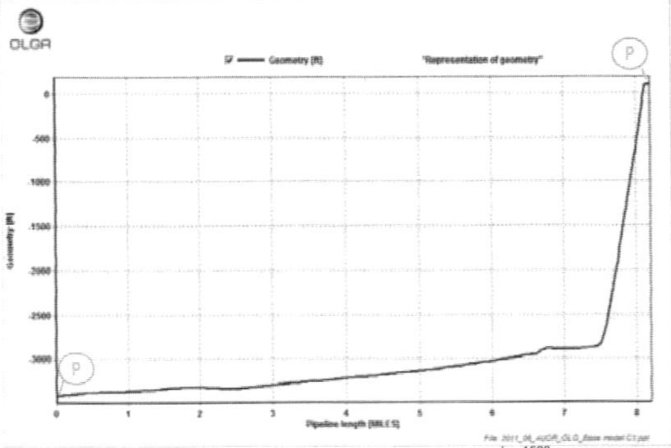

- Limited by single gauges at manifold and riser top over a long offset.
- Friction dependent in flowline; gravity dependent in the riser.

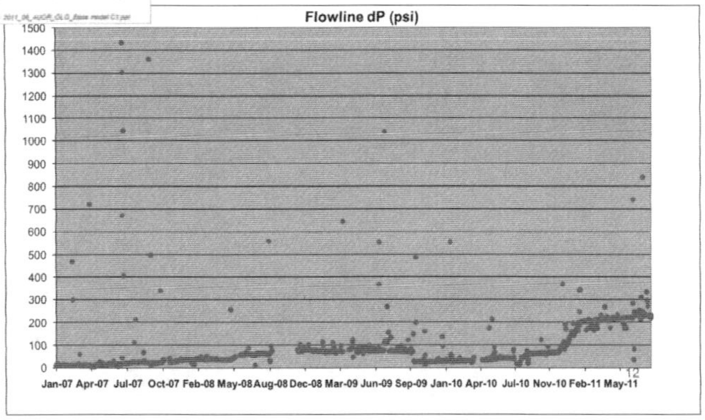

- Rate component required in analysis to offset natural decline.

Surveillance Methods – ΔP in Subsea Flowline

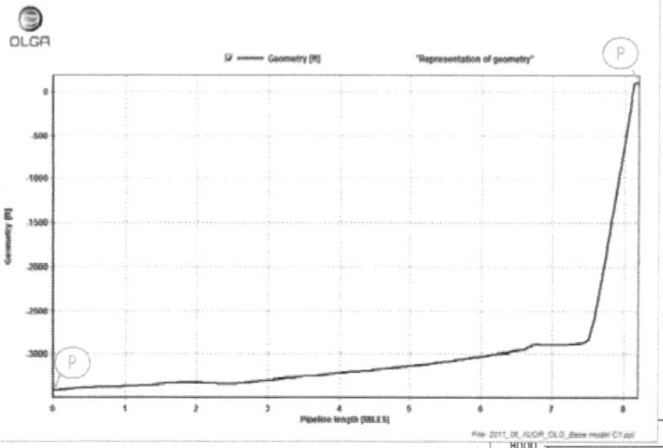

- Limited by single gauges at manifold and riser top over a long offset.
- Friction dependent in flowline; gravity dependent in the riser.

- Metric to assess ΔP changes independent of rate required ($\Delta P/Q^2$).

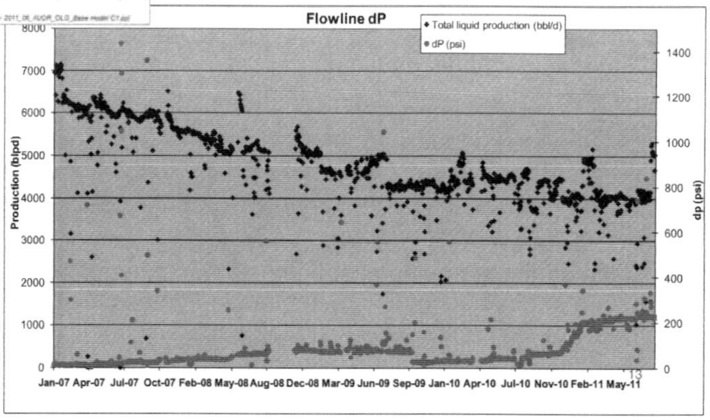

Surveillance Methods – ΔP in Subsea Flowline

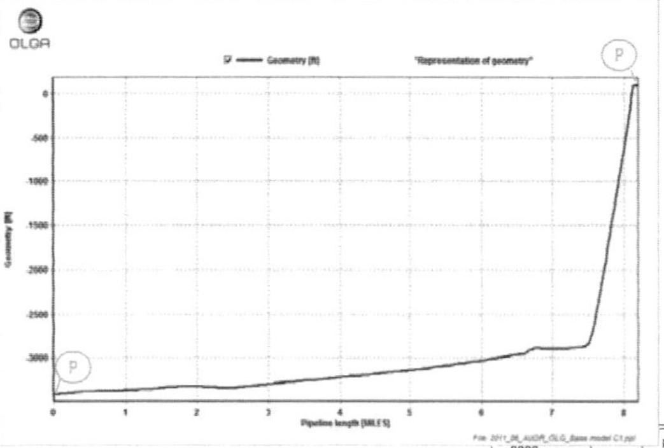

- Limited by single gauges at manifold and riser top over a long offset.
- Friction dependent in flowline; gravity dependent in the riser.

- Indicates deposition *could* be a factor.
- $\Delta P/Q^2$ less reliable in multiphase flow.
- Less reliable at low rates.
- Further analysis is required to assess degree of deposition.

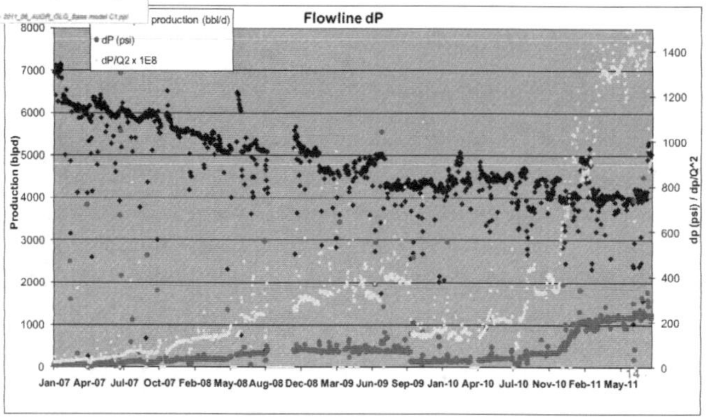

Surveillance Methods – Arrival Temperature

- Monitoring CWDT vs. temperature indication of deposition risk in riser.
- Temperatures decrease over time with production rates.
- Even if temperatures are below precipitation point, deposition is not inevitable.
- Often, J-T cooling in the riser obscures flowline ΔT.

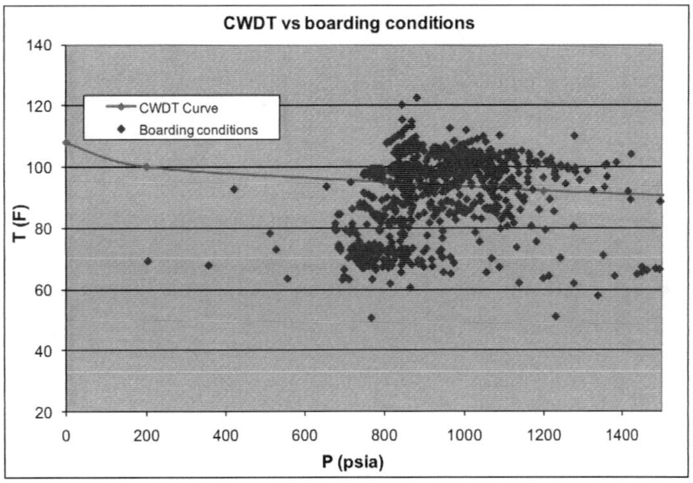

15

Surveillance Methods – Process Modeling

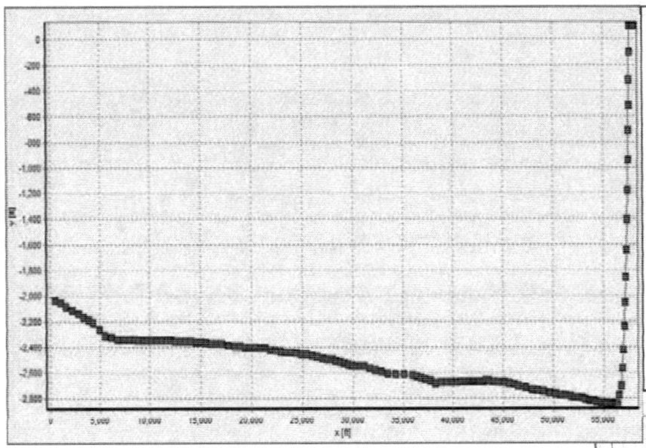

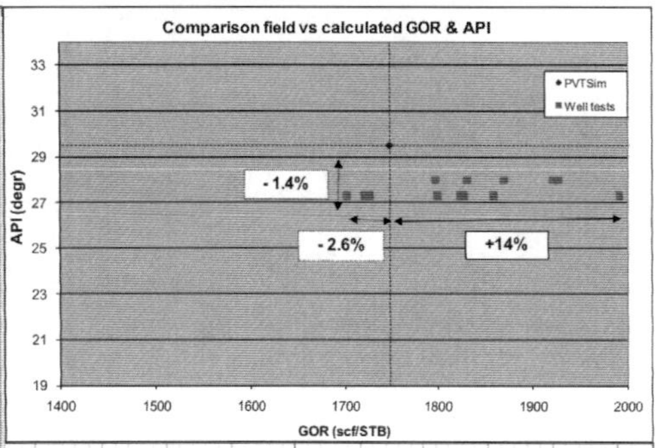

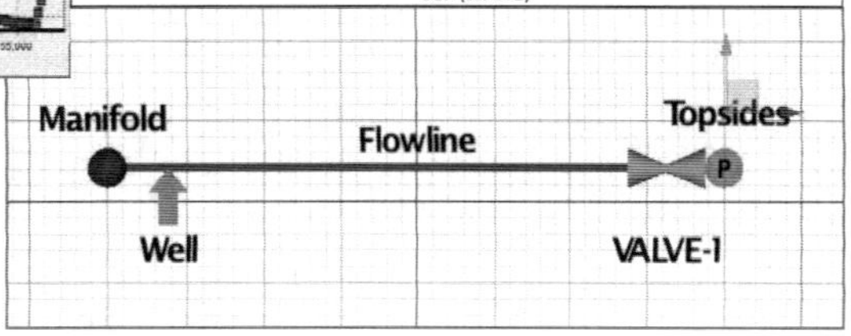

- Build flowline model that incorporates as-built coordinates with thermodynamic properties and PVT analysis for fluid definition.
- Better analyzes ΔP and thermal performance for multi-phase flow.
- High deviation an indication of deposition.

16

Surveillance Methods – Process Modeling

- Look back at Field S flowline after wax plug.
- Clear sign of changes in hydraulic performance.

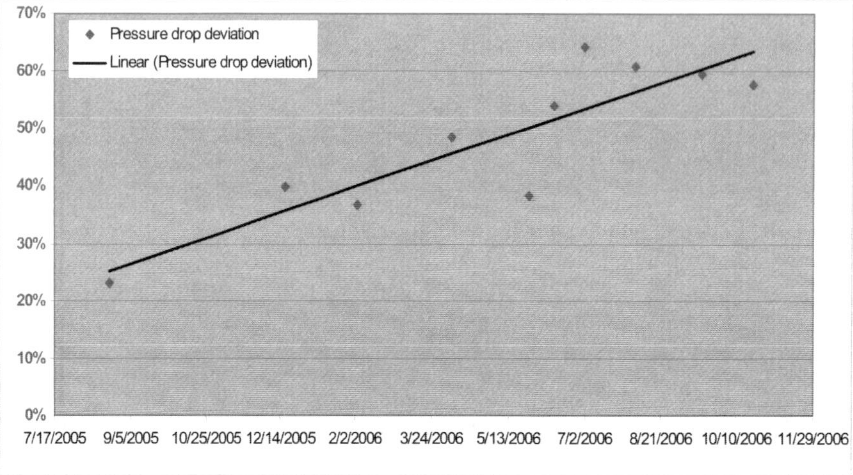

17

Surveillance Methods – Process Modeling

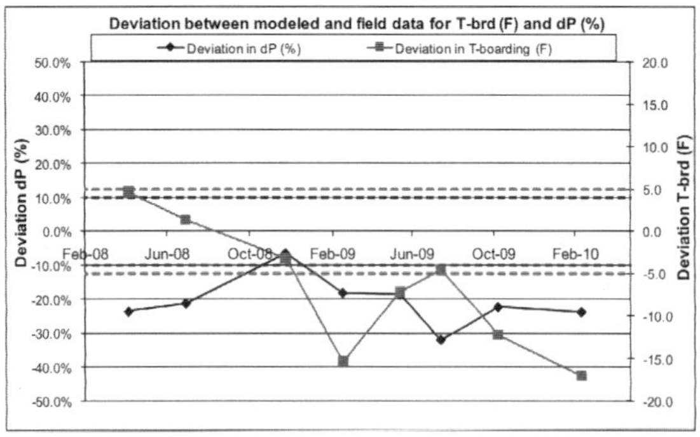

- Inaccurate benchmarking could be false indication of deposition.
- Fluid tuning and characterization with PVT data will improve benchmarking.
- Accuracy of temperature transmitters may vary.

- Tune fluids.
- Change temperature monitoring point.

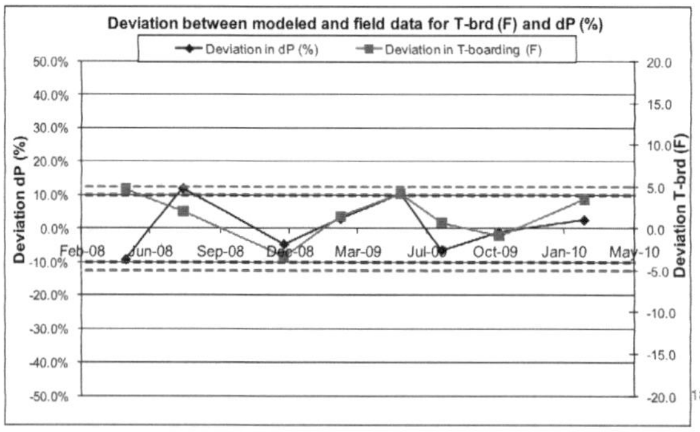

18

Asphaltenes - Temperature Trending (OTC 21587)

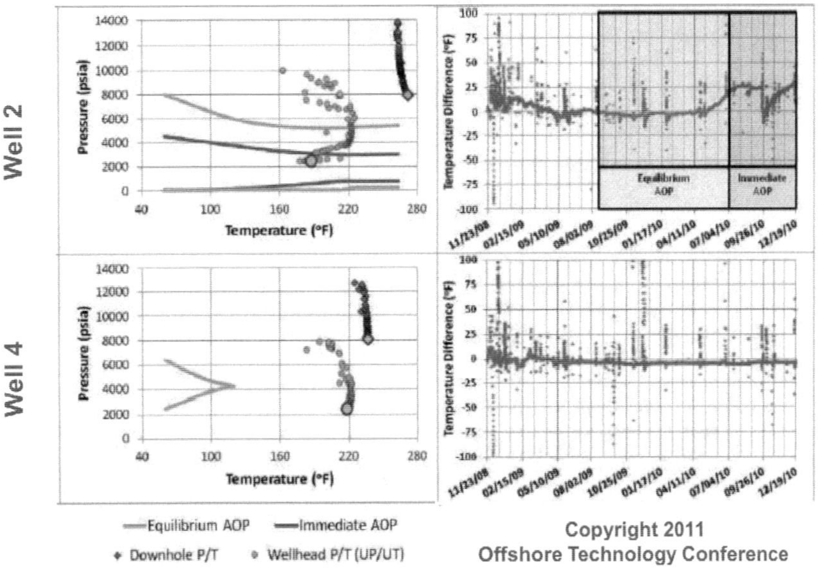

Asphaltene Management in GOM DW Subsea Development

Alberto Montesi, Chevron North America Exploration and Production Company, Robert Pinnick, Chevron North America Exploration and Production Company, Siva Subramanian, Chevron Energy Technology Company, Jianxin Wang, Chevron Energy Technology Company, Jeff Creek, Chevron Energy Technology Company

19

Chemical Tracer Testing (OTC 16690)

- Evaluated a range of tracers and chose MeOH as safe, available and detectable.
- Field A: Dual PIP 2.5" flowlines tied back 4.5 miles
  - 8.4 bbl wax removed by hot-oiling 310 bbl system
- Field B: Dual PIP 8" flowlines tied back 10 miles
  - Tracer showed no significant deposits.
  - Asphaltene inhibitor turned off
- Field M: Dual PIP 6" flowlines tied back 12 miles
  - Tracer showed 150 bbl deposit
  - 30 bbl removed by hot oiling before platform trip
  - No tracers have subsequently been run

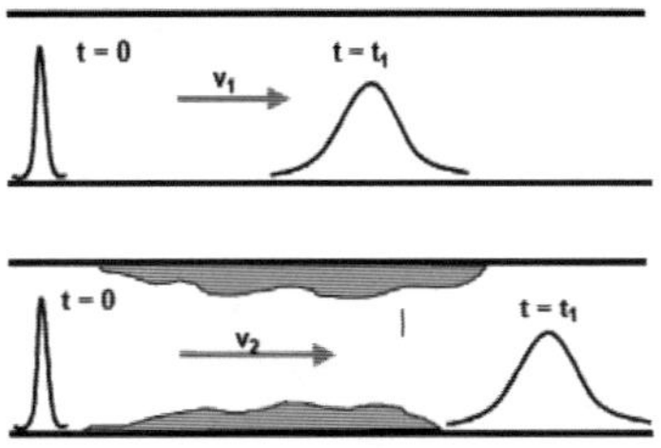

Figure1. Principal of the tracer test: at constant rate the higher average velocity caused by deposits results in the tracer arriving sooner or traveling further in the same time.

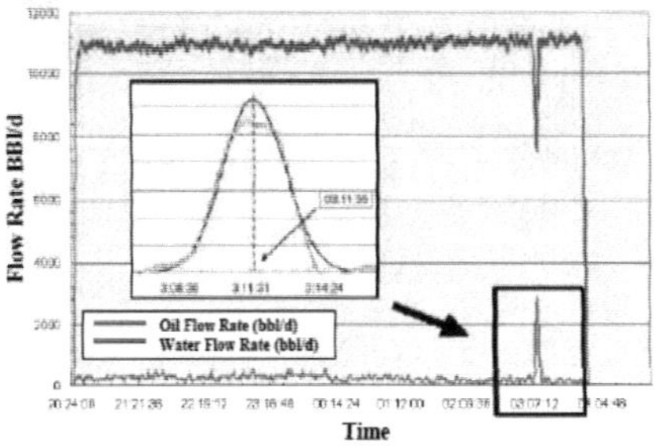

Figure 7 Example of Multi-Phase Meter detection of a methanol slug during a tracer test.

Flowline Deposition Assessment Using Chemical Tracers

William J.R. Nisbet, Kevin Cooper, Philip Wollenberg,
Shell E&P Company,
Moye Wicks, Szabolcs Balkanyi
Shell Global Solutions (US) Inc.

20

Gaps in Prevention/Surveillance Methods

Instrumentation and Models

- Limited pressure and temperature transmitters in system. Lack of redundancy.
- Tools equipped to detect larger deposits, but light amounts go undetected.

Direct Measurement Methods

- Not all systems have the capability to circulate pigs or tracers

Recommended Improvements

- Riser-base gauges would reduce uncertainty significantly
  - Friction vs. gravity ΔP
  - JT cooling in risers
- Accurate subsea multiphase flow meters (fluid changes, multi-well production)
- Distributed sensors
  - Enormous data capacity
  - How do you use all the data efficiently?

21

Remediation

Maintenance/Remediation Strategies

Strategy	Method	Wax	Asph	Comments
Maintenance	Standard pigging	✓	?	Proven approach if frequent enough; deferred production in subsea loops
	Hot oiling	✓	×	Can be effective for insulated systems
	Batch chemical treatment	✓	✓	OPEX & deferment
Remediation	Progressive pigging	✓	?	Slow process with high production deferment subsea
	Chemical Cleaning	✓	✓	Most effective with large volumes
	Coiled tubing	✓	✓	Proven; limited by weight/footprint, reach distance
	Burial	✓	×	One-time OPEX; can significantly increase heat retention in late life
	Exothermic heating	✓	×	Commercially available; HSE and logistical considerations
	Autonomous Devices	✓	?	"Jet pig", "crawler pig"; concerns about slow, untethered devices

23

Field C Wax Remediation by Burial

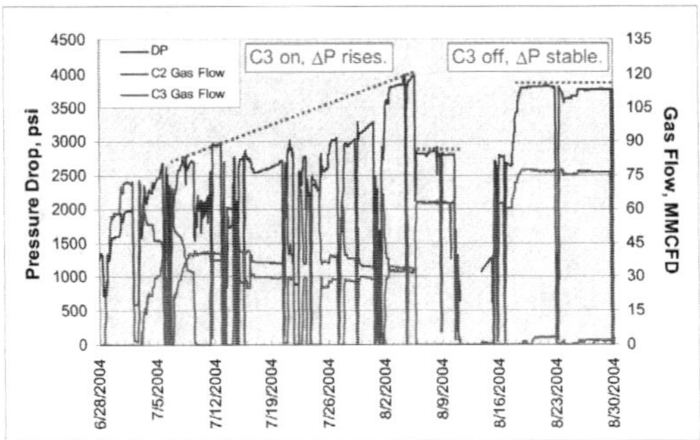

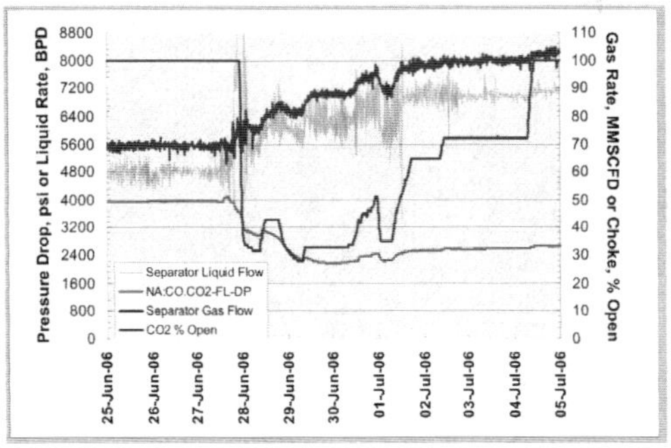

- Severe ΔP increase over very short time.
- Traced to waxy properties of 2nd well.
- Marginally effective management by shutting in 2nd well, toggling wells on and off.
- Decided to bury flowline
- Within a week, Q was up by 50% and ΔP was down by 35%.

24

System Cleanup

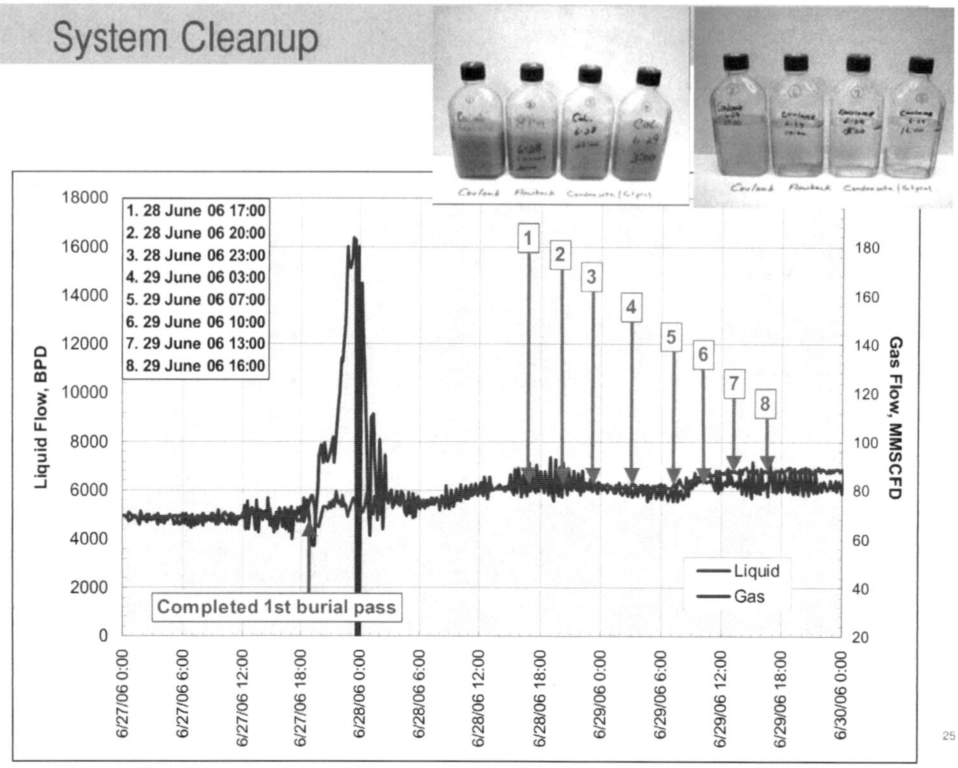

25

Field N Wax Remediation by Chemical Circulation

- 17-km dual 6" tieback, trenched and rock-dumped in shallow water.
- Wax identified when a subsea tie-in was carried out.
- Estimated that up to 510 bbl of wax was in the flow line.
- Numerous methods were considered and rejected over a period of years: progressive pigging, coiled tubing, novel devices.
- Chemical circulation was finally chosen. Terpene-based product was chosen to balance performance and HSE concerns.
- 3800-bbl system filled and circulated with chemical, followed by a gel pig.
- An estimated 370 bbl of wax was removed from the system.
- An additional 3000 BPD of oil was produced.

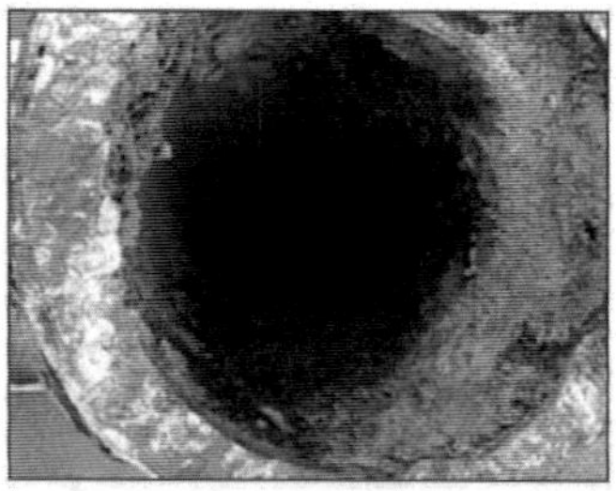

Field G: Wellbore ΔP Build-up

- From start-up, gradual increase in rate-adjusted wellbore ΔP

- FTP above bubble point through July '06
- Late '05: Asphaltene inhibitor increased significantly (150 ppm -> 1000 ppm) and shut-off over several day periods
- Variations in asphaltene inhibitor injection rates had no effect on increase in wellbore dP

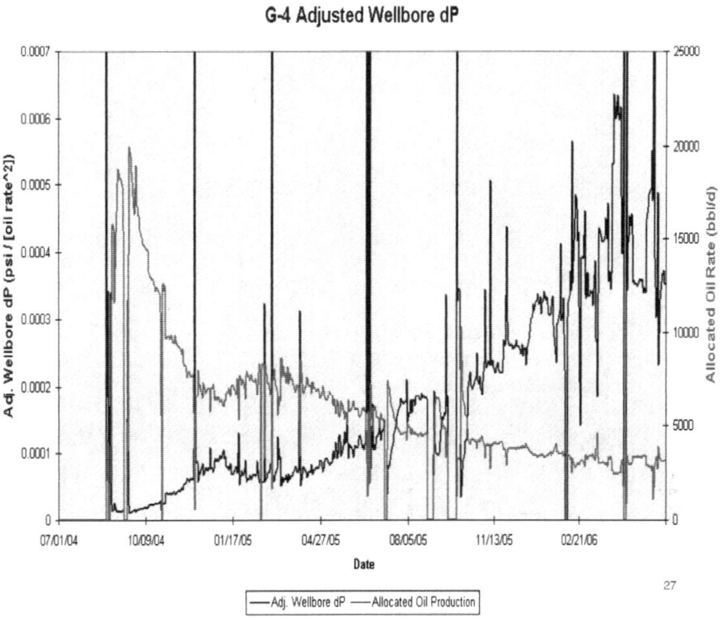

27

Well 4 Xylene Soak Results

- 1st soak completed May 17th & 18th, 2006
- First xylene treatment delivered to a subsea well from the host via existing umbilical system in the GoM
- Flush successful → 600 psi decrease in wellbore ΔP following soak, no associated topsides issues during flowback:

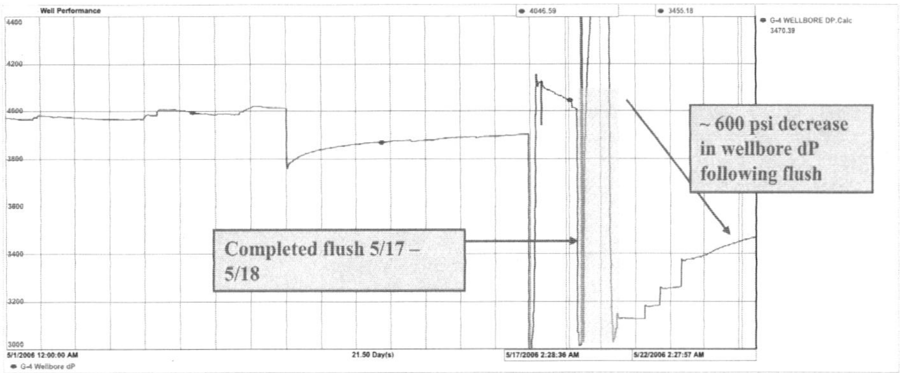

28

Observations and Conclusions

- It is impossible to anticipate all problems at the **design** phase.
- Even if we did, we could not afford to prevent them all.
- We must rely on **surveillance** to alert us to the *critical* issues – the *earlier* the better.
  - Fluids, instrument data, process models, tracers
- There are *many* available **remediation** tools
- There are *many* constraints
  - System design – access points, weight/footprint
  - Cost – especially deferred production vs. vessels/rigs
  - Logistics, availability & schedule
- Remediation would be greatly improved by reliable **deposit profiling** and **blockage location**
- We must continue to **innovate** both new tools and how we use them.

29

Acknowledgments

The authors would like to thank our respective companies for permission to present these data.

We would also like to thank several colleagues for consultation, data and reviews:

- William Nisbet, SEPCo
- Zhongxin Huo, Karthik Ramanathan & Nancy Utech, SIEP Inc.

30

31

Paraffins or Waxes
"The Cholesterol of the Petroleum Industry"

Costs the industry billions of dollars annually

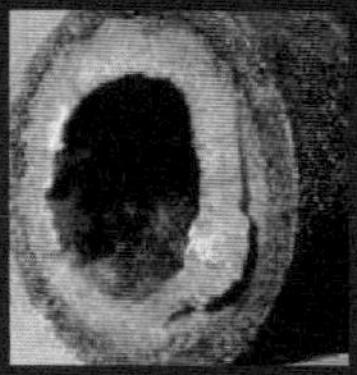

Wells Productivity
- Lower production
- Downtime during wax remedial jobs
- Expensive wax chemicals

Flowlines Management
- Extra insulation on flowlines
- Dual lines to enable round trip pigging
- Downtime during pigging
- Cost of chemical program

Greater challenge for deepwater projects (marginal fields)
- Unattractive economics for long tie-backs
- Need novel technologies

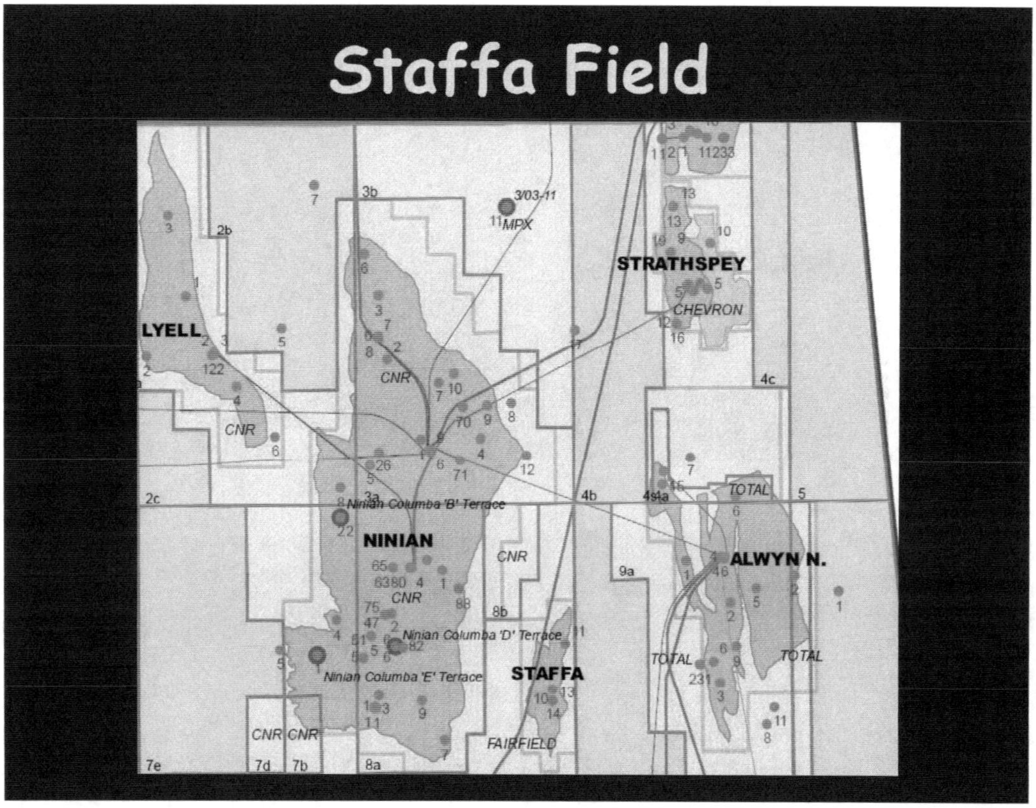
Staffa Field
LYELL
NINIAN
STRATHSPEY
CHEVRON
ALWYN N.
STAFFA
FAIRFIELD
TOTAL
CNR
Ninian Columba 'B' Terrace
Ninian Columba 'D' Terrace
Ninian Columba 'E' Terrace
3/03-11
MPX
2b
2c
3b
3a
4b
4a
4c
5
7e
7d
7b
8a
8b
9a

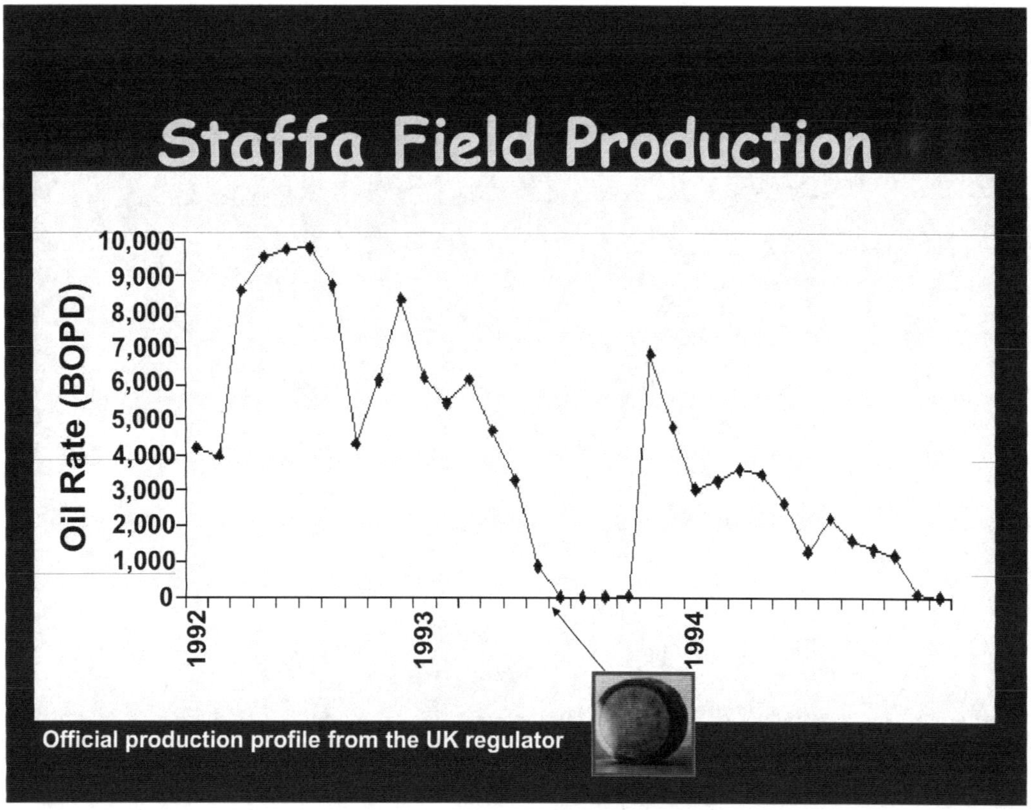
Staffa Field Production
Oil Rate (BOPD)
10,000
9,000
8,000
7,000
6,000
5,000
4,000
3,000
2,000
1,000
0
1992
1993
1994
Official production profile from the UK regulator

Badami Development

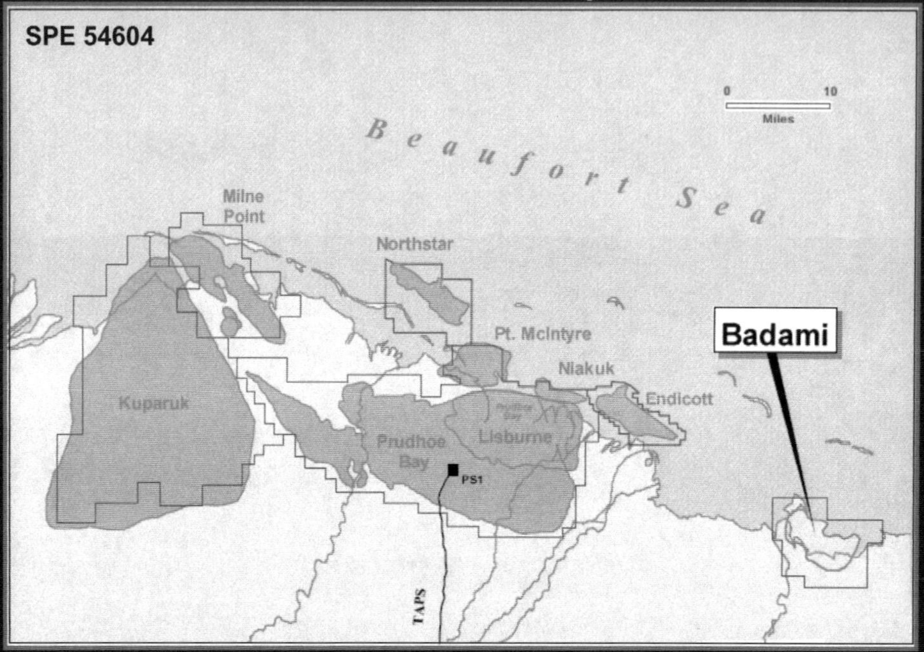

Badami Development

- Discovered by Conoco in 1990 (initial DST >4,000 BOPD)
- Conoco sold its share to BP
- Showcase project for BP Management
- Design included well pad and a processing facility
- 35 mile long production pipeline and a 6-in Miscible Injectant pipeline for WAG
- Expected recovery 120 MMbbl with 35,000 BOPD peak rate
- First oil in mid 1998 with peak prod of ~17,000 BOPD
- By late 1998, production dropped to < 5000 BOPD
- Suspended in 2003 with the rate of ~1300 BOPD
- During 2005-2007, produced seasonally with total of 850,000 Barrels
- Savant and ASRC have joined BP and restarted the production in late 2010

Managing Paraffins for Flow Assurance

What do we know

- What are paraffins or waxes?
- Paraffins phase behavior
- When and why do paraffins deposit?
- Prediction of paraffin deposition
- Current methods of paraffin control

What are the challenges

- Challenges in deepwater and arctic
- Potential new paraffin technologies
- Summary

What are Paraffins or Waxes?

- ***Saturated component of a crude oil – crystallizes upon cooling***
- ***Structure***

  $CH_3(CH_2)_nCH_3$

  n > 20 Petroleum Wax

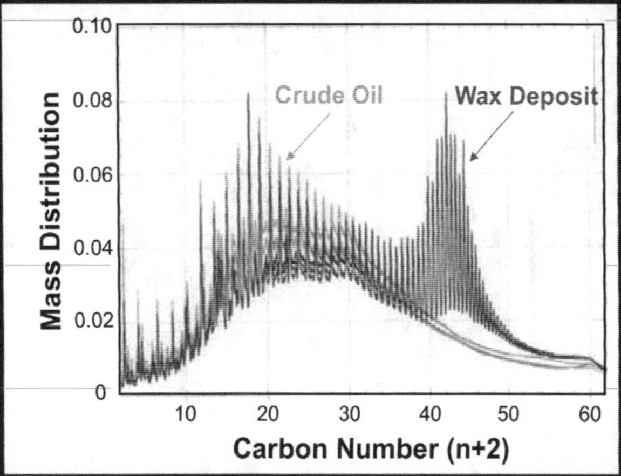

- ***Field Definition***

  A low melting point soft solid deposit that forms on cold walls of well tubing, flowlines and oil transport pipelines

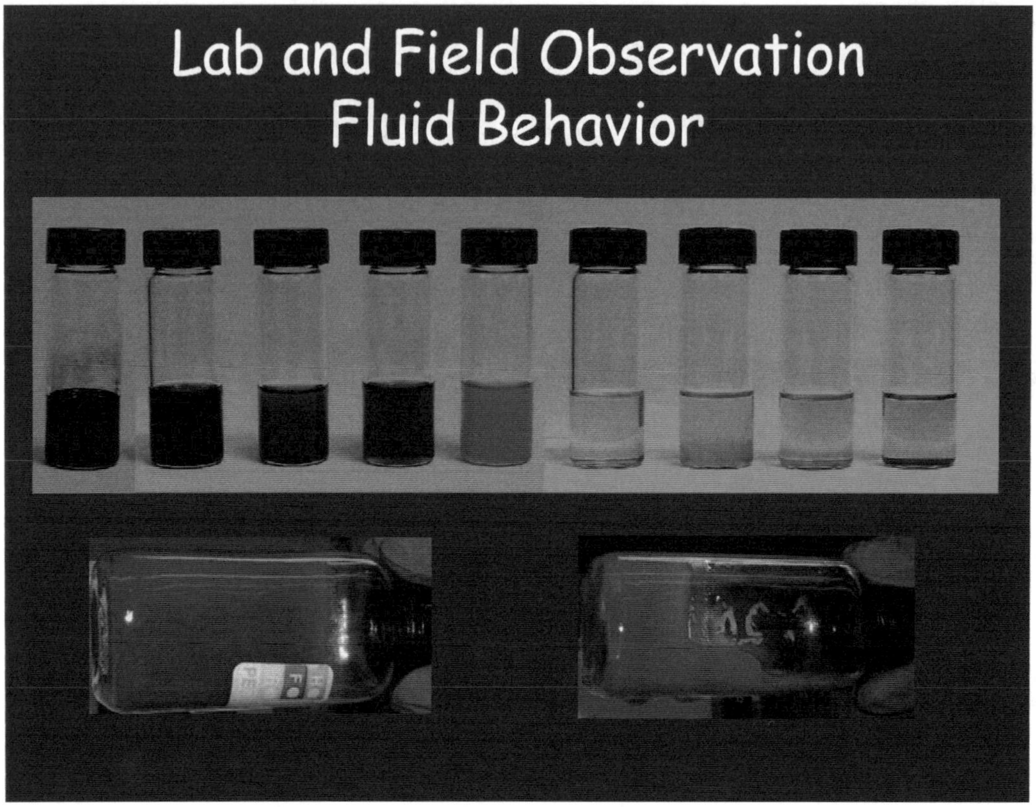
Lab and Field Observation
Fluid Behavior

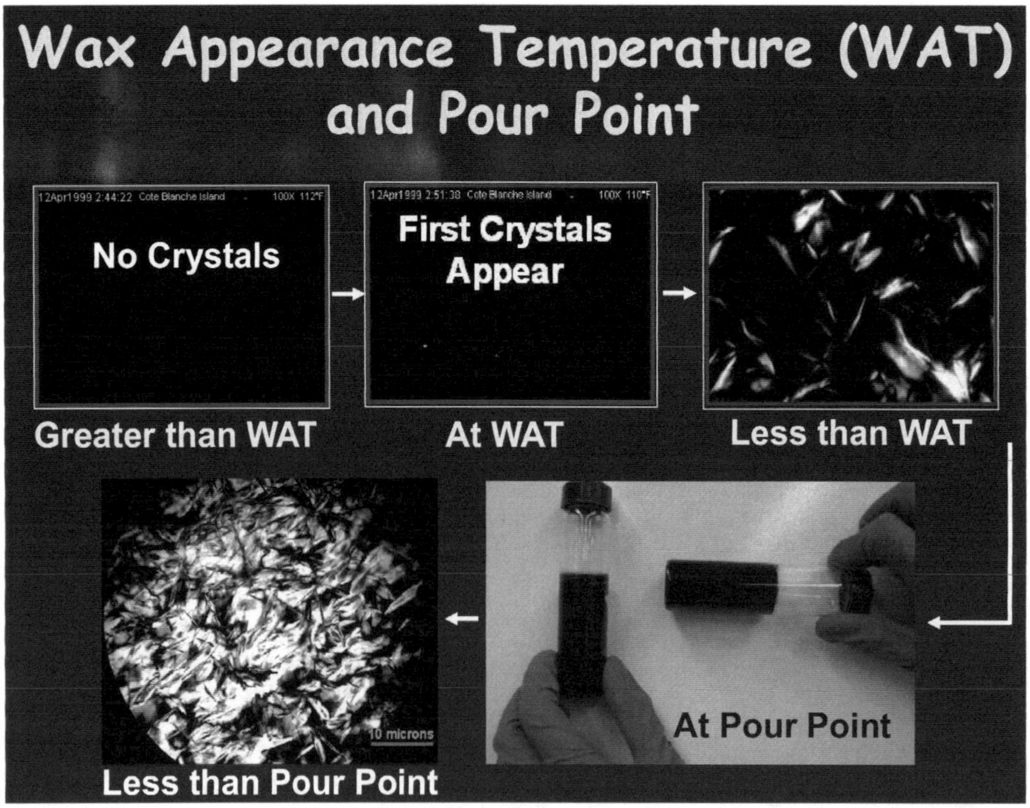
Wax Appearance Temperature (WAT)
and Pour Point
12Apr1999 2:44:22 Cote Blanche Island 100X 112°F
No Crystals
12Apr1999 2:51:38 Cote Blanche Island 100X 110°F
First Crystals
Appear
Greater than WAT
At WAT
Less than WAT
10 microns
Less than Pour Point
At Pour Point

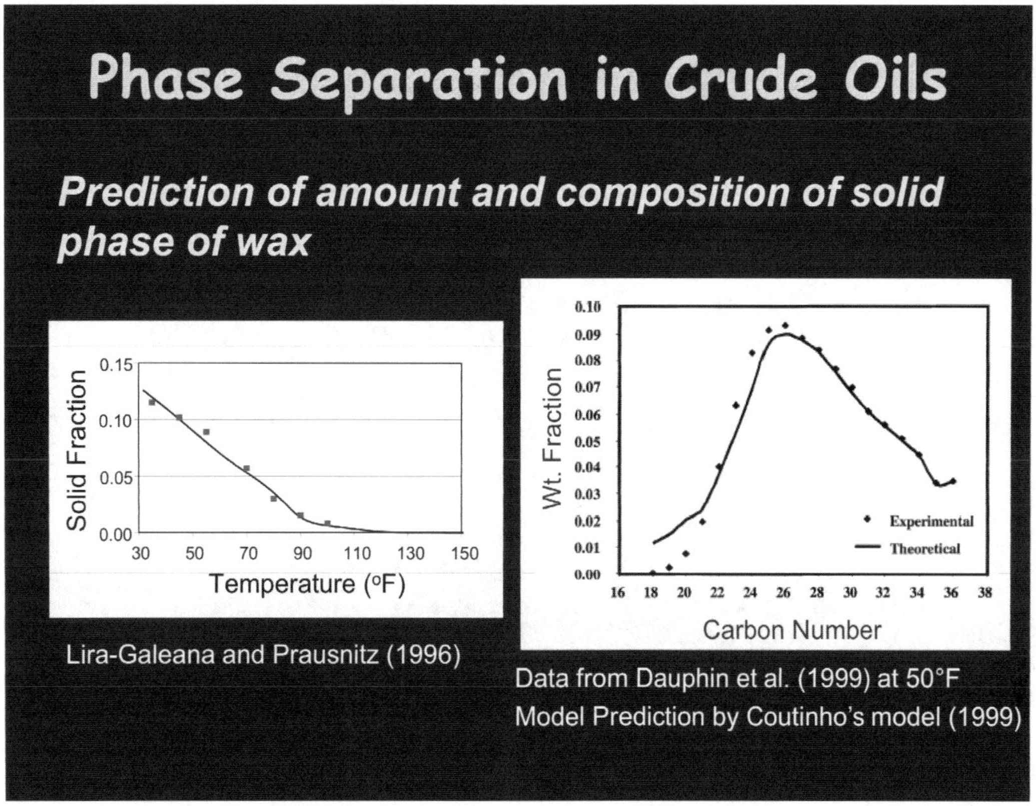
Phase Separation in Crude Oils
Prediction of amount and composition of solid phase of wax
Solid Fraction
0.15
0.10
0.05
0.00
30 50 70 90 110 130 150
Temperature (°F)
Lira-Galeana and Prausnitz (1996)
Wt. Fraction
0.10
0.09
0.08
0.07
0.06
0.05
0.04
0.03
0.02
0.01
0.00
16 18 20 22 24 26 28 30 32 34 36 38
Experimental
Theoretical
Carbon Number
Data from Dauphin et al. (1999) at 50°F
Model Prediction by Coutinho's model (1999)

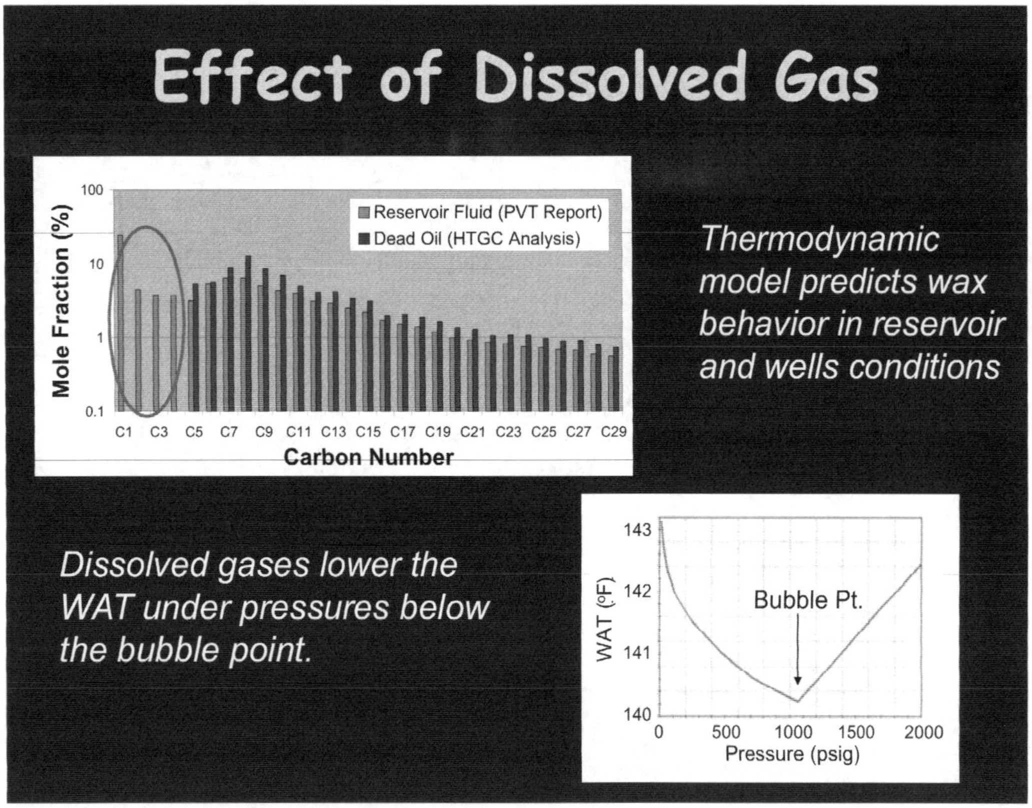
Effect of Dissolved Gas
Mole Fraction (%)
100
10
1
0.1
Reservoir Fluid (PVT Report)
Dead Oil (HTGC Analysis)
C1 C3 C5 C7 C9 C11 C13 C15 C17 C19 C21 C23 C25 C27 C29
Carbon Number
Thermodynamic model predicts wax behavior in reservoir and wells conditions
Dissolved gases lower the WAT under pressures below the bubble point.
WAT (°F)
143
142
141
140
Bubble Pt.
0 500 1000 1500 2000
Pressure (psig)

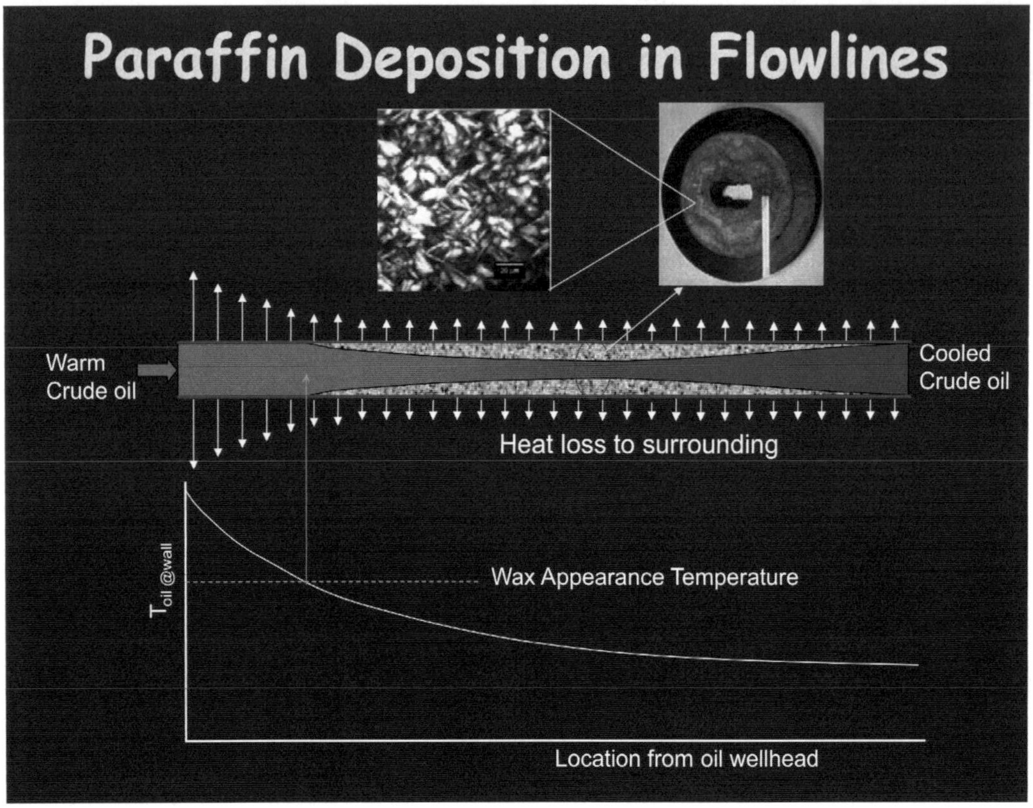
Paraffin Deposition in Flowlines
Warm
Crude oil
Cooled
Crude oil
Heat loss to surrounding
T oil @wall
Wax Appearance Temperature
Location from oil wellhead

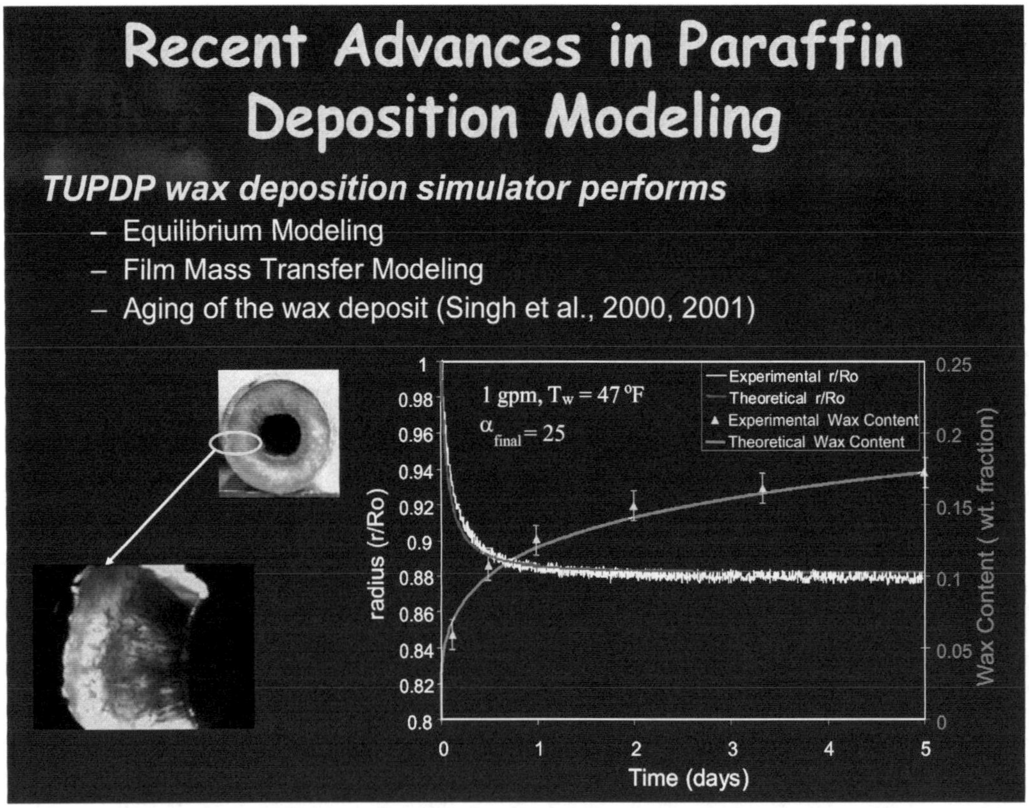
Recent Advances in Paraffin Deposition Modeling
TUPDP wax deposition simulator performs
– Equilibrium Modeling
– Film Mass Transfer Modeling
– Aging of the wax deposit (Singh et al., 2000, 2001)
1 gpm, Tw = 47 °F
α final = 25
Experimental r/Ro
Theoretical r/Ro
Experimental Wax Content
Theoretical Wax Content
radius (r/Ro)
Wax Content (wt. fraction)
Time (days)
1
0.98
0.96
0.94
0.92
0.9
0.88
0.86
0.84
0.82
0.8
0.25
0.2
0.15
0.1
0.05
0
0
1
2
3
4
5

Current Methods of Paraffin Control

- Chemical Inhibition
- Thermal Insulation
- Hot Solvent Treatment
- Mechanical Removal

Chemical Inhibitors for Paraffin Control

- Chemical performance is crude specific
- Need a rigorous laboratory testing program to qualify a chemical
- Screening of wax inhibitors using cold fingers or flow loops

Screening of Wax Inhibitors Using Cold Fingers

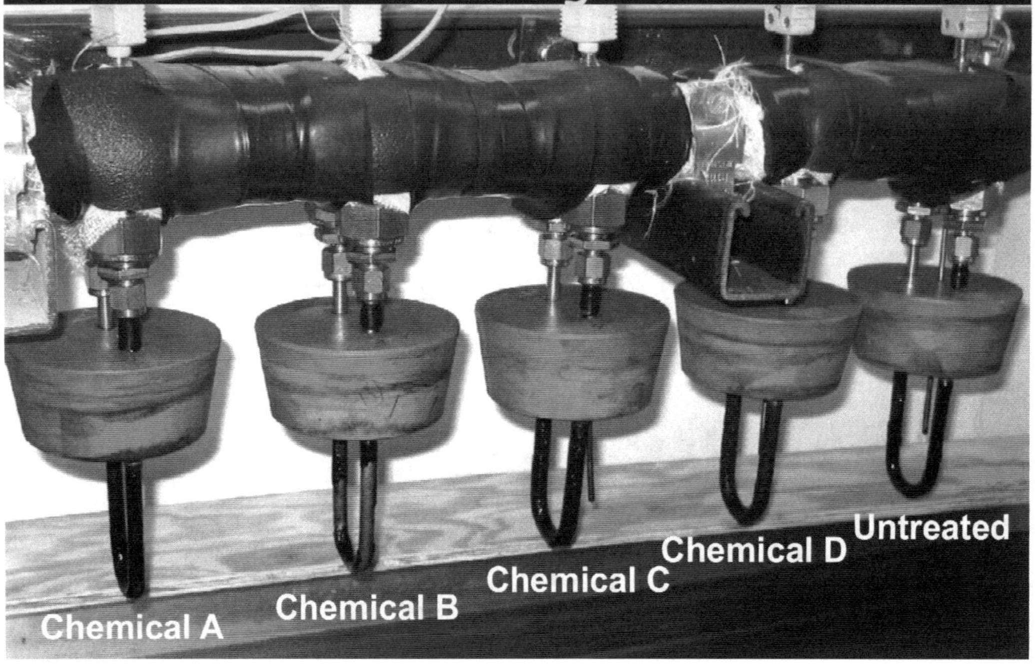

Paraffin Control for Production Wells

- *Wax inhibitors, hot solvent / dispersants*

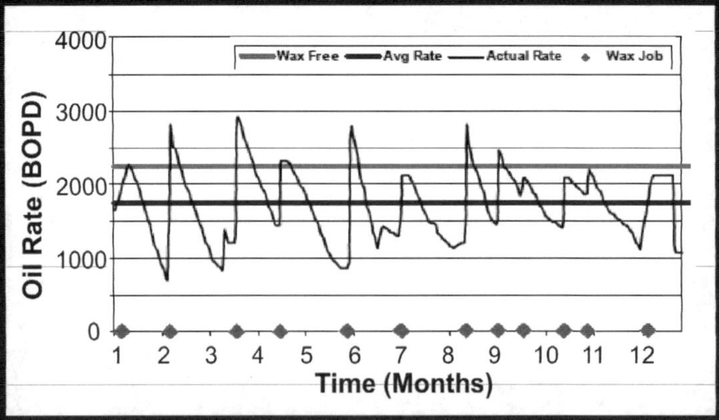

Singh et al., SPE Drilling and Completions, 2007

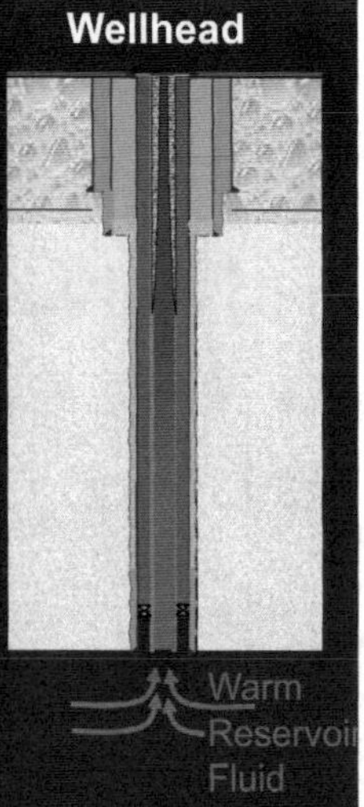

- *Coiled tubing access and wire line*
- *Heat retention using Vacuum Insulated Tubulars (VIT)*
- *Heating cable strap onto tubing string*

Heat Retention in the Well using VIT

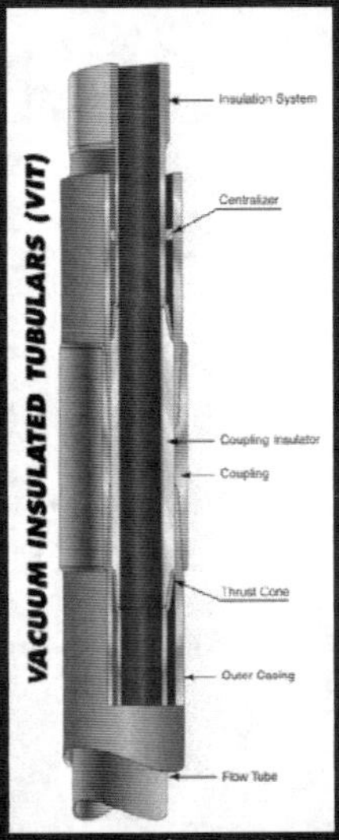

Courtesy of grantprideco

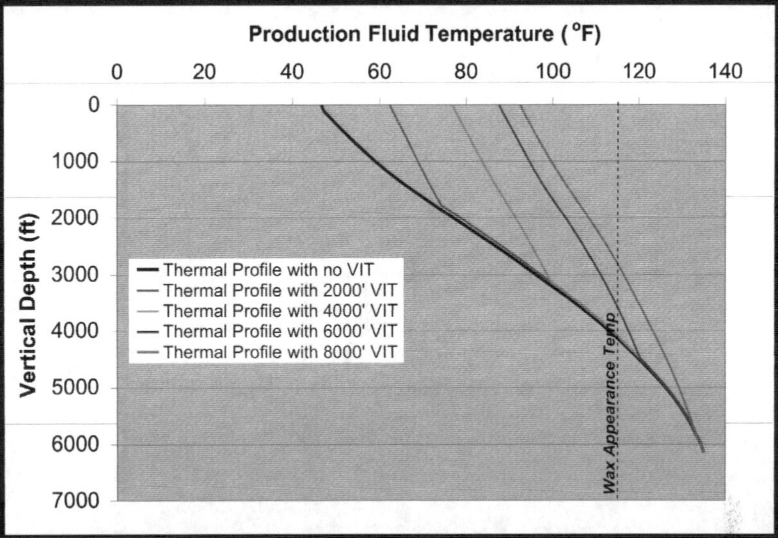

Singh et al., SPE Drilling and Completions, 2007

Lowered wax remediation job requirement to 1 per year

Offshore Indonesia Location

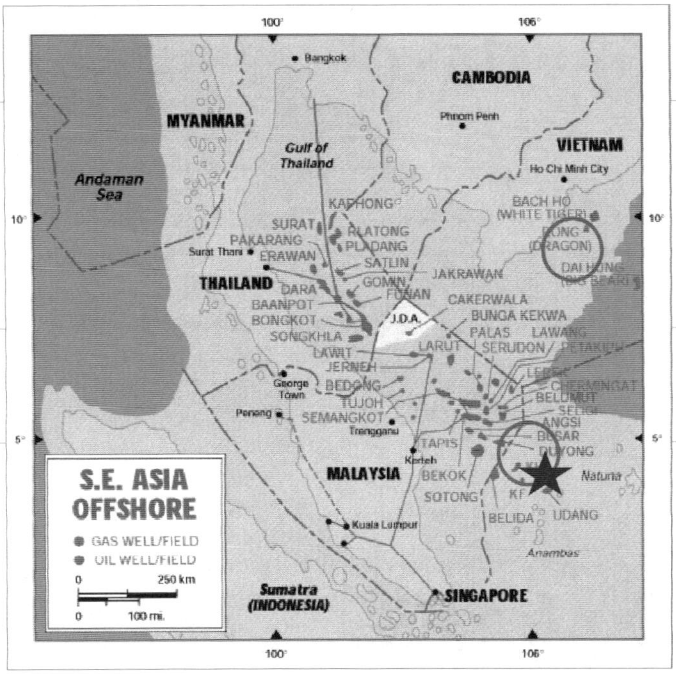

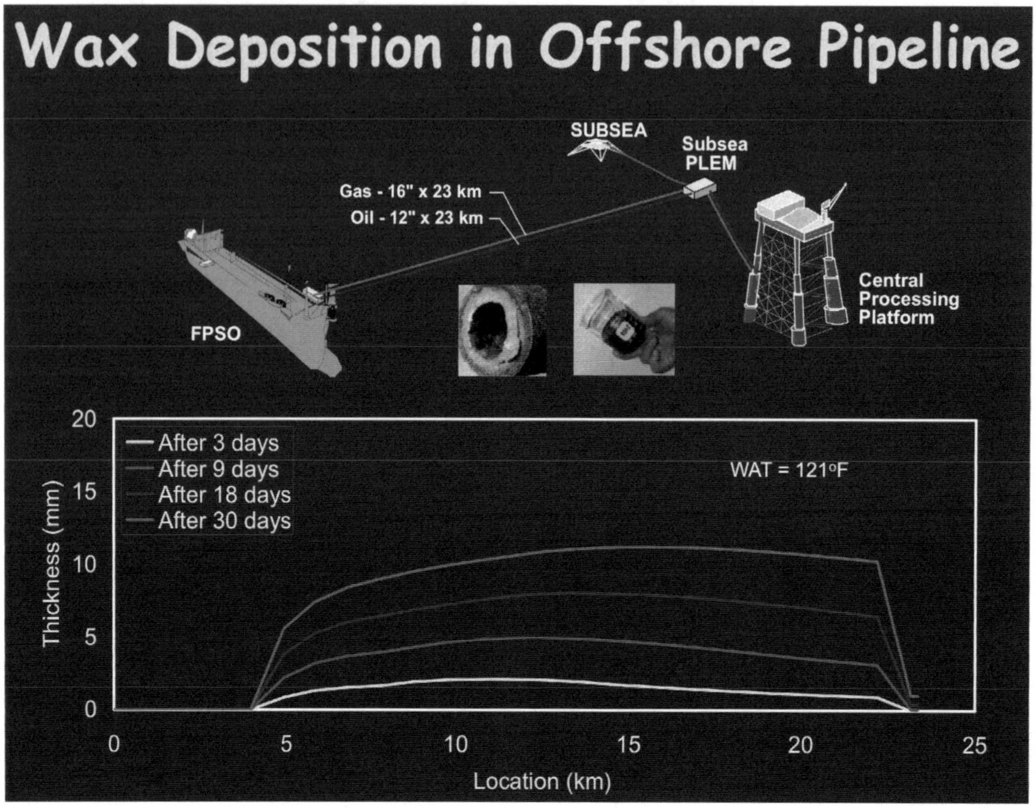
Wax Deposition in Offshore Pipeline
SUBSEA
Subsea PLEM
Gas - 16" x 23 km
Oil - 12" x 23 km
FPSO
Central Processing Platform
After 3 days
After 9 days
After 18 days
After 30 days
WAT = 121°F
Thickness (mm)
Location (km)
0 5 10 15 20 25

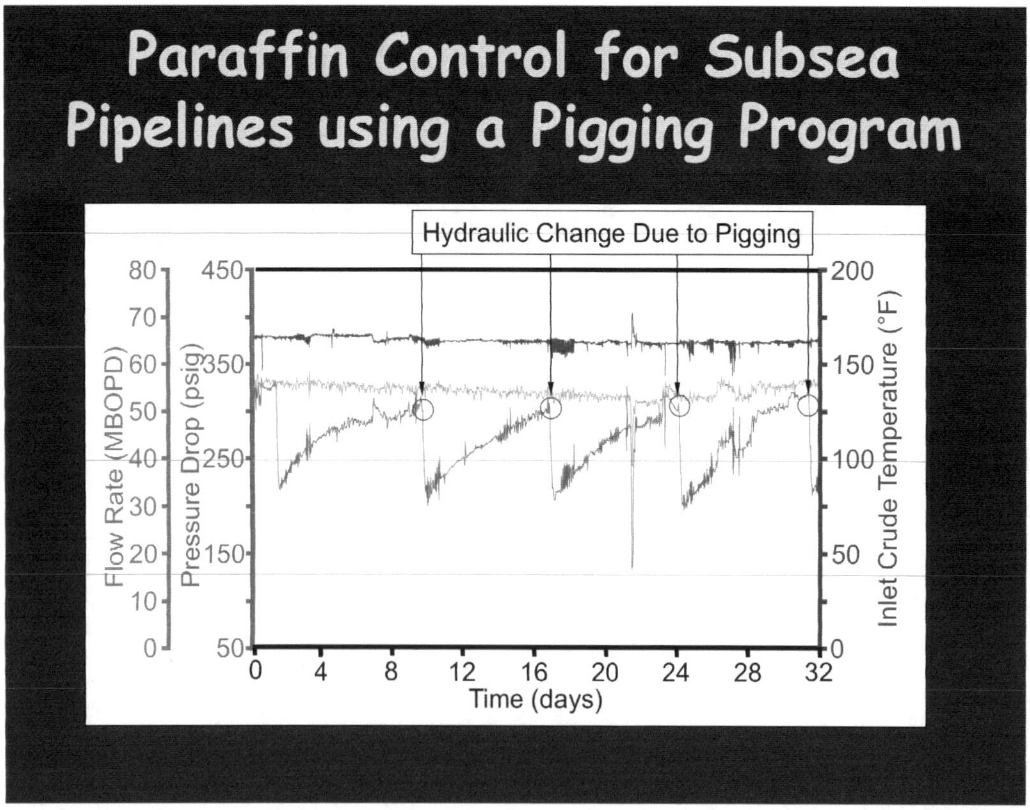
Paraffin Control for Subsea Pipelines using a Pigging Program
Hydraulic Change Due to Pigging
Flow Rate (MBOPD)
Pressure Drop (psig)
Inlet Crude Temperature (°F)
Time (days)
0 4 8 12 16 20 24 28 32

Pigging Requires Forethought

Mechanical Methods for Wax Remediation

- *Pigging – Hard pigs, Scraper pigs, By-pass pigs, Multi-diameter*
- *Coiled tubing – limited reach*

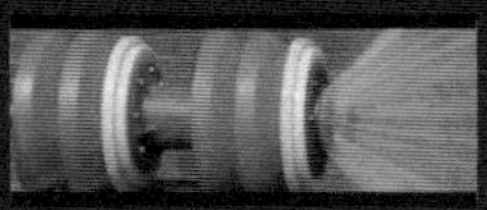

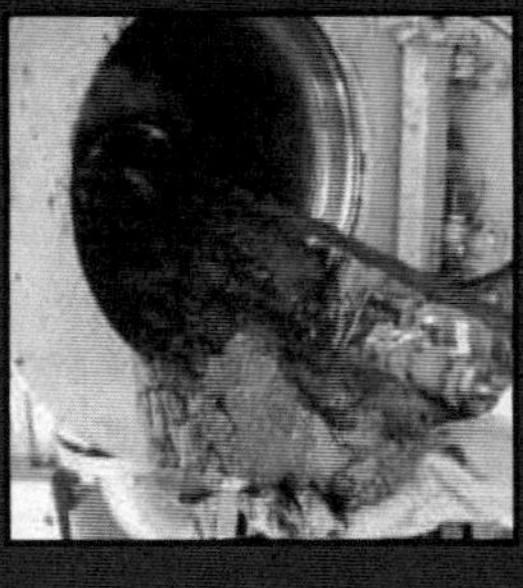

Improved pig design to lower the stuck pig risk

Preventing Wax Deposition using Thermal Management

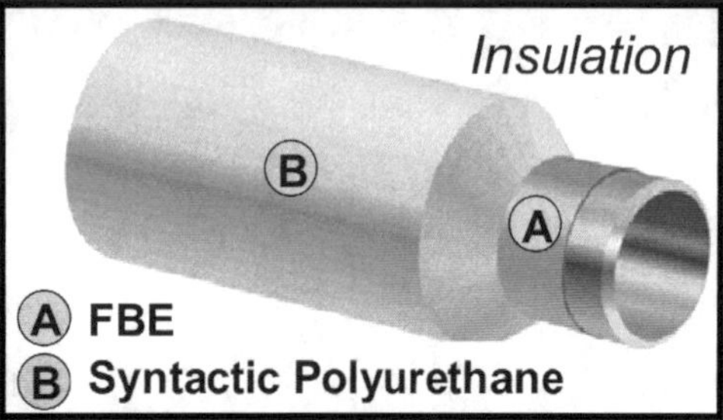

Courtesy of Bredero Shaw

Subsea tie-back options - insulation or dual pipeline?

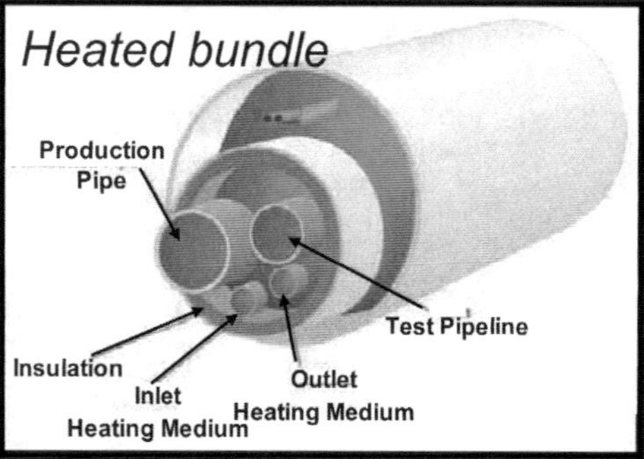

Courtesy of Subsea7

Challenges in Deepwater

- ***Longer distances***

  from wellheads to host facilities

- ***Concept selection***

  dry tree vs. subsea tie-back

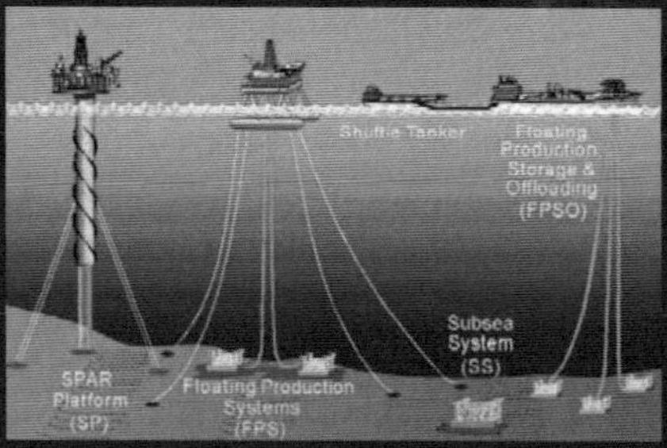

Courtesy of MMS

- ***Traditional dual pipelines for round-trip pigging***

  not attractive for developments with marginal reserves

- ***Chemical delivery from host platforms***

  high viscosity of the wax inhibitors especially at low temperatures and high pressures

Challenges in Arctic

- ***Well production loss***
  - Hot oil and scraper jobs
  - Wax inhibitor via lift gas or string
  - Arctic spec wax inhibitors
  - Vacuum Insulated Tubing

TRANS ALASKA PIPELINE SYSTEM

- ***Pipeline management***
  - Crudes form strong gels under arctic conditions
  - Displace the crude during shut-in or treat with PPD prior to shut-in
- ***Traditional by-pass pigging***
  - Not that effective for low flow rate scenarios

- ***Downhole chemical delivery***
  - High viscosity and pour point of the wax inhibitors especially at low temperatures

Potential Paraffin Technologies

- ***Reliable subsea pig launchers***
  - for routine single-trip pigging operations
- ***Crawler pigs***
  - capable of traveling upstream of the flow and return
- ***Cold flow technology***
  - demonstrating cold slurry flow
- ***Low viscosity wax inhibitors***
  - for effective pumping from host platforms via long distance umbilical lines

Summary

What do we know already

- Characterization of paraffin or wax
- Thermodynamics of paraffin in hydrocarbon fluid system
- Physics of paraffin deposition
- Several methods of paraffin control

What are the challenges in deepwater

- Long subsea tie-backs for marginal fields
- Round trip pigging a very expensive option
- Long umbilical distances to pump inhibitor chemical

What needs to be developed for deepwater

- Demonstrate novel ways to control wax such as cold flow
- Develop low viscosity easy pump-able wax inhibitors
- Field trial for subsea pig launchers and crawler pigs

Asphaltene Nanoscience and Downhole Fluid Analysis for Reservoir Evaluation

Oliver C. Mullins
Schlumberger

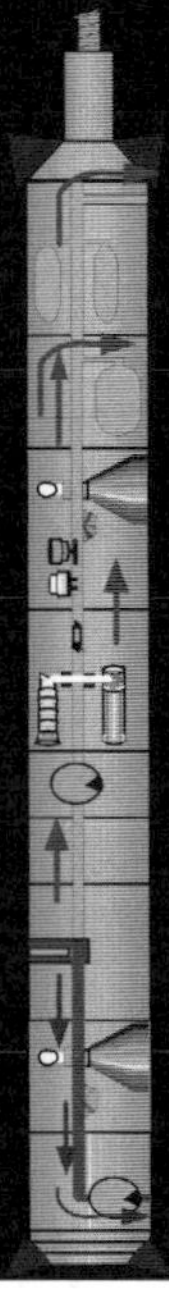

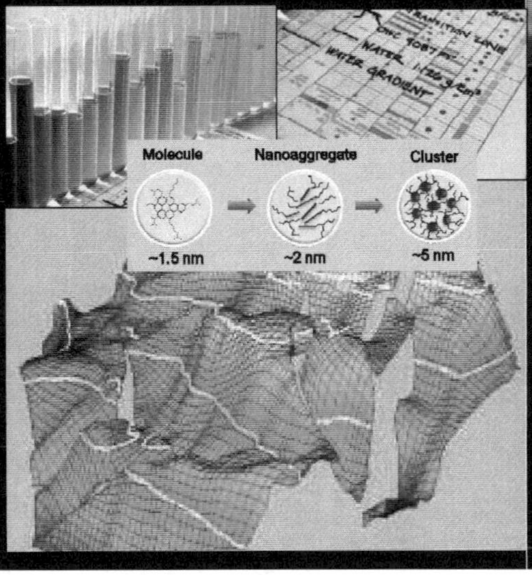

Downhole Fluid Analysis (DFA)

Outline:

1) Key Reservoirs Issues.

2) DFA & Compositional Variations & Compartments.

3) Reservoir Evaluation
   New Technology. DFA
   New Asphaltene Science
   New Work Flows

4) Conclusions

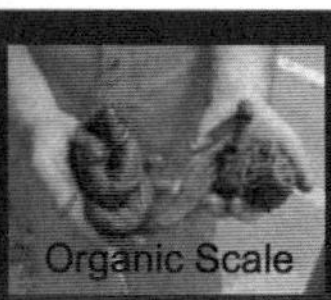

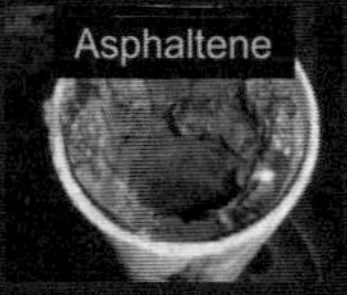

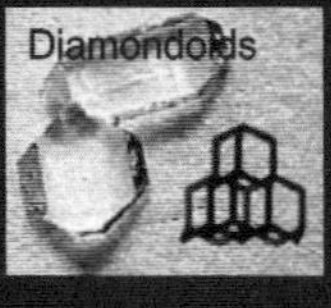

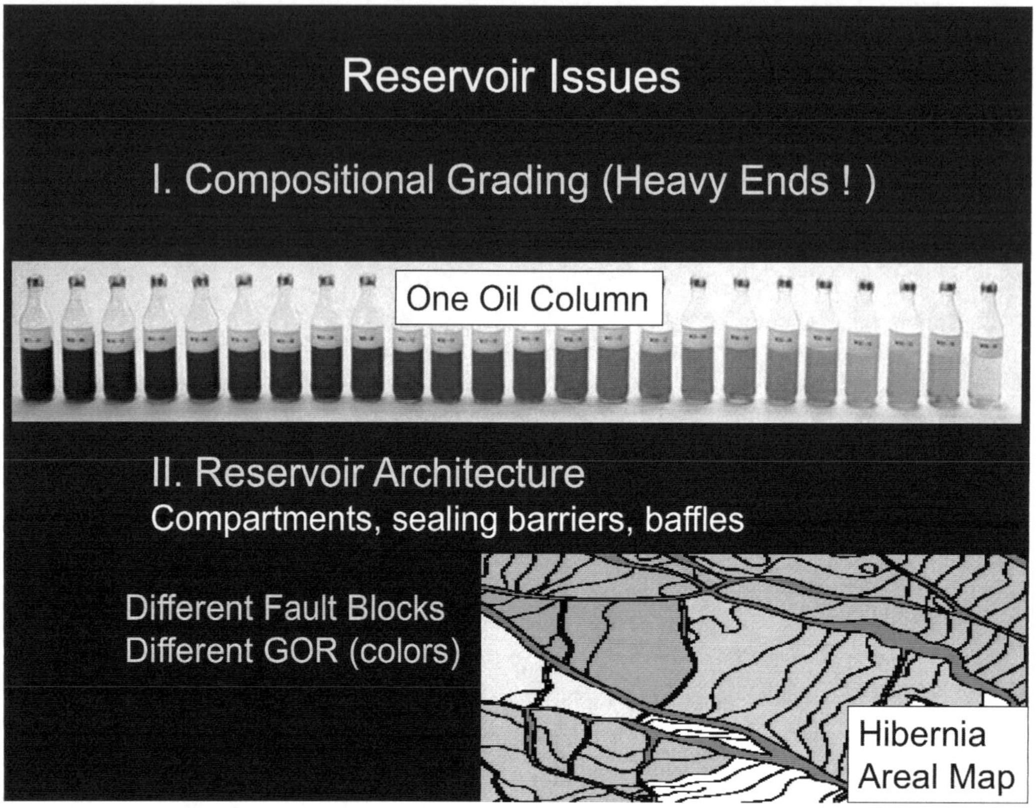
Reservoir Issues
I. Compositional Grading (Heavy Ends !)
One Oil Column
II. Reservoir Architecture
Compartments, sealing barriers, baffles
Different Fault Blocks
Different GOR (colors)
Hibernia
Areal Map

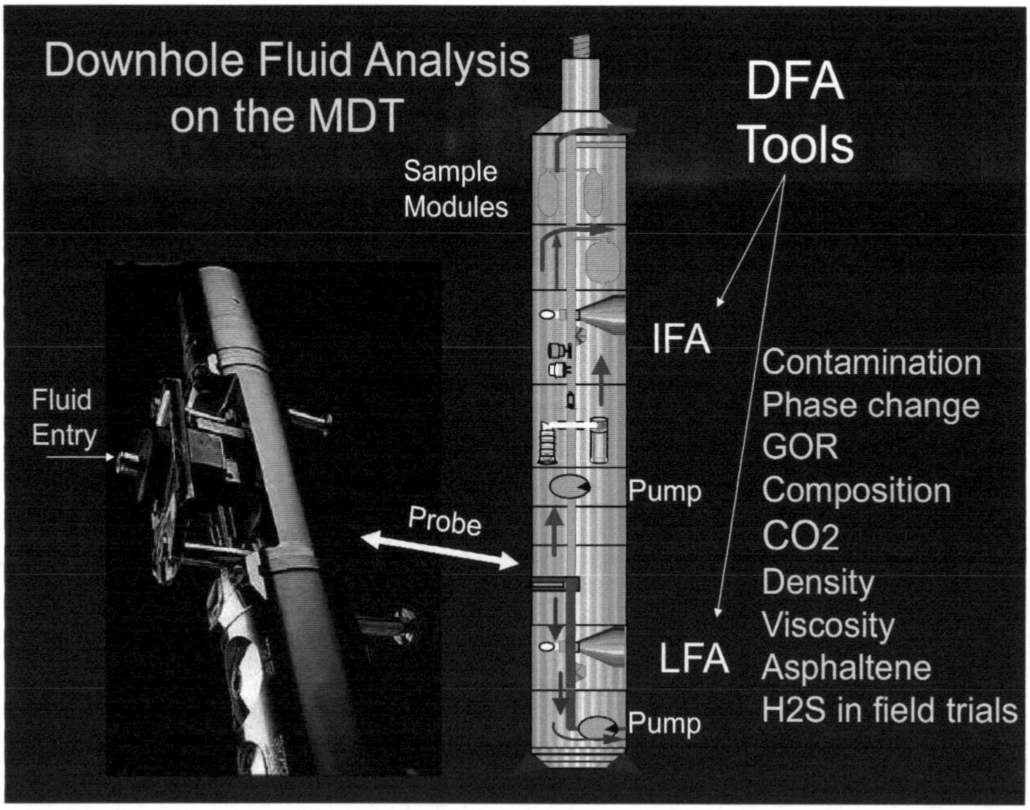
Downhole Fluid Analysis
on the MDT
DFA
Tools
Sample
Modules
IFA
Fluid
Entry
Pump
Probe
LFA
Pump
Contamination
Phase change
GOR
Composition
CO2
Density
Viscosity
Asphaltene
H2S in field trials

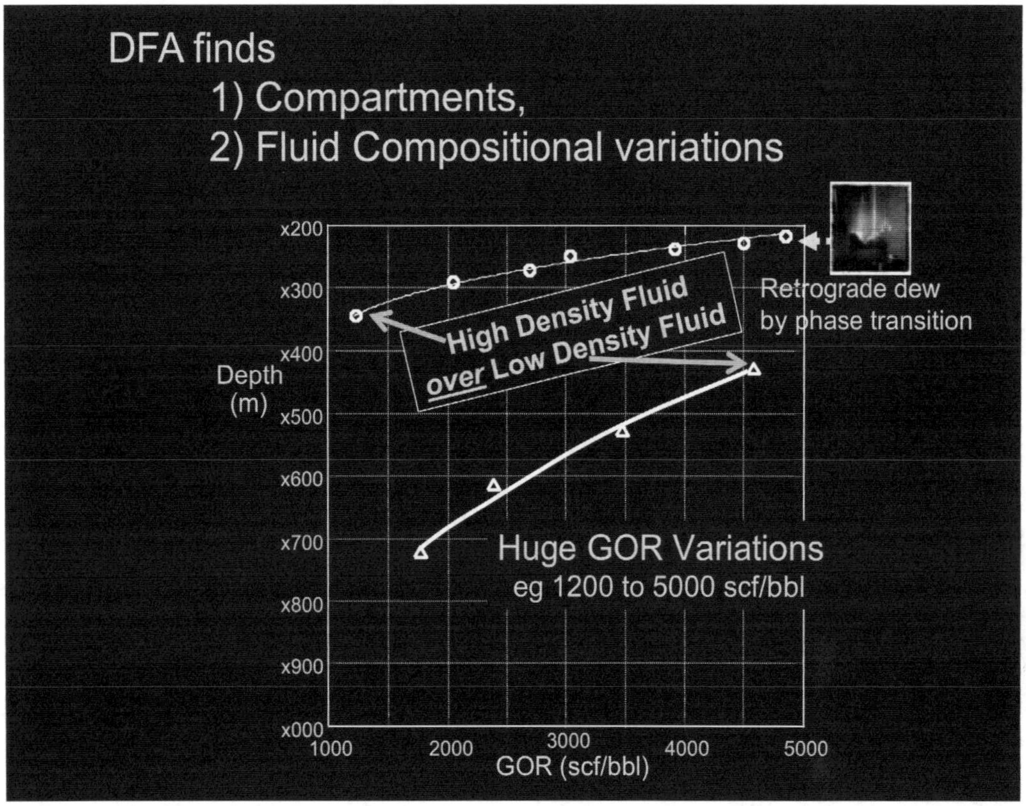
DFA finds
1) Compartments,
2) Fluid Compositional variations
Retrograde dew by phase transition
High Density Fluid over Low Density Fluid
Depth (m)
x200
x300
x400
x500
x600
x700
x800
x900
x000
Huge GOR Variations
eg 1200 to 5000 scf/bbl
1000
2000
3000
4000
5000
GOR (scf/bbl)

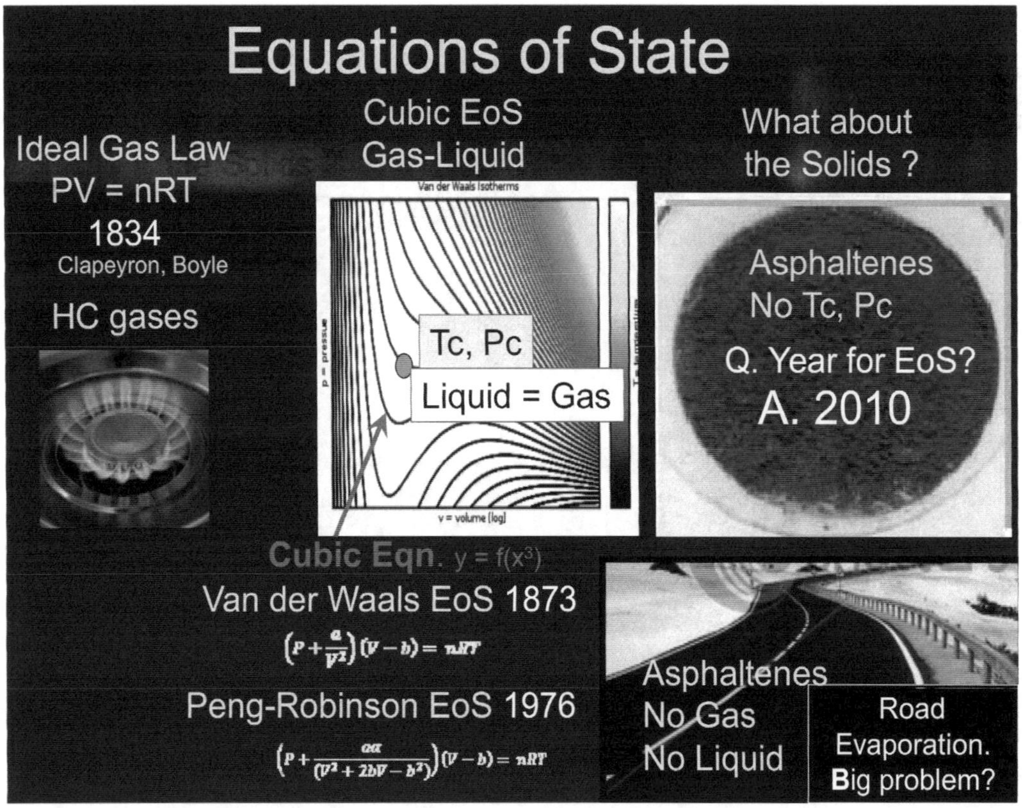
Equations of State
Ideal Gas Law
PV = nRT
1834
Clapeyron, Boyle
HC gases
Cubic EoS
Gas-Liquid
Van der Waals Isotherms
Tc, Pc
Liquid = Gas
p = pressure
v = volume (log)
What about the Solids ?
Asphaltenes
No Tc, Pc
Q. Year for EoS?
A. 2010
Cubic Eqn. y = f(x³)
Van der Waals EoS 1873
$\left(P+\frac{a}{V^2}\right)(V-b)=nRT$
Peng-Robinson EoS 1976
$\left(P+\frac{a\alpha}{(V^2+2bV-b^2)}\right)(V-b)=nRT$
Asphaltenes
No Gas
No Liquid
Road
Evaporation.
Big problem?

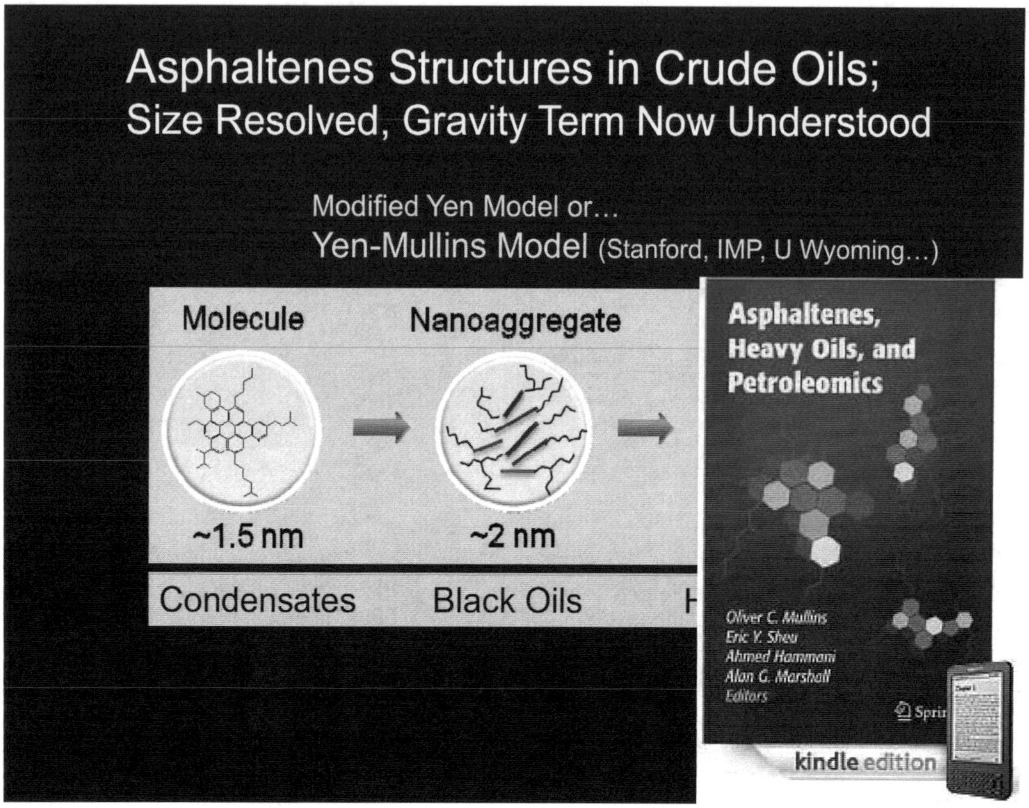

Asphaltenes Structures in Crude Oils;
Size Resolved, Gravity Term Now Understood
Modified Yen Model or…
Yen-Mullins Model (Stanford, IMP, U Wyoming…)
Molecule
Nanoaggregate
~1.5 nm
~2 nm
Condensates
Black Oils
Asphaltenes, Heavy Oils, and Petroleomics
Oliver C. Mullins
Eric Y. Sheu
Ahmed Hammami
Alan G. Marshall
Editors
kindle edition

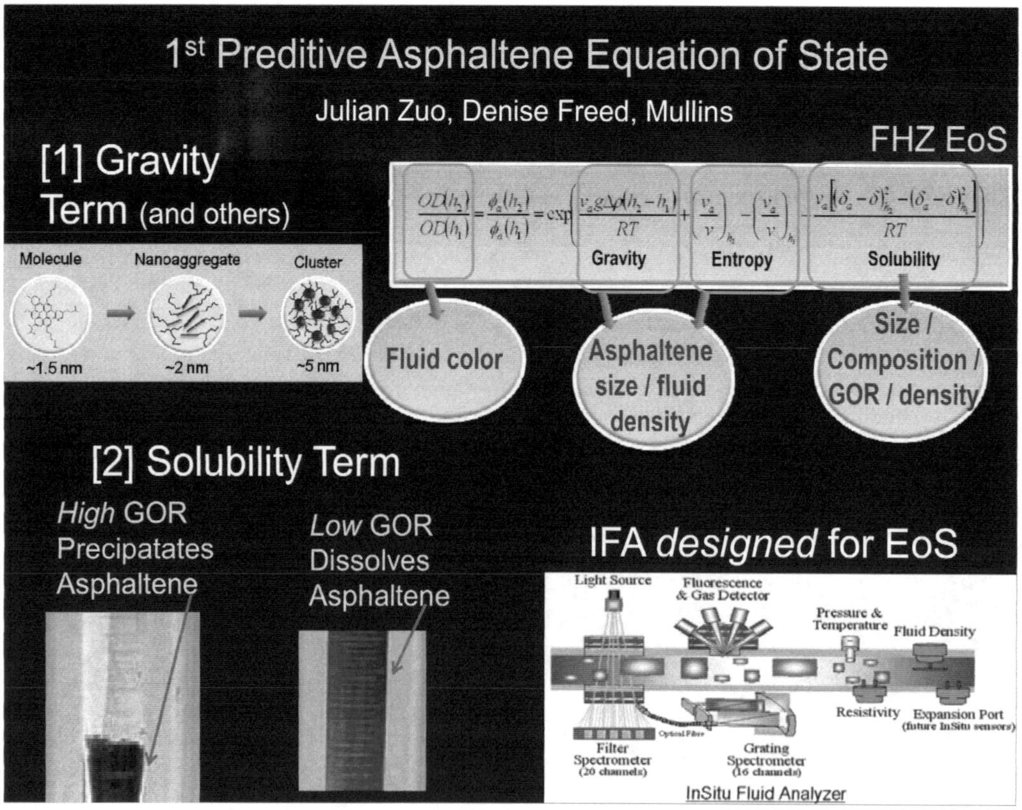

1st Preditive Asphaltene Equation of State
Julian Zuo, Denise Freed, Mullins
FHZ EoS
[1] Gravity Term (and others)
$\frac{OD(h_2)}{OD(h_1)} = \frac{\phi_a(h_2)}{\phi_a(h_1)} = \exp\left\{\frac{v_a g \Delta\rho (h_2 - h_1)}{RT} + \left(\frac{v_a}{v}\right)_{h_2} - \left(\frac{v_a}{v}\right)_{h_1} - \frac{v_a\left[(\delta_a-\delta)^2_{h_2} - (\delta_a-\delta)^2_{h_1}\right]}{RT}\right\}$
Gravity
Entropy
Solubility
Molecule
Nanoaggregate
Cluster
~1.5 nm
~2 nm
~5 nm
Fluid color
Asphaltene size / fluid density
Size / Composition / GOR / density
[2] Solubility Term
High GOR Precipatates Asphaltene
Low GOR Dissolves Asphaltene
IFA designed for EoS
Light Source
Fluorescence & Gas Detector
Pressure & Temperature
Fluid Density
Resistivity
Expansion Port (future InSitu sensors)
Filter Spectrometer (20 channels)
Grating Spectrometer (16 channels)
InSitu Fluid Analyzer

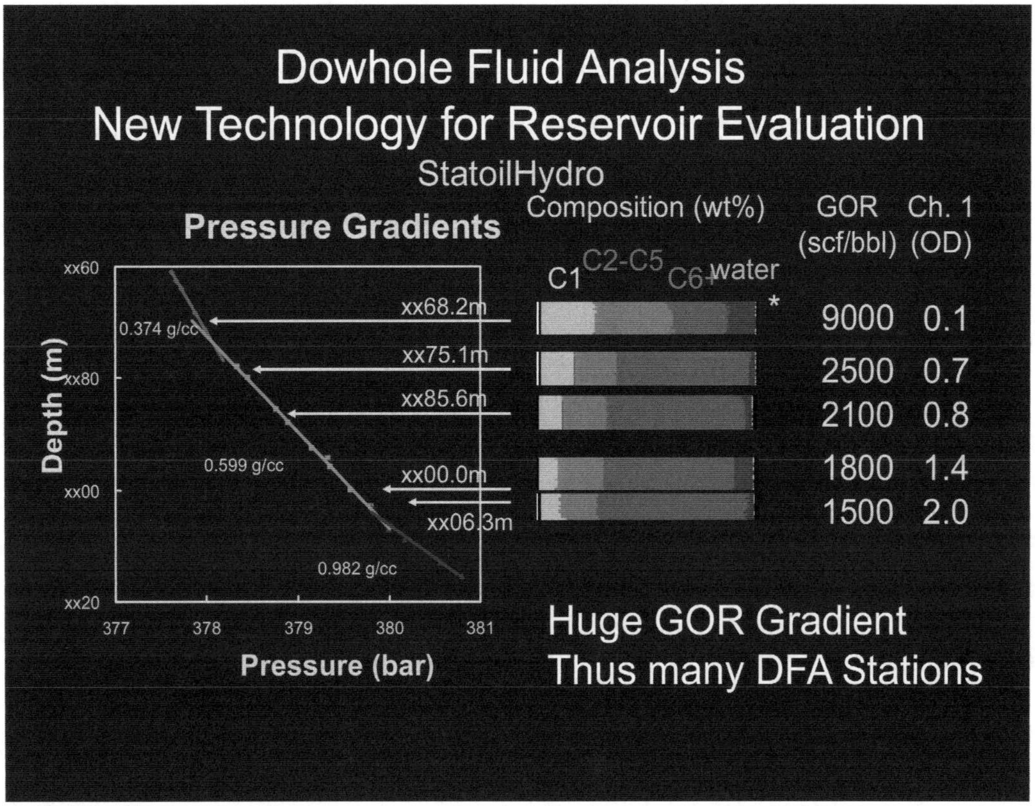
Dowhole Fluid Analysis
New Technology for Reservoir Evaluation
StatoilHydro
Pressure Gradients
Composition (wt%)
GOR (scf/bbl)
Ch. 1 (OD)
C1 C2-C5 C6+ water
xx60
xx80
xx00
xx20
Depth (m)
0.374 g/cc
0.599 g/cc
0.982 g/cc
xx68.2m
xx75.1m
xx85.6m
xx00.0m
xx06.3m
377 378 379 380 381
Pressure (bar)
9000 0.1
2500 0.7
2100 0.8
1800 1.4
1500 2.0
Huge GOR Gradient
Thus many DFA Stations

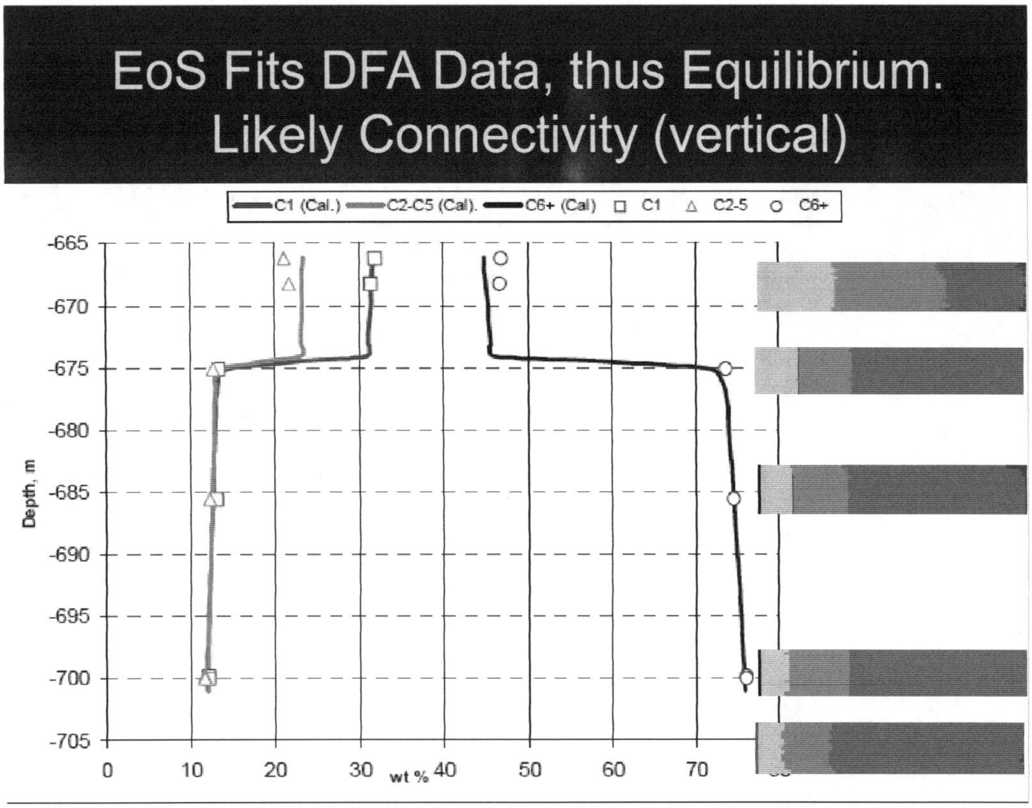
EoS Fits DFA Data, thus Equilibrium.
Likely Connectivity (vertical)
C1 (Cal.) C2-C5 (Cal). C6+ (Cal) C1 C2-5 C6+
-665
-670
-675
-680
-685
-690
-695
-700
-705
Depth, m
0 10 20 30 40 50 60 70
wt %

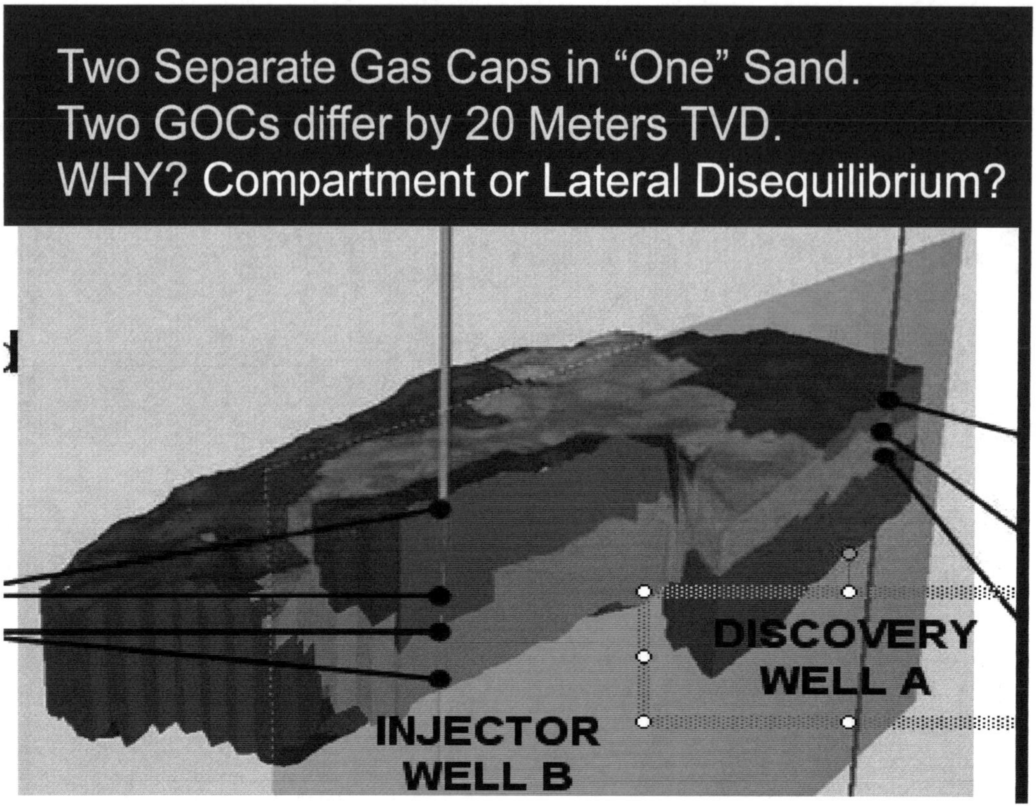
Two Separate Gas Caps in "One" Sand.
Two GOCs differ by 20 Meters TVD.
WHY? Compartment or Lateral Disequilibrium?
DISCOVERY WELL A
INJECTOR WELL B

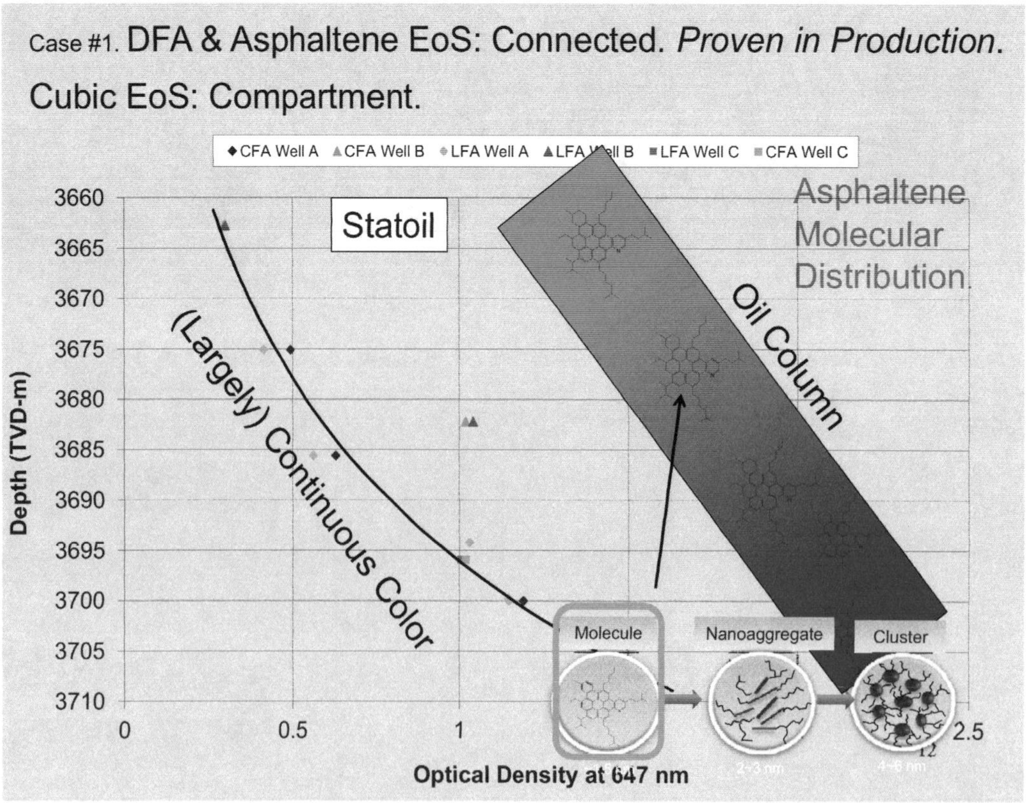
Case #1. DFA & Asphaltene EoS: Connected. *Proven in Production.*
Cubic EoS: Compartment.
◆CFA Well A ▲CFA Well B ◆LFA Well A ▲LFA Well B ■LFA Well C ■CFA Well C
Statoil
(Largely) Continuous Color
Oil Column
Asphaltene Molecular Distribution
Molecule
Nanoaggregate
Cluster
Depth (TVD-m)
3660
3665
3670
3675
3680
3685
3690
3695
3700
3705
3710
0
0.5
1
2.5
Optical Density at 647 nm

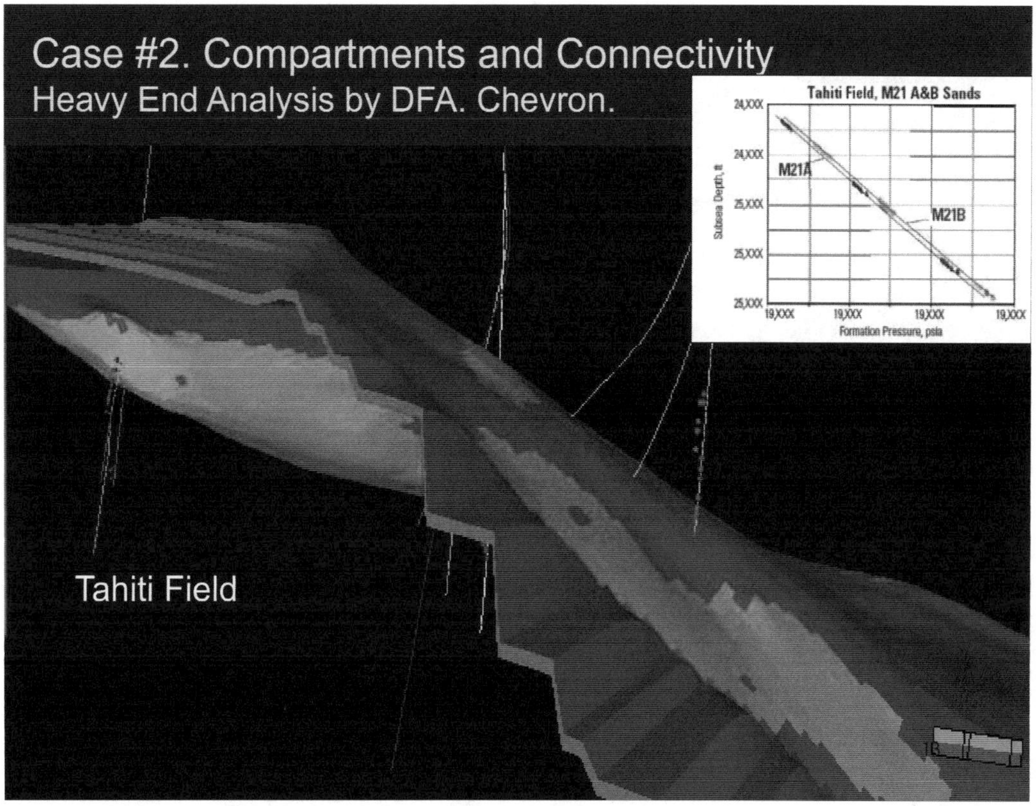

Case #2. Compartments and Connectivity
Heavy End Analysis by DFA. Chevron.
Tahiti Field, M21 A&B Sands
M21A
M21B
Subsea Depth, ft
Formation Pressure, psia
Tahiti Field

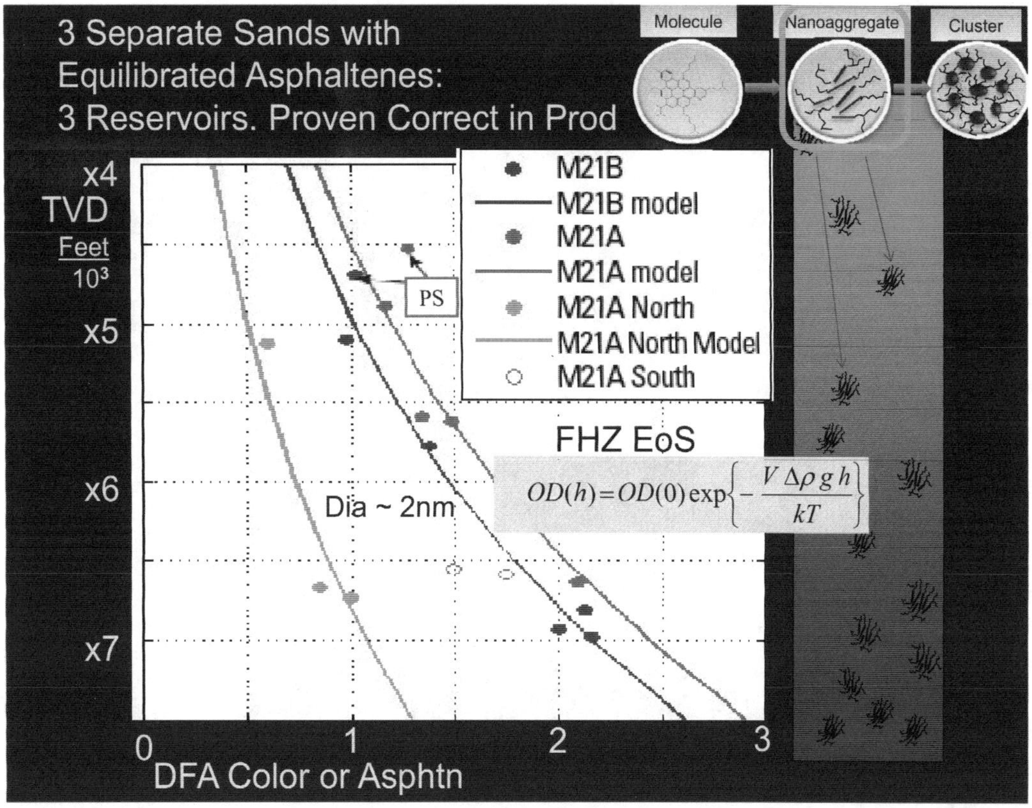

3 Separate Sands with
Equilibrated Asphaltenes:
3 Reservoirs. Proven Correct in Prod
Molecule
Nanoaggregate
Cluster
TVD
Feet
10^3
x4
x5
x6
x7
M21B
M21B model
M21A
M21A model
M21A North
M21A North Model
M21A South
PS
FHZ EoS
$OD(h) = OD(0)\exp\left\{-\frac{V\Delta\rho g h}{kT}\right\}$
Dia ~ 2nm
0
1
2
3
DFA Color or Asphtn

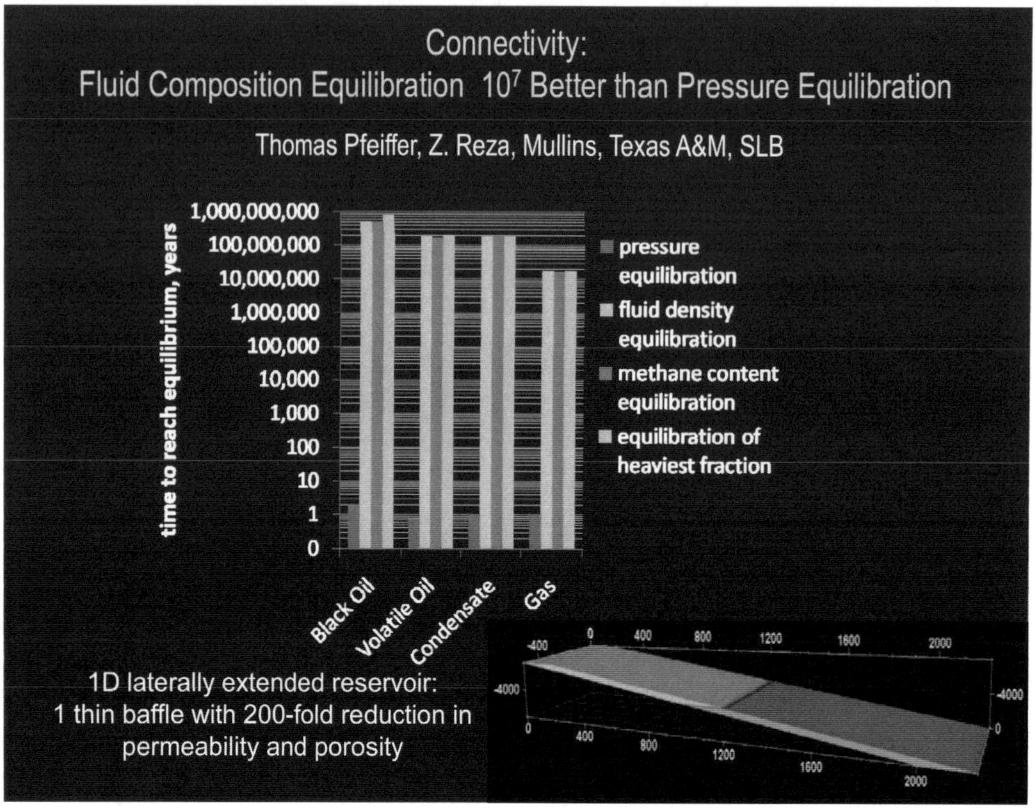
Connectivity:
Fluid Composition Equilibration 10^7 Better than Pressure Equilibration
Thomas Pfeiffer, Z. Reza, Mullins, Texas A&M, SLB
time to reach equilibrium, years
1,000,000,000
100,000,000
10,000,000
1,000,000
100,000
10,000
1,000
100
10
1
0
pressure equilibration
fluid density equilibration
methane content equilibration
equilibration of heaviest fraction
Black Oil
Volatile Oil
Condensate
Gas
1D laterally extended reservoir:
1 thin baffle with 200-fold reduction in permeability and porosity

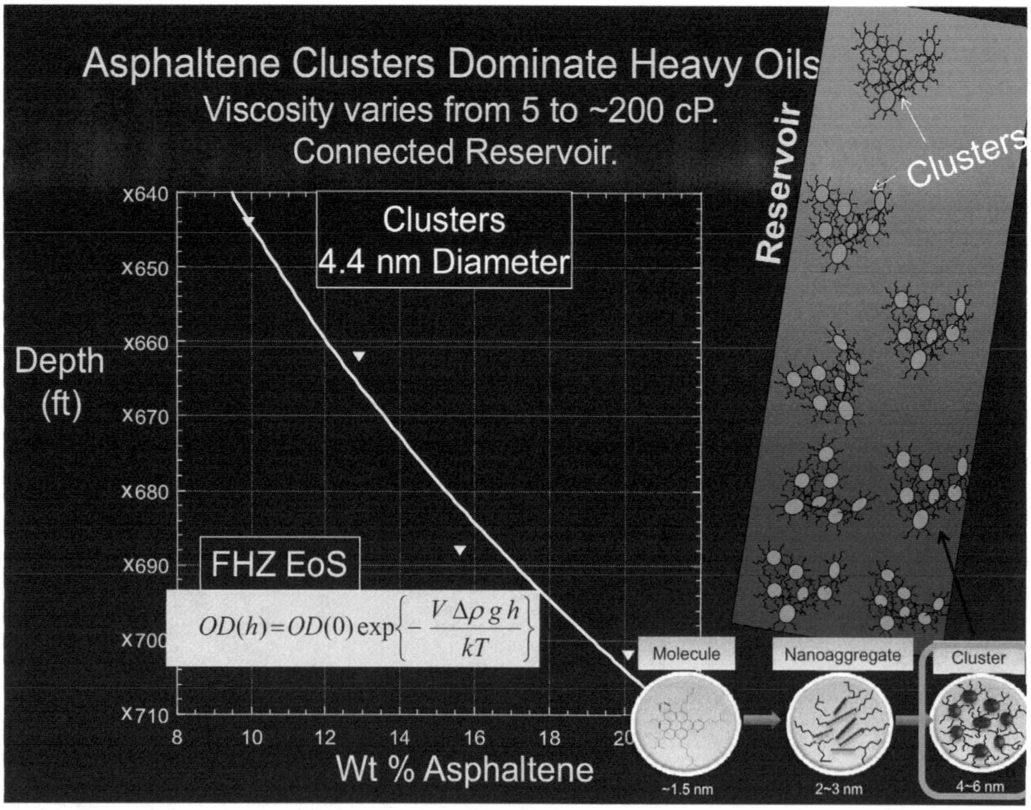
Asphaltene Clusters Dominate Heavy Oils
Viscosity varies from 5 to ~200 cP.
Connected Reservoir.
Clusters
4.4 nm Diameter
Depth (ft)
X640
X650
X660
X670
X680
X690
X700
X710
FHZ EoS
$OD(h)=OD(0)\exp\left\{-\frac{V\Delta\rho g h}{kT}\right\}$
8
10
12
14
16
18
20
Wt % Asphaltene
Reservoir
Clusters
Molecule
Nanoaggregate
Cluster
~1.5 nm
2~3 nm
4~6 nm

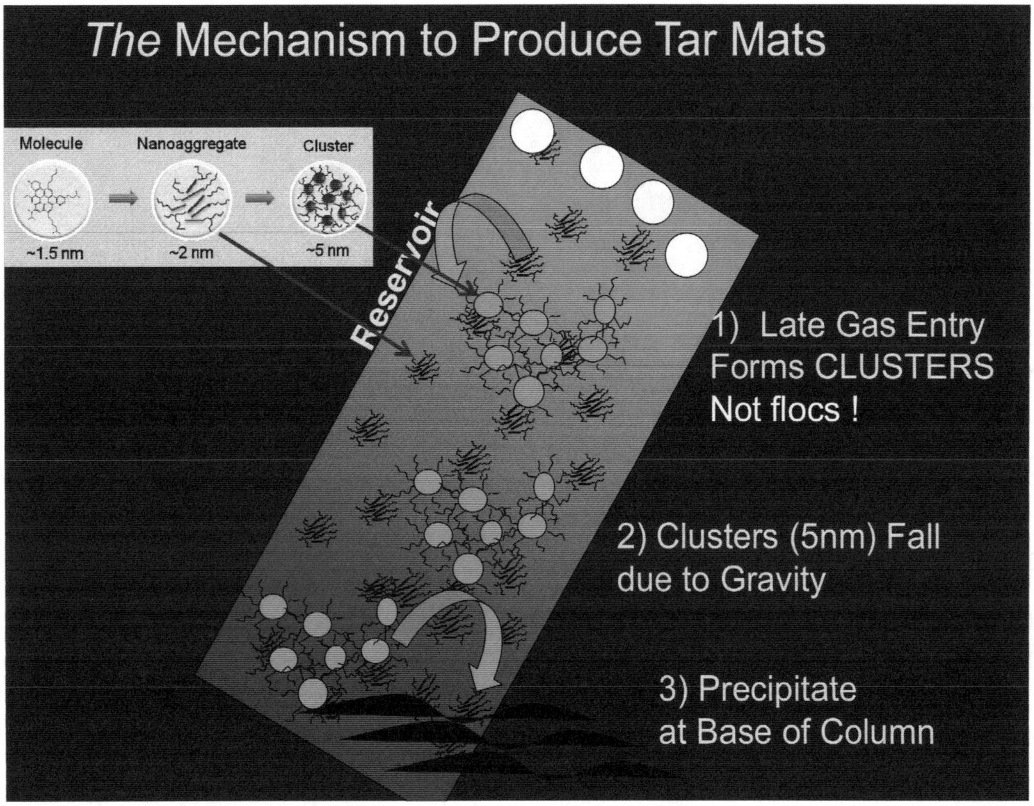
The Mechanism to Produce Tar Mats
Molecule
Nanoaggregate
Cluster
~1.5 nm
~2 nm
~5 nm
Reservoir
1) Late Gas Entry
Forms CLUSTERS
Not flocs !
2) Clusters (5nm) Fall
due to Gravity
3) Precipitate
at Base of Column

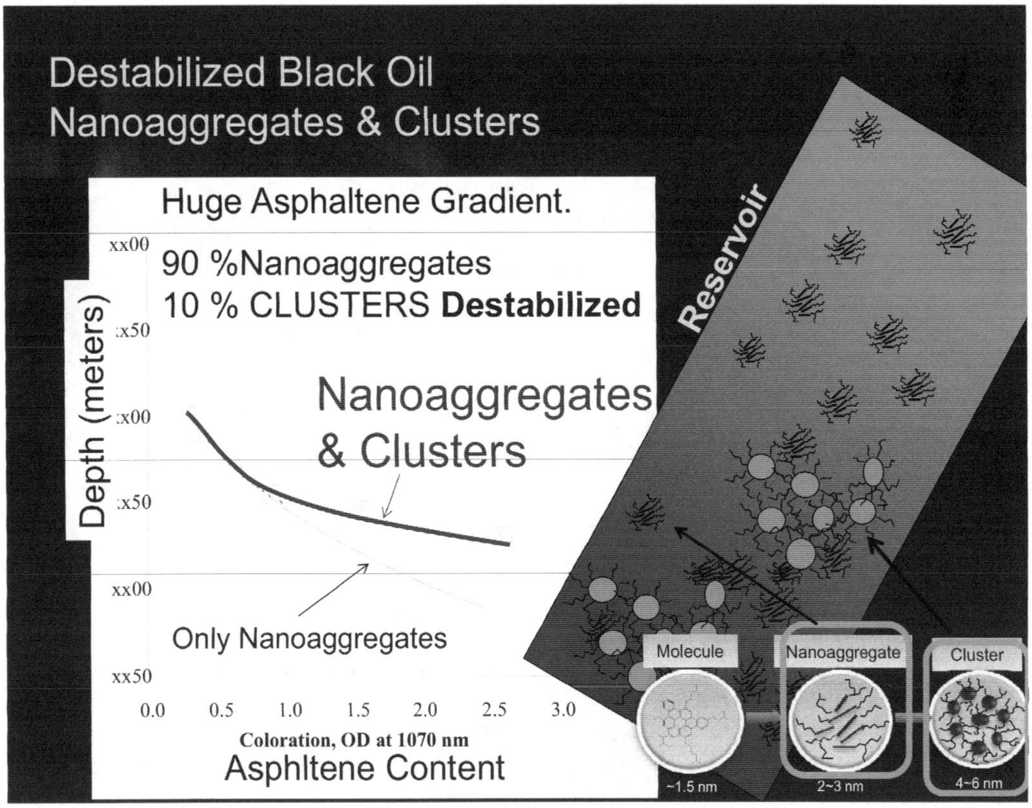
Destabilized Black Oil
Nanoaggregates & Clusters
Huge Asphaltene Gradient.
90 %Nanoaggregates
10 % CLUSTERS **Destabilized**
Nanoaggregates
& Clusters
Only Nanoaggregates
Depth (meters)
xx00
:x50
:x00
:x50
xx00
xx50
0.0
0.5
1.0
1.5
2.0
2.5
3.0
Coloration, OD at 1070 nm
Asphltene Content
Reservoir
Molecule
Nanoaggregate
Cluster
~1.5 nm
2~3 nm
4~6 nm

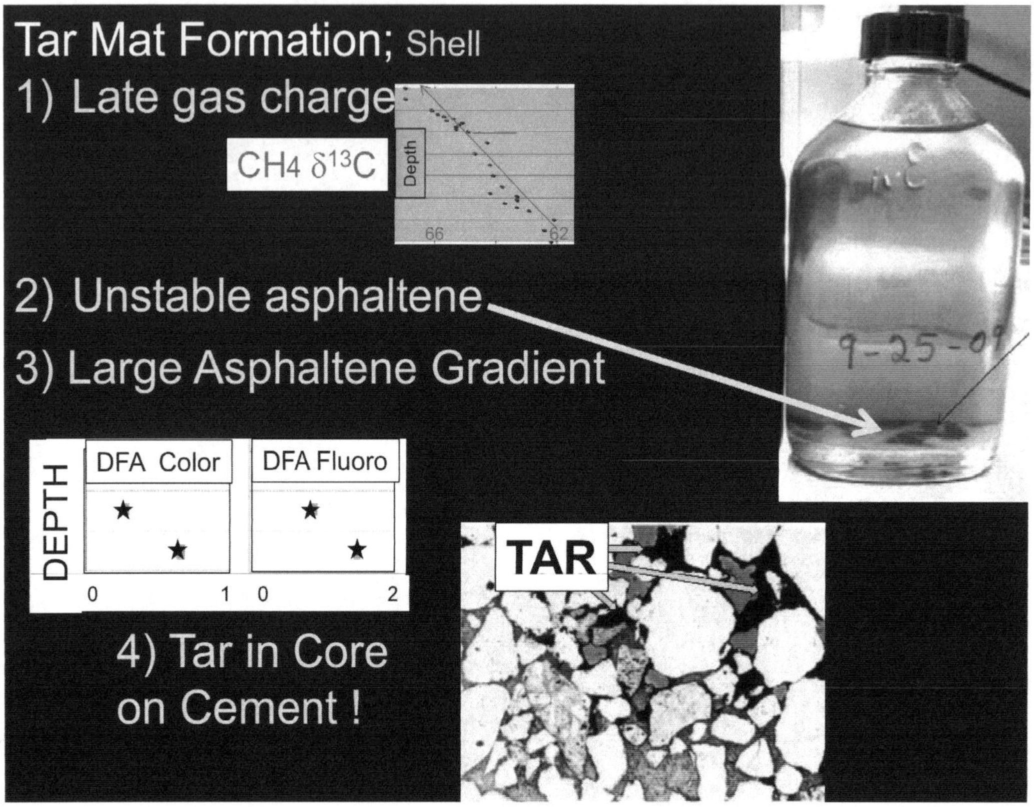

Conclusions:

1) Fluid complexities and compartmentalization are large unresolved issues in reservoir management.

2) DFA is required to optimize finding *fluid compositional variation*

3) Asphaltene Nanoscience enabled Development of the Flory-Huggins-Zuo Eos, the 1st Predictive Asphaltene EoS for Gradients

4) Reservoir Evaluation: New Science & DFA Technology
Connectivity, Heavy Oils, Tar Mats, Disequilibrium

Asphaltene Grading, Flow Assurance and Tar Mat Formation in Hydrocarbon Reservoirs

Julian Y. Zuo

Schlumberger

AIChE/SPE Joint Workshop-Challenges in Flow Assurance
Sept 26-28,2011, Houston

Outline

- Reservoir and fluid issues
- FHZ EoS for asphaltene and new workflow
- Applications
  - Equilibrium fluid
  - Non-equilibrium fluid with a late stage of gas charge
  - Heavy oil
- Conclusions

Schlumberger

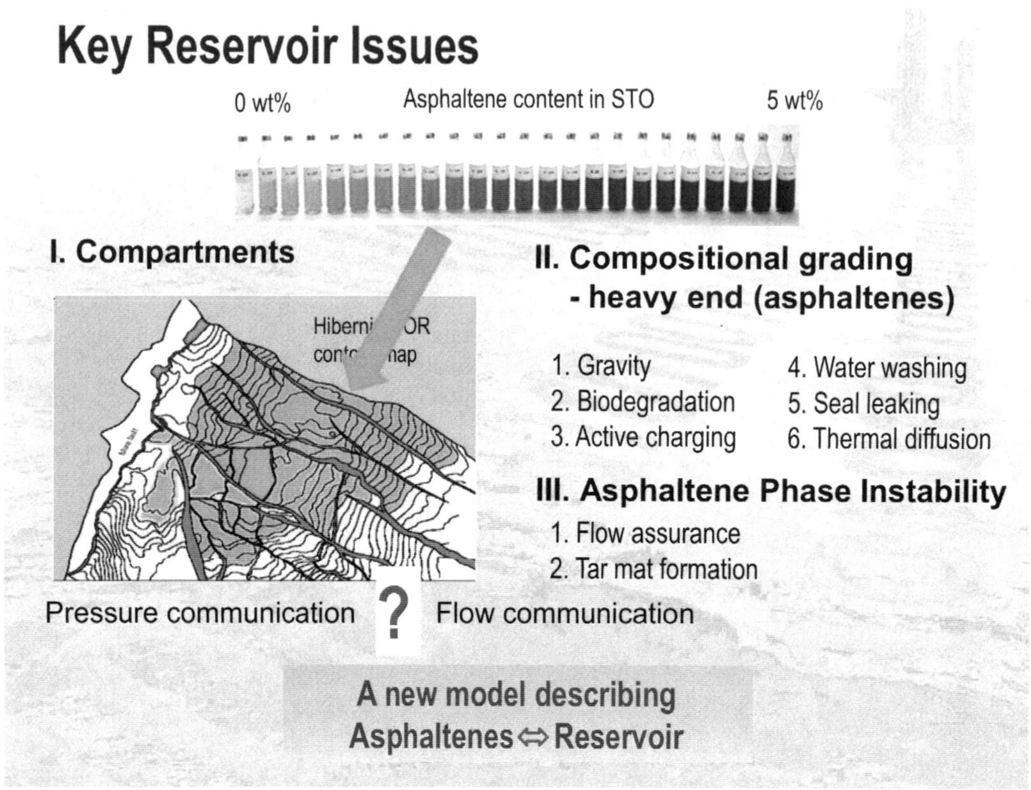
Key Reservoir Issues
0 wt%
Asphaltene content in STO
5 wt%
I. Compartments
II. Compositional grading
- heavy end (asphaltenes)
1. Gravity
2. Biodegradation
3. Active charging
4. Water washing
5. Seal leaking
6. Thermal diffusion
III. Asphaltene Phase Instability
1. Flow assurance
2. Tar mat formation
Pressure communication
?
Flow communication
A new model describing
Asphaltenes⇔Reservoir

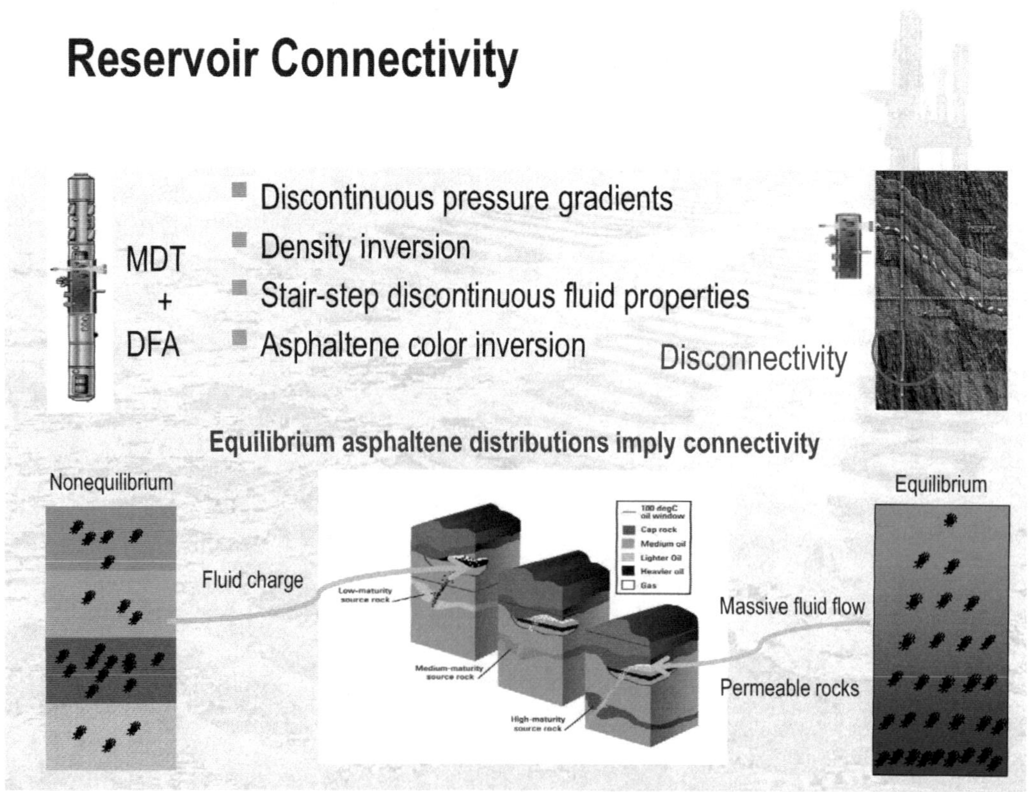
Reservoir Connectivity
MDT
+
DFA
Discontinuous pressure gradients
Density inversion
Stair-step discontinuous fluid properties
Asphaltene color inversion
Disconnectivity
Equilibrium asphaltene distributions imply connectivity
Nonequilibrium
Equilibrium
Fluid charge
Massive fluid flow
Permeable rocks
Low-maturity source rock
Medium-maturity source rock
High-maturity source rock
100 degC oil window
Cap rock
Medium oil
Lighter Oil
Heavier oil
Gas

Equations of State

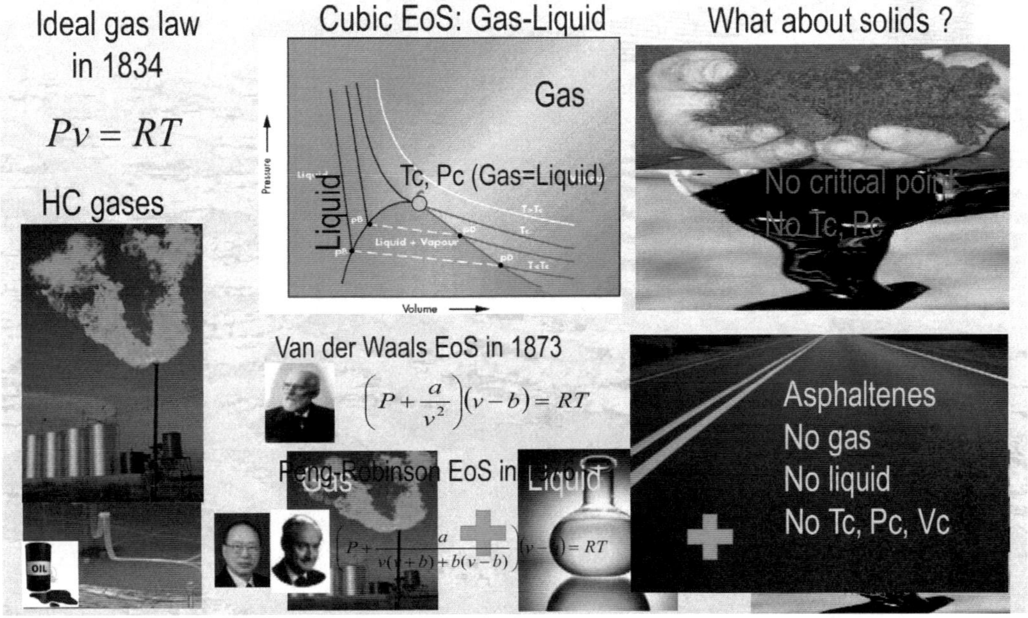

Gravity Plays a Big Role in Oil Columns

We are dealing with oil reservoirs with large vertical depths, so gravity matters. And for gravity, SIZE counts.

Although Hirschberg (1988) suggested to use the Flory-Huggins theory to model asphaltene gradients, however, no GOR variations were considered and could not be used to heavy oil. Most importantly, asphaltene size was not understood throughout in crude oil at that time.

Nobody knew the size of the asphaltenes in crude oil.
No predictive models for asphaltene gradients.

Hails fall through the air due to gravity

Small fogs can remain suspended in the air

Size of asphaltenes in crude oil now determined (Year 2010)

Asphaltene size is resolved and codified in the Yen-Mullins model.

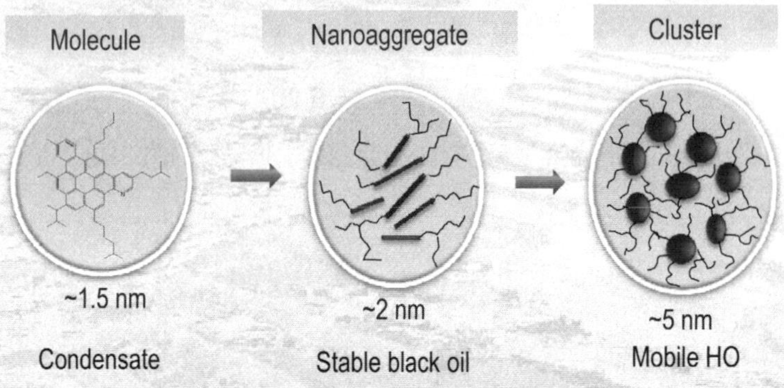

Yen-Mullins model

With the size known, the gravity term is now resolved

Flory-Huggins-Zuo (FHZ) Model

Paul John Flory

Model = Gravity Term + Solubility Term + Entropy Term

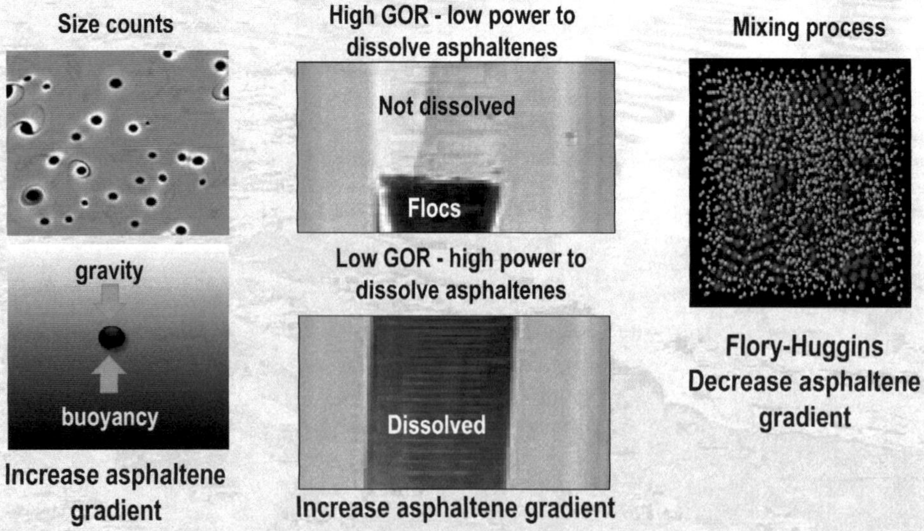

Asphaltene Grading FHZ Model (cont'd)

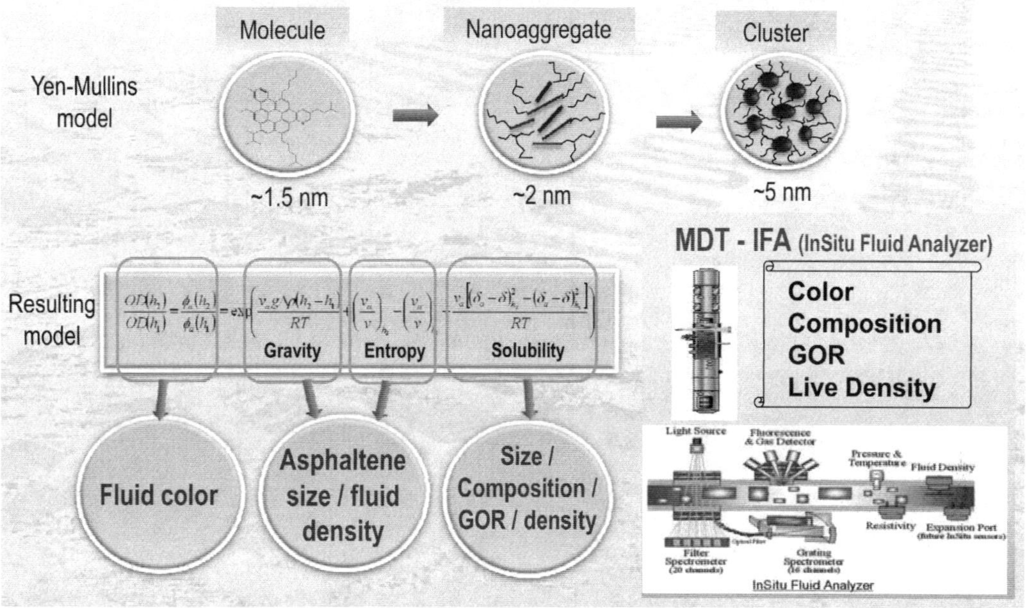

New Workflow of DFA Profiling

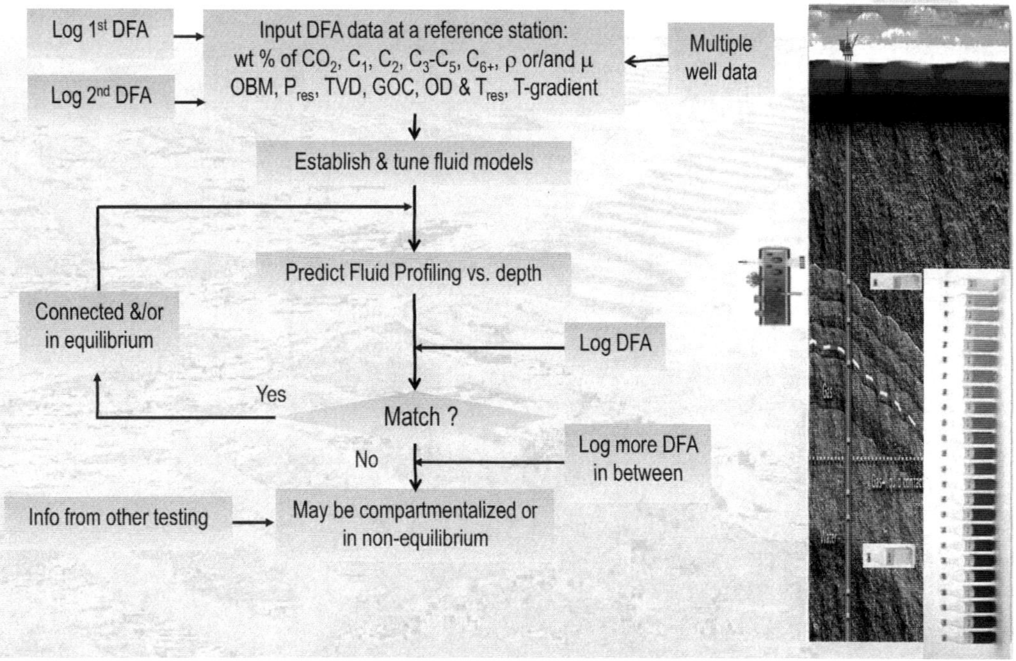

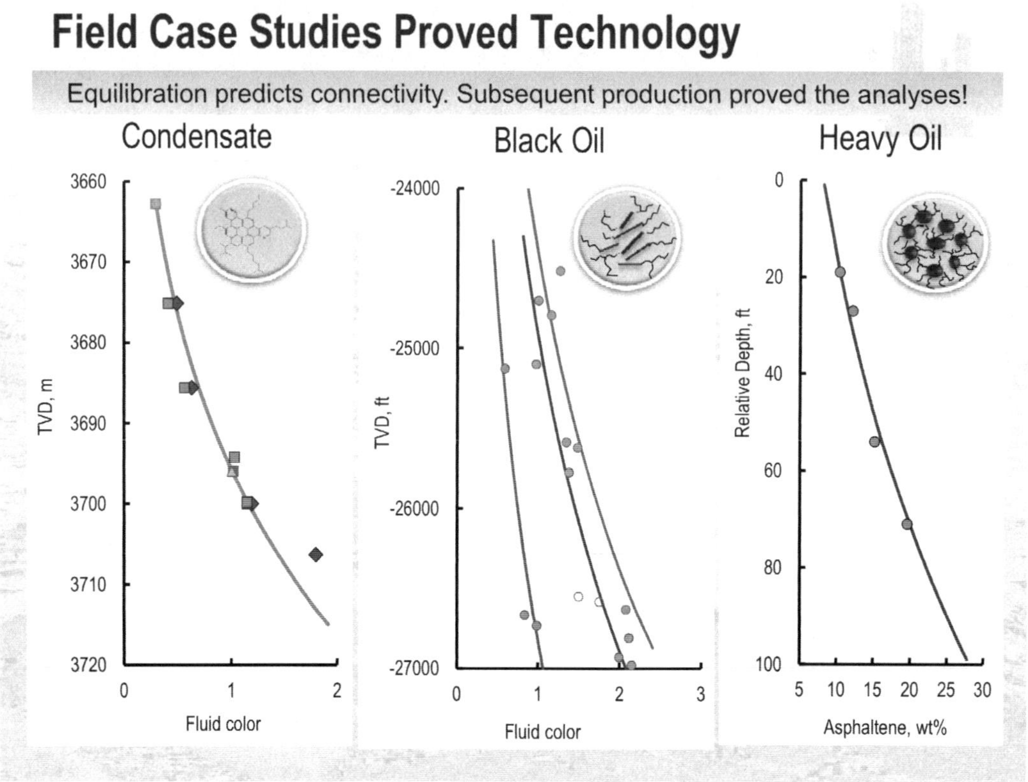
Field Case Studies Proved Technology
Equilibration predicts connectivity. Subsequent production proved the analyses!
Condensate
Black Oil
Heavy Oil
TVD, m
Fluid color
TVD, ft
Fluid color
Relative Depth, ft
Asphaltene, wt%

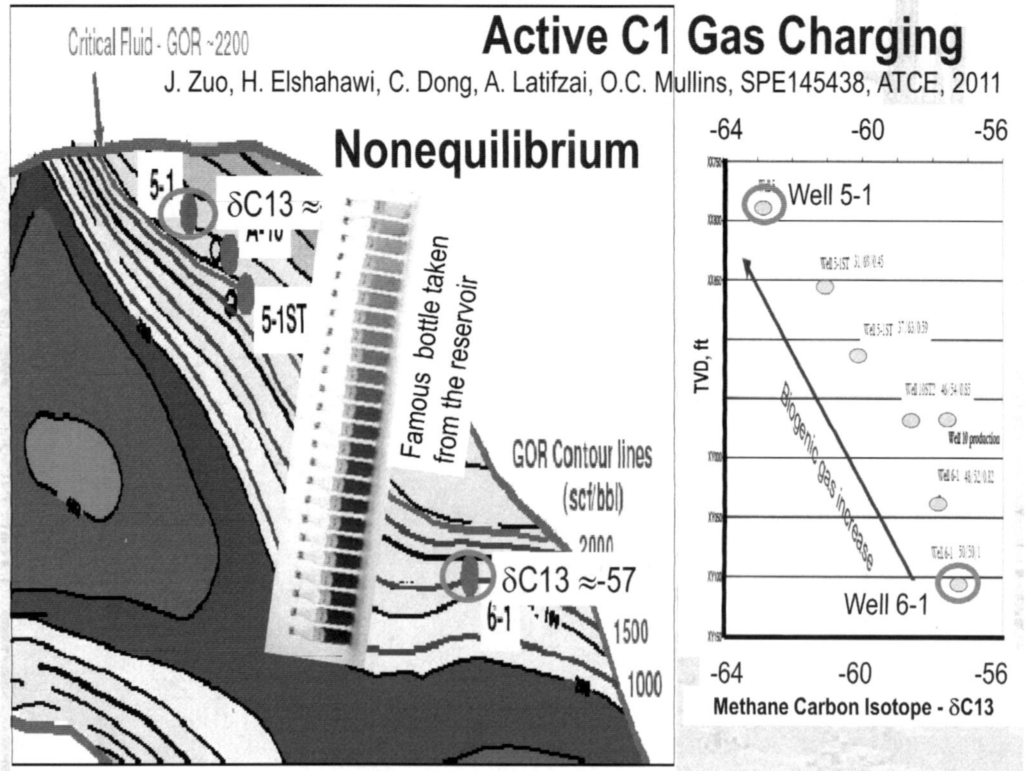
Critical Fluid - GOR ~2200
Active C1 Gas Charging
J. Zuo, H. Elshahawi, C. Dong, A. Latifzai, O.C. Mullins, SPE145438, ATCE, 2011
Nonequilibrium
5-1
δC13 ≈
5-1ST
Famous bottle taken from the reservoir
GOR Contour lines (scf/bbl)
2000
δC13 ≈-57
6-1
1500
1000
Well 5-1
TVD, ft
Biogenic gas increase
Well 6-1
Methane Carbon Isotope - δC13

Proposed Mechanism for a Large Color Variation

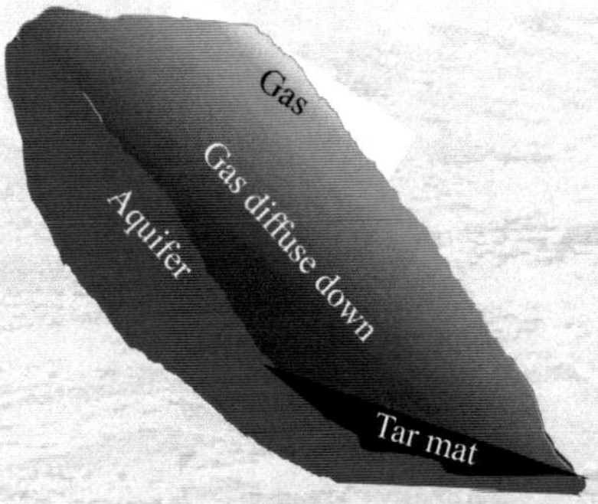

Active Late Gas Charge

High GOR fluid reduces solvation power for asphaltenes and push asphaltenes out

Gas charge did not diffuse fully down to the bottom and asphaltenes are concentrated at the base

Tar mat may form if asphaltenes are destabilized

Can we model this non-equilibrium process using the FHZ model?

Pressure Gradients by Cubic EoS

Equilibrium model does not work

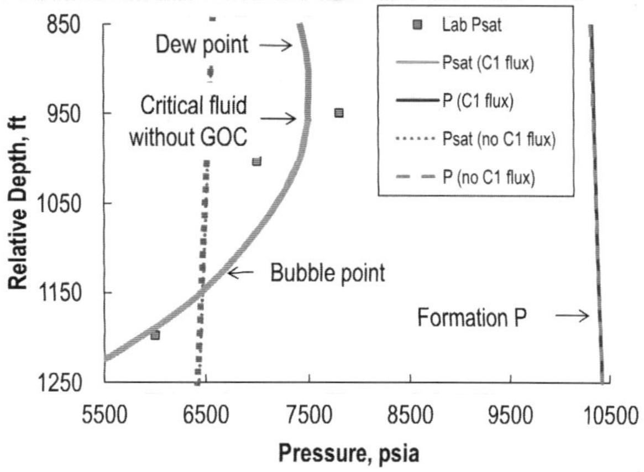

Solids curves are non-equilibrium with methane influx = 0.78 $Stdm^3/m^2/MY$

GOR & API Gravity Gradients by Cubic EoS

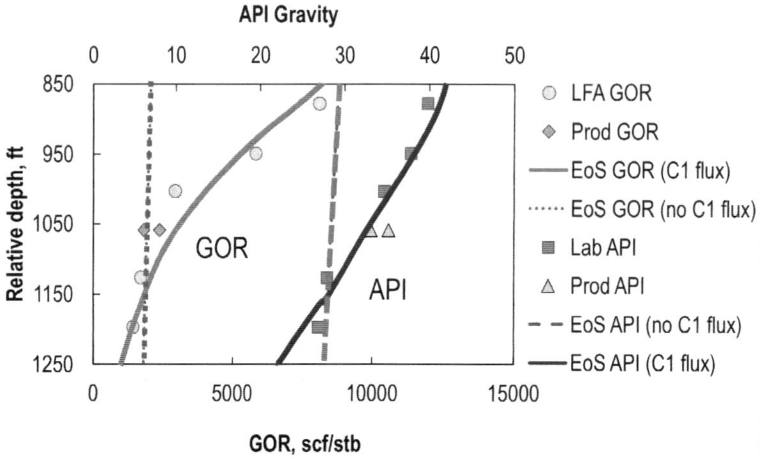

Solids curves are non-equilibrium with methane influx = 0.78 $Stdm^3/m^2/MY$

Predicted Asphaltene/Color Gradients

Asphaltene nanoaggregates
Diameter (d) = 2 nm

1. Continuous disequilibrium distribution of asphaltenes implies connectivity!
2. C1 influx also indicates connectivity!

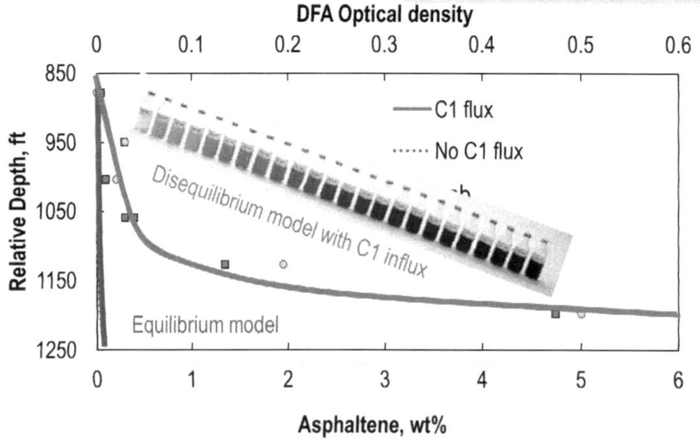

The FHZ model coupled with GOR and density gradients gives right color variation.
Local asphaltenes are equilibrated with local GOR at each depth.

Possible asphaltene phase instability

Equilibrium criteria

$$x_i^{oil}\gamma_i^{oil} = x_i^{Asph}\gamma_i^{Asph}$$

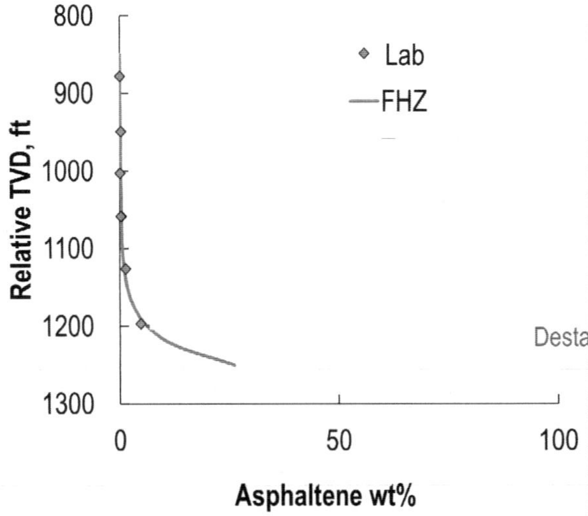

Destabilized asphaltenes may form a tar mat

Commingling production from top and bottom may result in severe flow assurance problems

Flow assurance - asphaltene phase instability

Field Case shows asphaltenes destabilization

- Geochem – gas washed
- $P_{onset} = P_{formation}$
- Bitumen in core
- Flow assurance issue

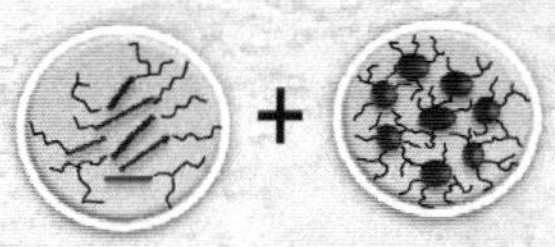

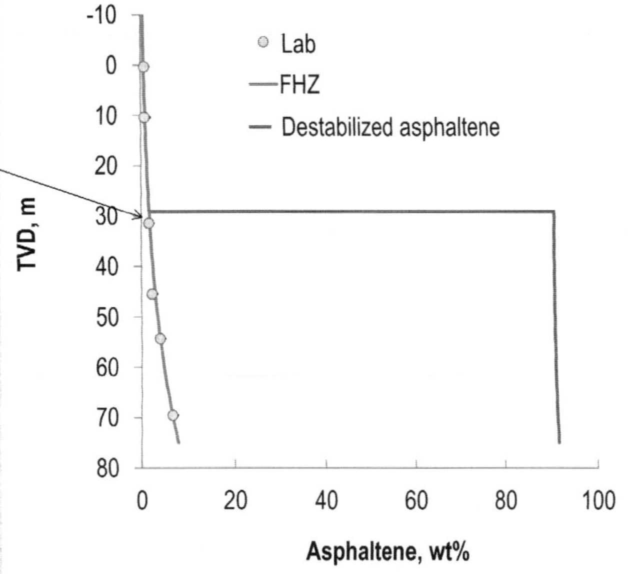

Heavy Oil – Change from Mobile to Immobile

T = 200 F, API = (13 – 21), GOR = < 90 scf/bbl

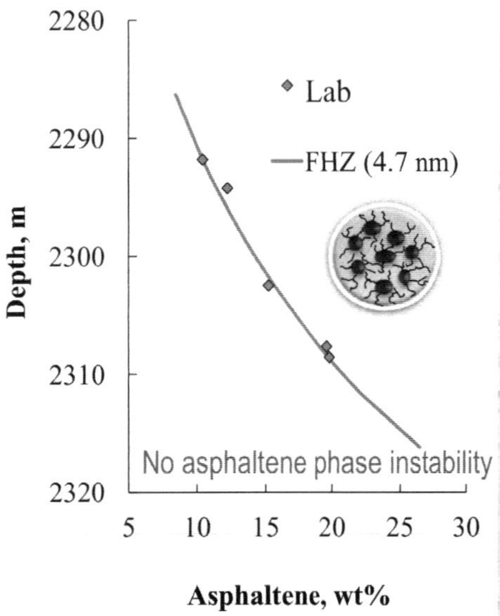

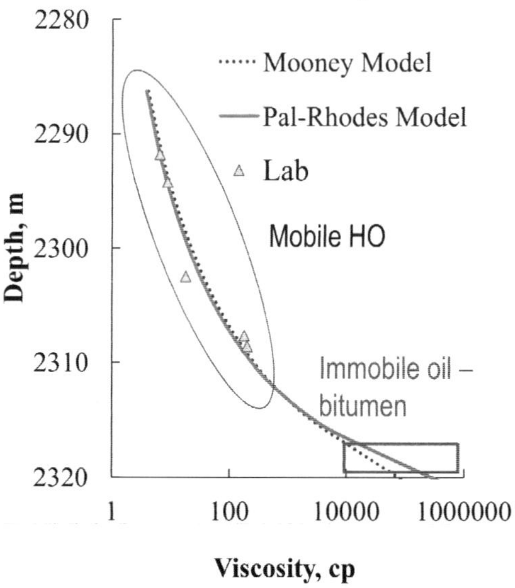

Conclusions

- Fluid complexities, compartmentalization, tar mat formation are large unresolved issues in reservoir management.
- Integration of the Yen-Mullins model, FHZ EoS and new workflow with DFA is a powerful approach to find fluid compositional grading, compartments/connectivity & asphaltene phase instability (tar mat formation, flow assurance issues).

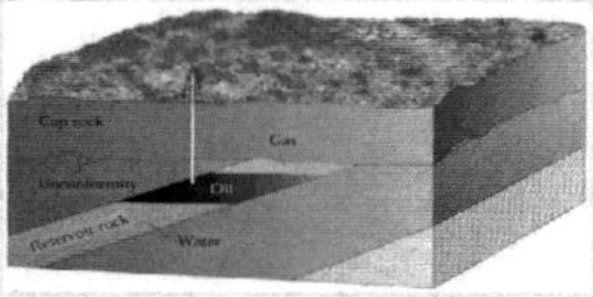

Oil-field Chemicals in the Changing Face of the Industry

Allenson, S., NALCO

To meet the ever-increasing energy demands, the oil industry continues to push the limits in deep and ultra-deep oil and gas exploration. Crude oil production in deep/ultra-deep water is an extremely capital intensive process in often hostile environments. Ensuring smooth production is immensely challenging due to remoteness of the locations, longer distances and subsea environment, greater depths and extreme temperature, high pressure and corrosive conditions, and problems such as hydrates plugging, wax deposition; foaming and emulsions; or topside operational shut-downs such as from slugging. Intervention and remediation costs for such hindrances can be significantly high, along with substantial losses in revenue, and may often threaten the viability of the project. Thus, with the changing face of the oil and gas industry, flow assurance currently encompasses a wide spectrum of responsibilities such as reliable system deliverability, insulation, operability and performance; production chemistry and system integrity along with safe operations, all of which involve tremendous obligations throughout the entire life cycle of the hydrocarbon exploitation. Achieving maximum production and simultaneously ensuring unimpeded flow through tubings and pipelines can be attained by increasingly robust system designs that however raise overall system economics to staggering heights, not always feasible for especially smaller fields.

With the boom in deep-water production and growth in developing regions such as Latin America, Asia-Pacific and Africa, the global market for oilfield chemicals is forecasted to reach US$31 billion by the year 2015. Oil field chemicals are gaining increasing importance such that chemical usage strongly influences flow system design, operating procedures and implementation of a successful flow assurance strategy. Chemicals indeed provide cost effective and invaluable solutions to various problems that plague the oil and gas production systems from drilling and completion, through fluid transportation to topside operations. Controlling the deposition of hydrates, waxes, asphaltenes, mineral scales, and solids to ensure smooth flow of hydrocarbons through the flow-lines or maintaining the structural integrity of the operator's assets to improve system performance and longevity and ensuring safe operations during production and transportation along with environmental compliance are overriding concerns that can be effectively managed with chemicals, while striking a balance between cost and system performance. In the harsh subsea environment, where errors can have unfortunate consequences, the oil industry is constantly pushing oilfield chemical companies to develop innovative products that meet a wide range of challenges in terms of quality, reliability, pumpability, deliverability and environmental compliance besides meeting the required performance standards. To rise up to this challenge, the oil chemicals industry continues to put arduous efforts to tailor new-generation chemistries that can sustain the extremes of temperatures and pressure variations in terms of stability and viscosity of the products, are compatible with the materials used in the chemical delivery system, besides successfully meeting stringent chemical performance and delivery requirements. This talk will discuss the transitioning impingent on the oil field chemical industry as the global oil and gas industry shifts to deeper offshore production.

Advancing Reservoir Performance

Flow Assurance Chemical Solutions:
" What Can be Done to Promote Increased Application of Specialty Chemicals for Deepwater Developments? "

David W. Jennings
Baker Hughes
Petrolite Technology

September 28, 2011

BAKER HUGHES

Introduction

- Flow Assurance Challenges are Handled Through a Combination of System Design & Operating Procedures
  - Mechanical, Thermal, and Chemical Means
  - Or Combinations Thereof
- Intent of this Presentation is to Stimulate Discussion on Chemical Applications. In Particular, on Ways to Improve and Increase Chemical Treating
  - Briefly Overview Chemical Applications
  - Discuss Chemical Delivery & Environmental Regulations
- New or Improved Applications Provide Increased Options
  - More Optimal Flow Assurance Designs

Outline

- Current Status of Chemical Applications
  (Hydrates, Asphaltenes, Paraffin, & Scale)
  - Benefits
  - Limitations / Future Needs
- Chemical Delivery
  - Umbilical Delivery
  - Alternative Delivery
- Environmental Regulations
- Acknowledgements

Chemical Applications (Current Status)

Hydrate Applications - Hydrate Inhibitor Benefits

- Hydrates are Most Common & Biggest Flow Assurance Concern
  - Gas Systems: Treated with Continuous Injection of Inhibitor
  - Oil Systems: Steady State Designed to be Outside Hydrate Region. Shut-in and Restarts Transients Treated
    - Thermodynamic Hydrate Inhibitors, Low Dosage Hydrate Inhibitors (LDHIs), & Other

- LDHI Benefits:
  - Lower Volumes than Methanol / Glycol
  - Lower Capital Costs than Methanol / Glycol
    - Storage Tanks, Pumps, and Umbilicals

LDHIs: Limitations & Needs

- Limitations:
  - KHI: Time & Sub-cooling Limitations, Biodegradability, and Toxicity
  - AA: High Water Cut, Water Quality, Biodegradability, and Toxicity

  - Industry Reluctance

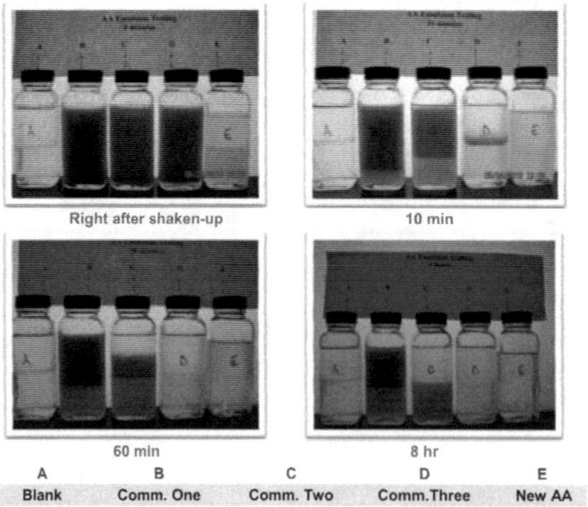

Test of Anti Agglomerate LDHI Water Quality

Aphaltene Applications: Asphaltene Inhibitor Benefits

- Asphaltenes are not a Problem for the Majority of Crude Oil Production. If Problems Occur in Deepwater Subsea Wells though, the Impact can be Catastrophic.
- Asphaltene Inhibitors Applied in Most Perceived Asphaltene Problem Wells
- Asphaltene Inhibitor Benefits:
  - Help Mitigate Deposition Which can be Devastating to Wells' Economics
  - Aid Mitigation of Topside Problems

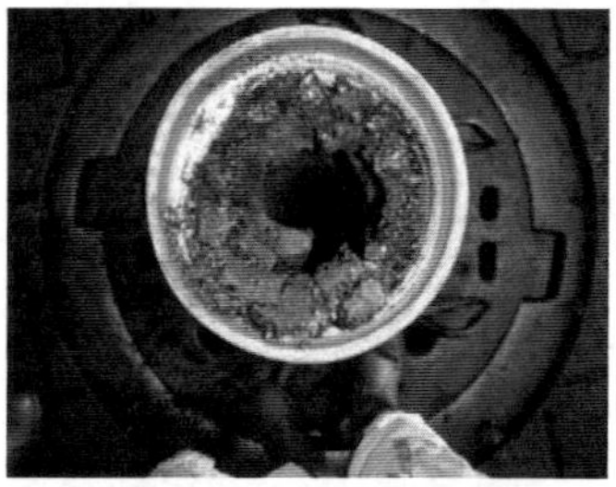

Asphaltene Well Deposition

Asphaltene Covered Subsea Completion (SPE 115338)

Asphaltene Inhibitors: Limitations & Needs

- Limitations:
  - Uncertainty in actual field deposition rates and amount of inhibition.
  - Injection location limited to depth of capillary string. Not into reservoir.
- Needs:
  - Improved performance
    - In Inhibition Effectiveness
    - In Assessment
      - On-shore well case study needed
  - Squeeze products
    - Allows Placement into Reservoir
    - Allows Treating Flexibility

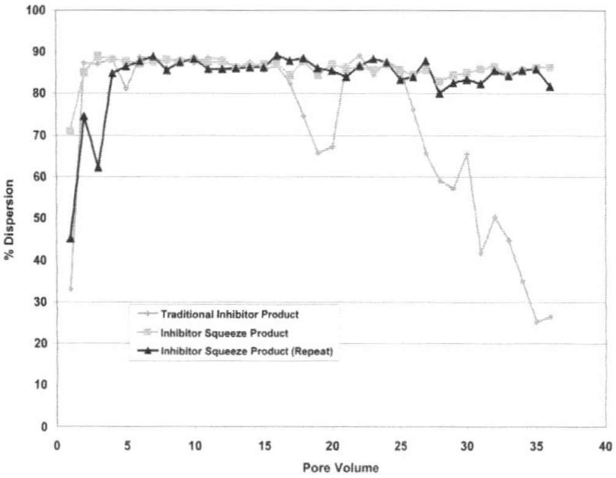

Asphaltene Inhibitor Retention Life

Paraffin Applications: PI / PPD/ Dispersant Benefits

- Paraffin Deposition Most Widely Handled with Pigging
  - Paraffin Inhibitors (PIs) Used Sometimes to Supplement Pigging & in Some Single Flowline Tiebacks
  - Dispersants for Cleaning for Deepwater Flowlines
- High Pour Point Crude Oil Production Handled with:
  - Pour Point Depressant (PPD)
  - Heat Retention
- Benefits:
  - PI: Reduced Deposition, Pigging Frequency, & Deferred Production
  - PPD: Allow Moderate-High Pour Point Crude Oil Production without Gelling Flowline or Provide Ability to Restart
  - Dispersant: Economic Means to Clean Severe Restrictions

PI / PPD: Limitations & Needs

- Limitations:
  - Effective Performance / Dose Rate
  - Delivery Through Umbilical
    - Diluted Product Activities
    - Elimination of Chemistries
- Needs:
  - Delivery of Higher Active Products

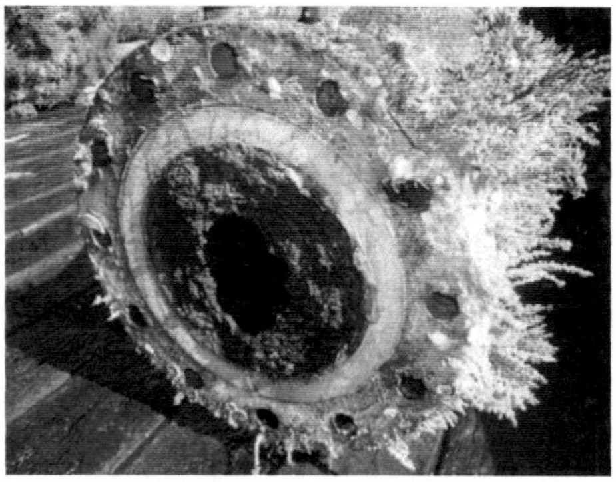

Paraffin Deposition in Subsea Line

Scale Applications: Scale Inhibitor Benefits

- Scale Deposition Predominately Handled Chemically with Scale Inhibitors. Membranes also Used for De-sulfating Seawater for Flooding.
- Scale Inhibitor Benefits:
  - Economical and Effective Treatment of Scale
  - Continuous Injection and Squeeze Applications
  - Low Minimum Effective Conc.
  - Long Squeeze Lives

Scale Deposition in Tubing

Scale Inhibitor: Limitations & Needs

- Needs:
  - Improved High TDS Tolerant Chemistries
  - Improved High Temperature Chemistries
  - Multiple Tagged Chemistries for Monitoring Residuals from Multiple Completions and/or Commingled Well Production

Chemical Delivery (Umbilical Systems)

Umbilical Delivery

- Umbilical Challenging Environment:
  - Low Temperature, High Pressure, Long Distance, Small Diameter, & Even Smaller Flow Paths (Check Valves, Orifices, etc ...)
  - Relatively In-accessible
  - Relatively Expensive
    - Few or No Spare Cores
- Critical that Umbilical Blockage Does not Occur
  - Flow Assurance may Depend on Chemical

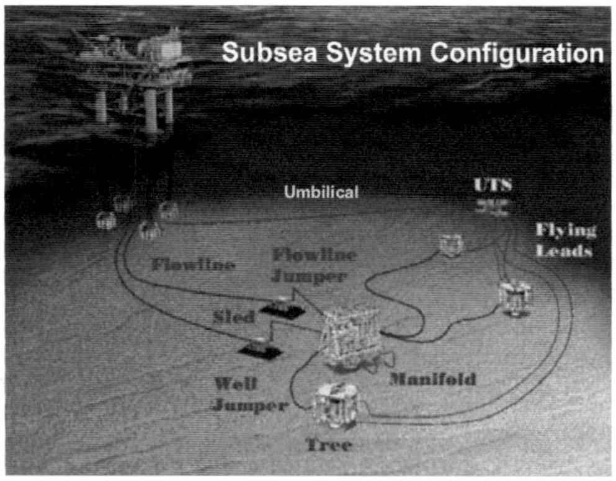

Umbilical Delivery

- Focus on Umbilical Blockage Needed:
  - BASICS JIP
    - Focus on Chemical Qualification & Recommended Practices
    - Good Starting Point
  - Need Similar Focus on Hardware
    - On Most Prone Plugging Locations
      - Addition of Redundancy
      - Ability to Treat and/or Replace
    - Most Prone Locations
      - At Line Ends
    - Most Common Causes
      - Backflow of Produced Fluids
        - Solids, Incompatible Fluids, & Hydrates
      - Solvent Permeation through Nylon
      - Unstable Fluids & Mixing Incompatible Fluids

Umbilical Delivery

- Don't Use Nylon
- Design Changes ??
  - Minimize Narrow Flow Paths Subsea
  - Redundant Dual Leads
    - For Deepwater By-Pass
  - Heating Capabilities
    - Electrical
    - Exothermic Reactions
  - Multiple Switching Flow Paths from UTS
    - Ability to Switch Umbilical Cores to Different Flying Leads
    - Establish Core Loop from Topside

Chemical Delivery
(Alternative Delivery)

Alternative Chemical Delivery

- Squeeze
  - Common Place for Scale Inhibition
  - Likely will be More Common for Asphaltene Inhibition
  - Design for Deployment from Host
- With Hydraulic Fracturing

Alt. Chemical Delivery - With Hydraulic Fracturing

- Controlled Release from within Proppant
  - Scale, Asphaltene, Paraffin, Corrosion, Biocide, & Combinations

- Relatively New, but 2000+ Applications
  - Majority Onshore Scale Applications
  - 20+ Offshore Applications
- Performance History very Good
  - Inhibition
  - Treatment Lifetime
  - Can "Re-Squeeze"

- Different Process / Cost Based On:
  - Confining Pressure
  - Loading

Controlled Release from Frac Treatments

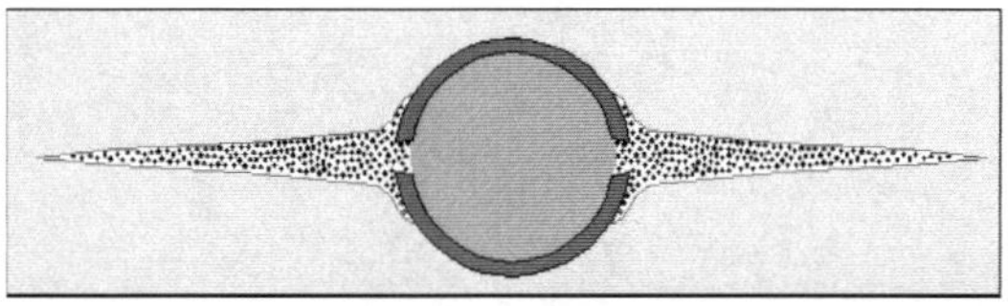

Environmental Regulations

Environmental Regulations

- Highly Effective, Cost Efficient Products with Low Environmental Impact & Risk are Endanger of Being Banned
  - Oil-Soluble Products
    - Partition into Oil
    - PPM Levels
    - Often HS&E of Products are No Worse than Crude Oil Constituents
  - De Minimis Quantity Limit
    - Threat to Ban Most Products

- Regulations Should be Based on HS&E Impact & Risks

Acknowledgements

- Baker Hughes Incorporated
  - For Permission to Participate in Workshop

- Baker Hughes Colleagues for Helpful Discussions in Preparing this Presentation
  - Mike Newberry, Steve Szymczak, Jun Tian, Tudor Ionescu, Marc Lehmann, Sam Asomaning, Geoff Leonard, Steve Jackson, & Vaithilingam Panchalingam

Guidelines for Subsea Hydrate Risk Management

Rama Alapati
Casey Rutkowski
Andrew Paille & Cristina Loya
Champion Technologies

Risk Assessment

Increasing Likelihood →

Severity Based on Consequences ↓

	A	B	C	D	E
0					
1					
2		Scale	Asset Integrity		Emulsions
3					
4				Wax Asphaltenes	
5		Water Quality			Hydrates

Low:	Medium:	High:
Manage for Continuous Improvement	Incorporate Risk Reduction Measures	Intolerable, Investigate Alternatives

Operator Goals

- Prevent or reduce loss of production revenue
- Eliminate hydrate blockage risk
- Most economic solution
- Meet water quality requirements

Current Solutions

- Methanol (MeOH)
- MEG (Mono Ethylene Glycol) regeneration systems
- Low Dose Hydrate Inhibitors (LDHI)– (AA)Anti-Agglomerate hydrate inhibitors

Application

Subcooling	Water Depth	Gas/Oil Well	GOR	Water-Cut	Injection
35°F	4,000 ft	Gas-Condensate	25,000 scf/bbl	50%	Continuous

- East Breaks Block - GoM
- Wells shut-in as MeOH pumps maxed out – 3500GPD
- LDHI trialed successfully:
  - MeOH completely eliminated
  - Excellent water quality
- LDHI in use since August 2010

Field Trial

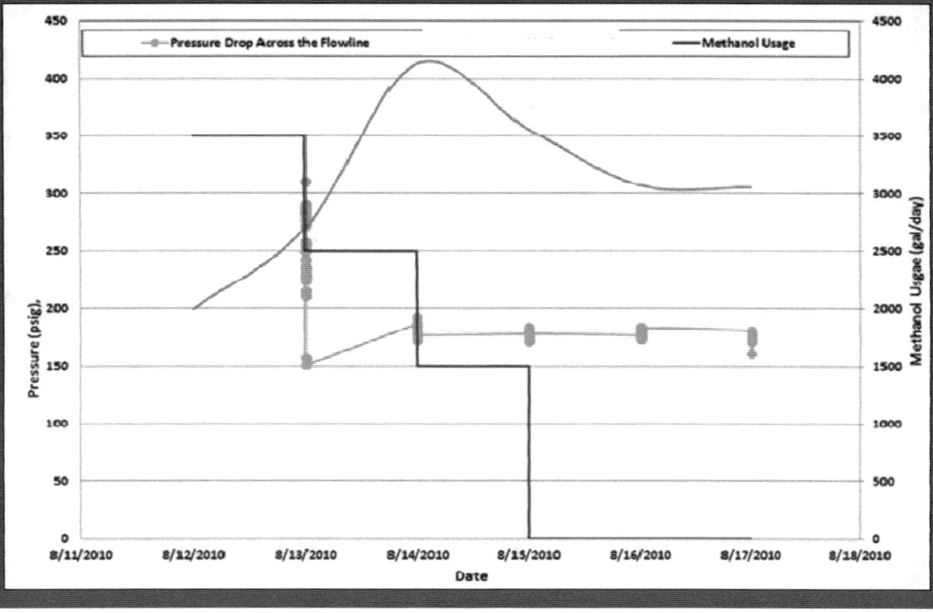

Success – No plugging

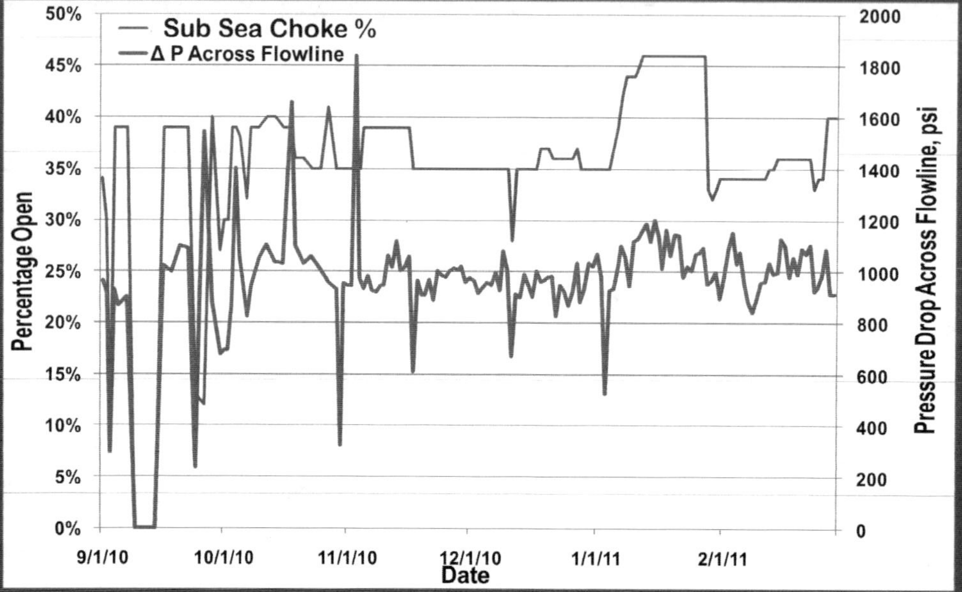

Success – Water Quality

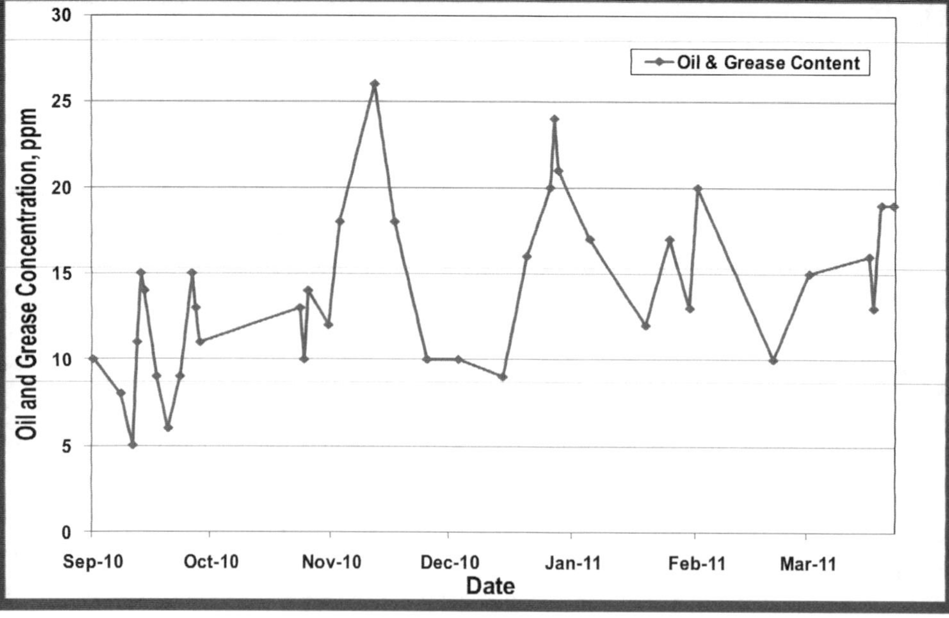

Application

SSTB Length	Water Depth	Gas/Oil Well	GOR	Water-Cut	Subcooling	Injection
8 miles	6350 ft	Gas-Condensate	24,000 scf/bbl	13%	35 °F	Continuous

- Mississippi Canyon Block
- Reduced operating pressure to lower subcooling
- Lost production revenue
- Product in use since March 2010.

Success – No plugging

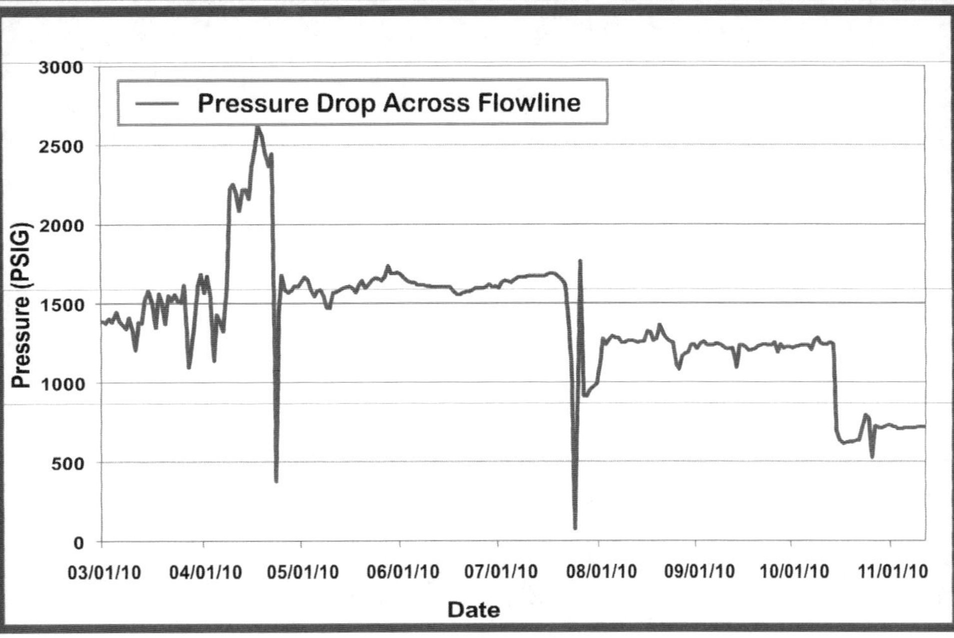

Success – Water Quality

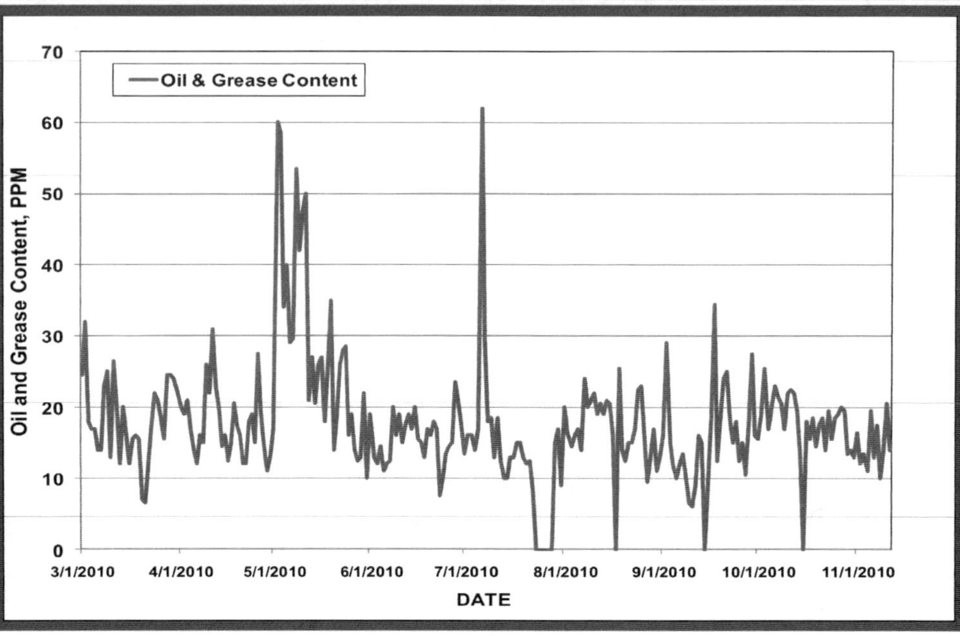

Success

- NO water quality issues
- NO increase in pressure drop
- NO MeOH slugging needed
- NO additional chemical treating costs

Success

- By-pass MEG. Reg. unit
- Increased production
- More than a year in application

Guideline -By Numbers

- Flowline length – 22miles
- Max water cut – 90%
- Highest GOR – 100,000
- Highest subcooling – 45°F

Guidelines

- **Low GOR Black Oil systems**
  - LDHI for planned Shut-in/restarts
  - Un-insulated Flowlines, steady state operation – LDHI application
  - Dose rate less than 1%
  - Unplanned shut-in/restarts needs – blowdown and bullhead MeOH
  - Verify LDHI temperature stability
  - Confirm LDHI materials compatibility

Guidelines

- **High GOR gas condensate systems**
  - 10% liquid hydrocarbon required
  - Phase behavior needs to be understood
  - Dose rate up to 2%
  - LDHI more cost effective than MeOH or MEG
  - Dual injection capability needed
  - Treatment needed for less than 5% water cut

Guidelines

- LDHI injection point
  - At the tree or above/below SCSSV
  - Line size less than 0.5”
- Methanol/MEG Injection point
  - Above/below SCSSV for gas condensate systems
  - Line size less than 1”

Acknowledgements

- Stacy Delafosse, Eric Vita, David Beard, Bruce Adams, Chad Bergeron and Elise Nguyen with Champion Technologies

Balancing Chemical Solutions and Engineering Solutions: A Holistic Approach to Flow Assurance Risk Management

Yuan, M., Chevron Energy Technology Company

On the one hand, although chemical additives are widely applied to prevent or manage flow assurance risks, it is not uncommon that 1) the applied chemicals fail to achieve the desired outcomes, or worse, 2) chemicals themselves have created additional production problems and/or risks to asset integrity or operation efficiency. On the other hand, there are instances that 1) engineering (i.e. non-chemical) solutions are simply preferred over chemical solutions, even though the latter may offer significant savings especially in capital expenditure (CAPEX), or 2) a production system was inadequately designed to allow effective chemical applications when a flow assurance problem has become apparent and a chemical solution becomes necessary.

This presentation attempts to strike a balanced view, where both the value of chemical solutions to flow assurance and its necessity in certain operating environments are duly recognized, but improvement of chemical applications and alternate options are also highlighted. The presenter believes that a holistic approach to flow assurance risk management will help reduce, if not eliminate, some missteps that may result in a flawed flow assurance strategy. This approach would recommend the following practices:

- Form a multi-disciplinary team covering a range of technical backgrounds and expertise, including both non-chemical and chemical subject matter experts (SME's), drawn from operator, design/engineering firm and chemical service company
- Identify and assess all the potential flow assurance risks that may occur in the early field life but also understand the evolving nature of flow assurance risks during the entire field life cycle
- Identify alternates (mechanical, thermal, electrical, operational and chemical) to prevent/manage the flow assurance risks
- Develop the preferred alternate which may be engineering solution only, or chemical solution only, or a combination of both
- Ensure proper documentation of decision rationales and knowledge -gained from development phase and its transition to the operating staff
- Allow flexibility in the production system to accommodate future, new chemical applications, if needed

In addition, this presenter will highlight several key obstacles/challenges that have been hindering a wider acceptance of chemical applications in the flow assurance area.

Application of ANSYS engineering simulation to flow assurance - with focus on multiphase flows

Massah, H., ANSYS

With increasing complexities to ensure safe and reliable transport of hydrocarbon between production to collection and processing points, detail engineering analysis including 2D-3D and transient simulations increasingly complement the current modeling capabilities for flow assurance. The applications include a range of physics including multiphase flows, heat and mass transfer, thermodynamics, phase change, structural integrity, electromagnetic and thermal analysis. In this presentation, we will review modeling of several flow assurance issues such as wax and hydrates, hydrodynamic and terrain slugging, gas lift, water and sand management, erosion, and hydrate melting to illustrate the use of ANSYS engineering simulation solution to flow assurance.

CD-adapco

STAR-OLGA coupled model for multiphase flows in pipeline and equipment

Sven Enger, Simon Lo, Mike Henneke

AIChE/SPE Workshop, Houston, September 2011

STAR-OLGA coupled model - Objective

To study 3D effects in in-line equipment: valve, junction, elbow, obstacle, jumper, separator, slug catcher, compressor, ...

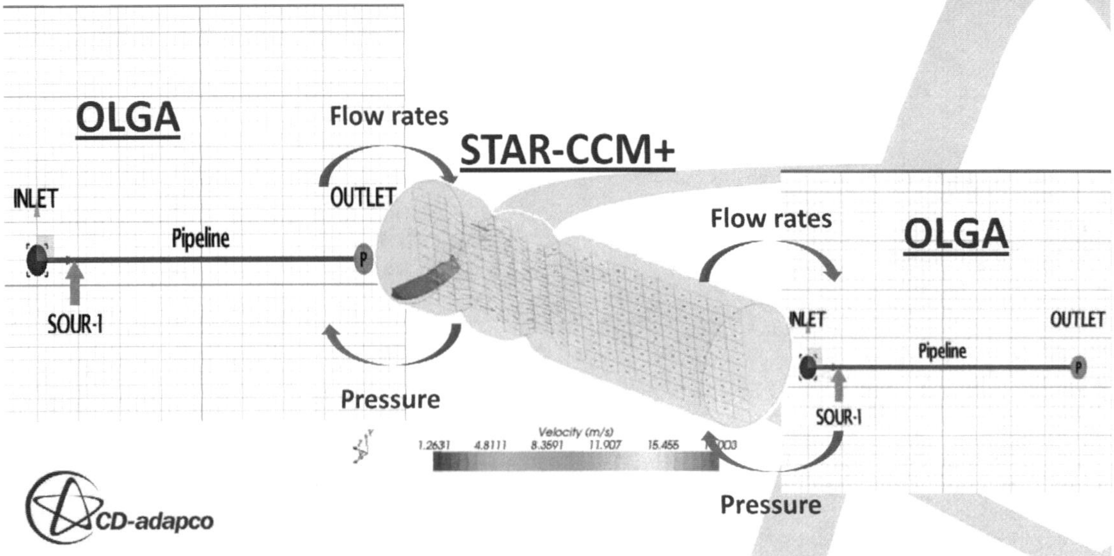

CD-adapco

CD-adapco – SPT partnership

- CD-adapco & SPT are partners in developing tools for Computational Flow Assurance
- Both companies fully committed from top management to technical engineers
- Very close working relationship between engineers
- CD-adapco and SPT are partners in TMF and FACE research consortia

CD-adapco

STAR-OLGA coupling – some details

- OLGA and STAR-CCM+ can run on any computer in the network
  - Example: we ran STAR-CCM+ in our Houston office and OLGA in our New York office
- User specify which STAR-CCM+ boundary connect to which OLGA zone
- All physical properties are interpolated from same OLGA pvt file to ensure consistency
- Stratified flow is assumed at STAR-CCM+ inlet

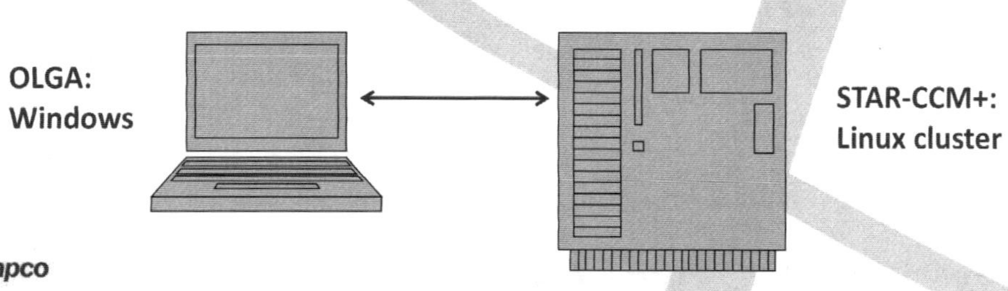

CD-adapco

STAR-OLGA coupling – time step controls

- Variable time step:
  - Allow OLGA to take *larger* time steps
  - Allow STAR-CCM+ to take *smaller* steps
  - STAR-CCM+ will interpolate OLGA results for intermediate steps

OLGA time step		STAR-CCM+ time step
O1		S1
		S2
		S3
O2		S4
		S5
		S6
O3		S7

CD-adapco

STAR-OLGA Coupling – Point & Click setup

Familiar 1D Code Interface

- New "Olga" node appears in Tools
- Port Number, Host Name, Input file, Property file etc. set as properties of this node
- Right Click allows Olga to be started

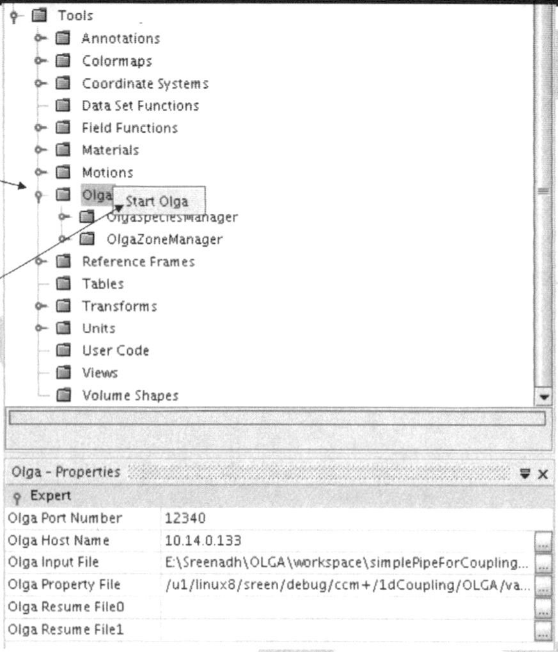

CD-adapco

STAR-OLGA Coupling – Clear association of model parameters

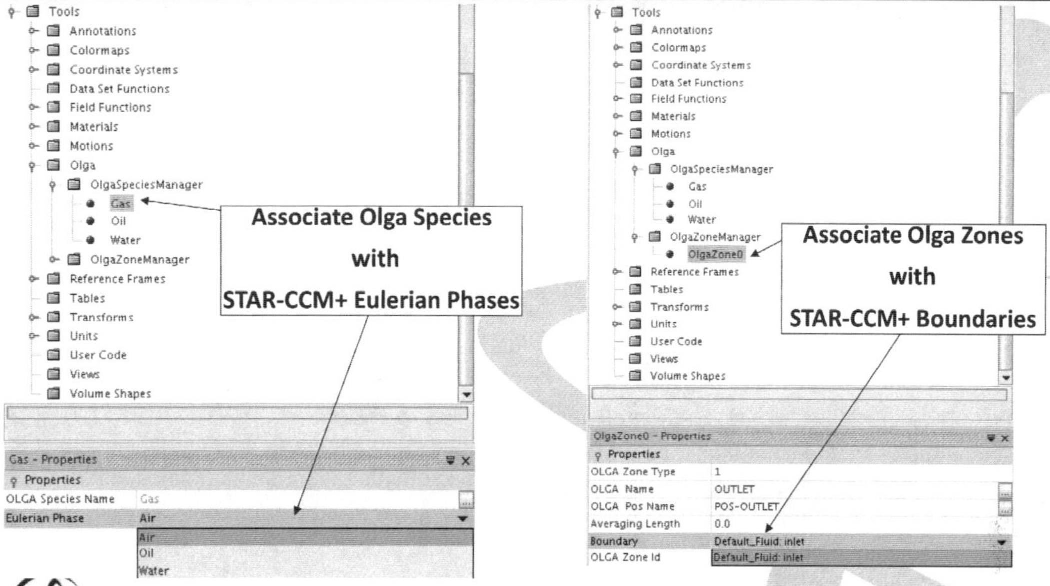

CD-adapco

Example: Cyclic forces on elbow in slug flows (Cranfield Univ.)

- Slug flow around pipe bend causes cyclic forces on the bend

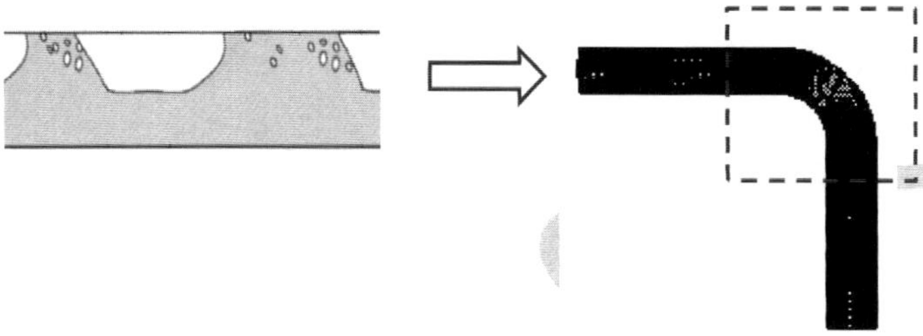

- Slug force: excessive movement; fatigue cracking; challenge to ensure the integrity of the pipeline

CD-adapco

Traditional "steady state" analysis

- **Most models are based on force balance analysis of steady state flow**

$$F_x = \boxed{\rho A U_x^2} + \boxed{(P_{in} - P_a)A}$$

$$F_y = \boxed{\rho A U_y^2} + \boxed{(P_{out} - P_a)A}$$

Momentum change
+
Pressure-area force

$$F_R = \sqrt{F_x^2 + F_y^2}$$

- **Industrial practice: impact (safety) factor '2' added to the momentum term**

CD-adapco

Test case: Tay's PhD thesis (2002)

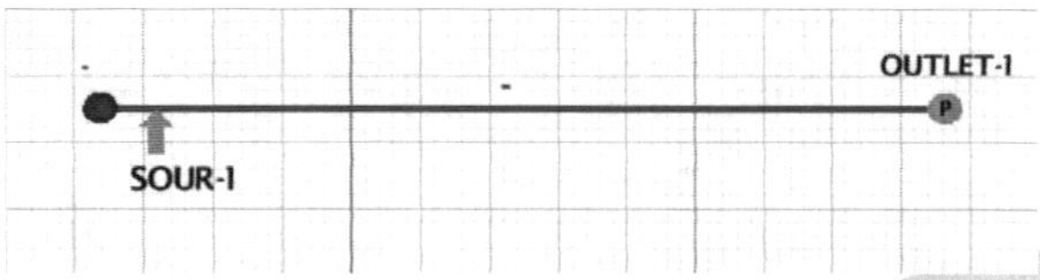

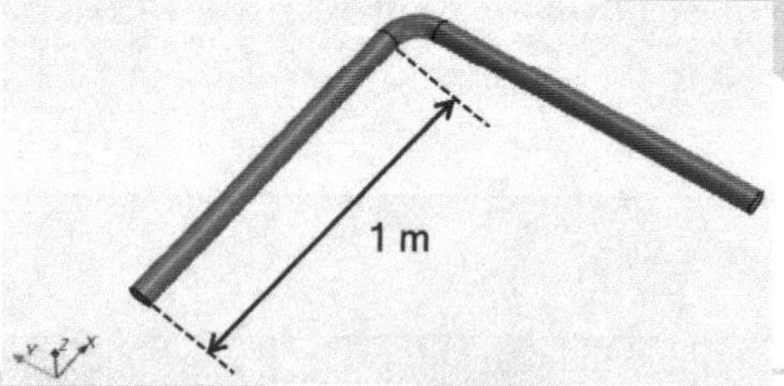

OLGA pipe: 20 m (+ slug tracking) **STAR pipe: 1m + bend + 1m**

- **Pipe diameter: 70 mm; 90° bend radius: 105 mm (horizontal)**
- **Inlet (SOUR-1): U_{SG} = 2.02 m/s and U_{SL} = 0.6 m/s (superficial)**
- **Outlet (bend outlet): P = atmospheric pressure**

CD-adapco

Model configuration and Data exchange

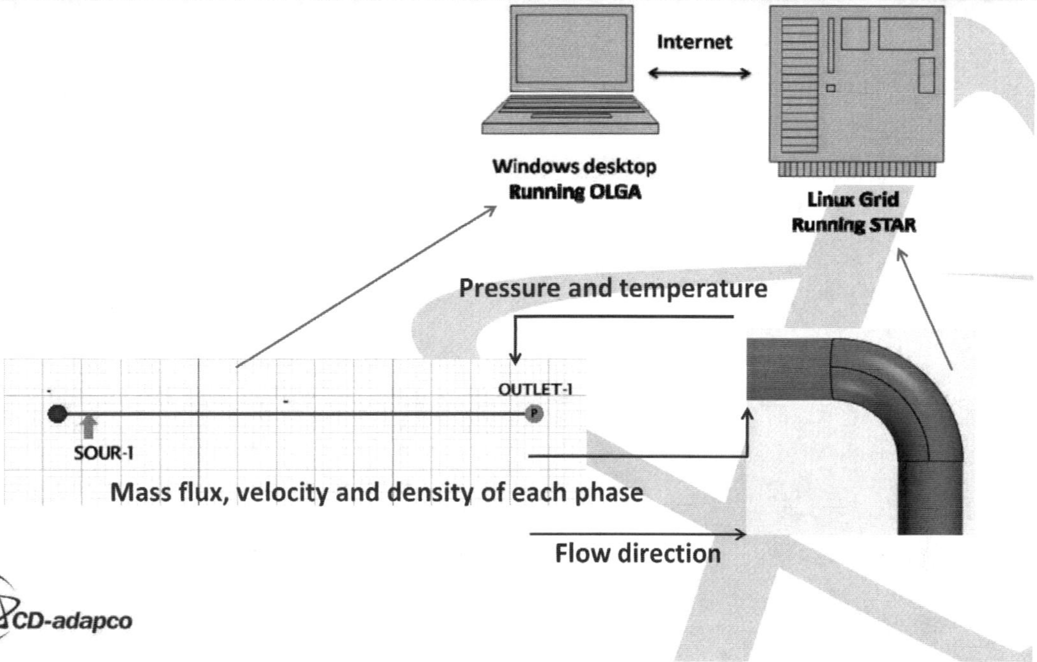

CD-adapco

Pressure variation due to slug flow pass elbow

Note the passing of liquid slug in "blue"

Note the increase in pressure as liquid slug passes

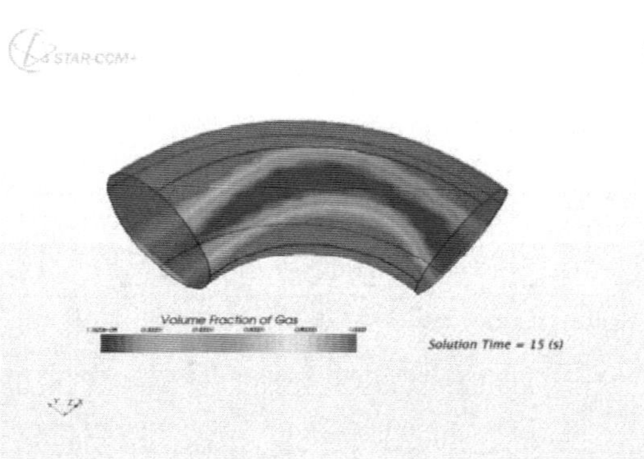

Gas volume fraction

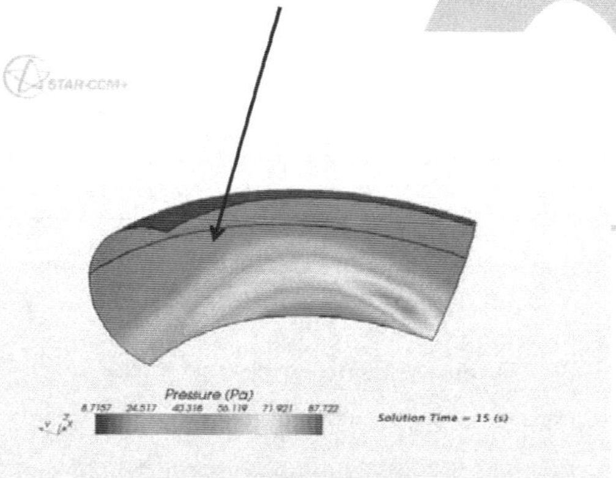

Pressure on the outer part

CD-adapco

Forces on the inner part

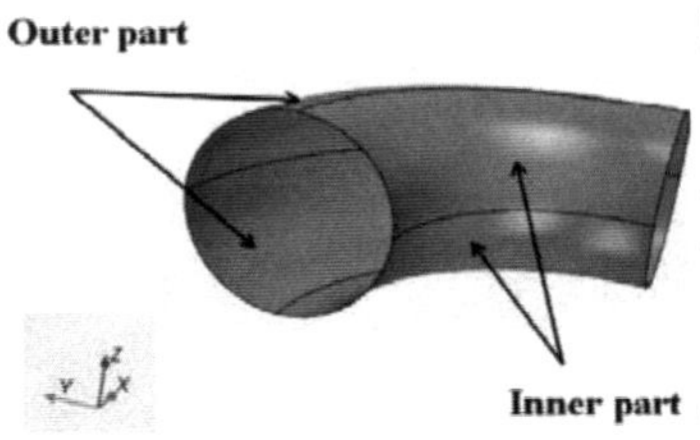

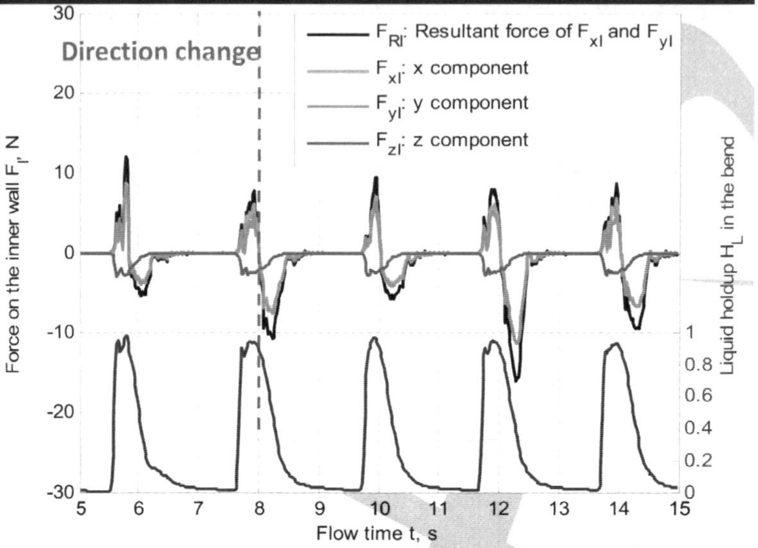

- 'Sine wave' behaviour
- Slug tail passing the bend centre

Resultant force and force components on the inner part

CD-adapco

Comparison

	Coupling model	Experiment
Slug frequency (Hz)	0.5	0.5
Slug velocity (m/s)	slug front: 2.8 to 3.6 slug tail: 3.0 to 3.5	3.6
Peak force on bend (N)	44 to 54	40 to 60
Maximum force on bend (N)	54	60

In industrial design with safety factor '2': maximum force 141 N

CD-adapco

Example: Multi-point STAR-OLGA coupling

Example Set-up

I1: Interface1; 1st coupling pt
OLGA Outlet – STAR-CCM+ Inlet

I2: Interface2; 2nd coupling pt
OLGA Outlet – STAR-CCM+ Inlet

I3: Interface3; 3rd coupling pt
STAR-CCM+ Pressure – OLGA Source

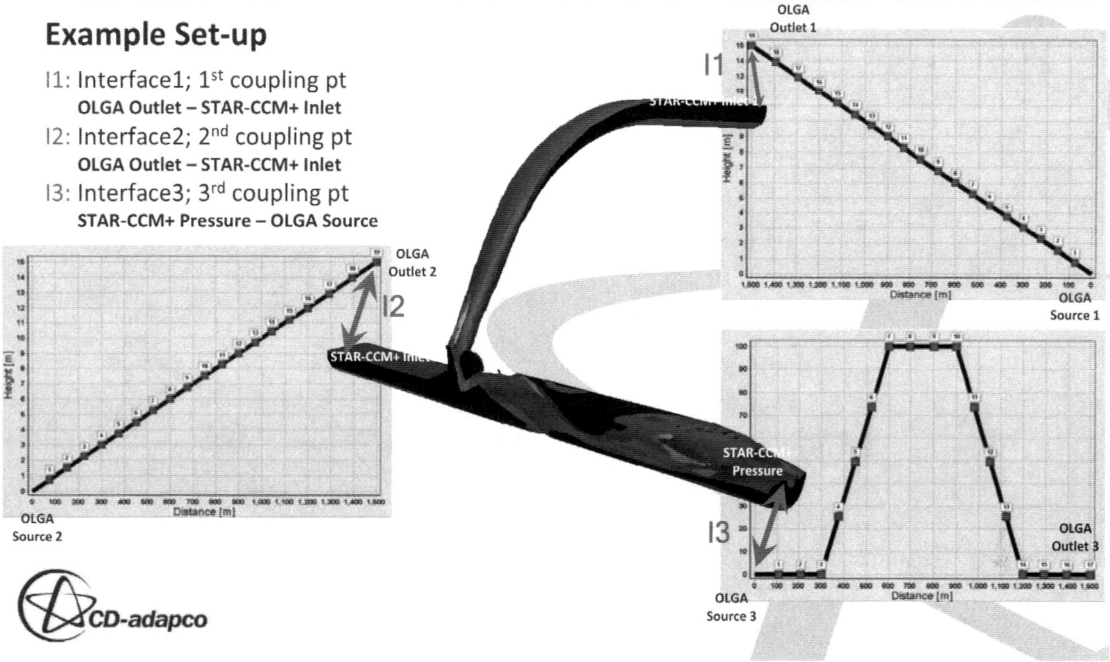

CD-adapco

T-junction + 90° side branch

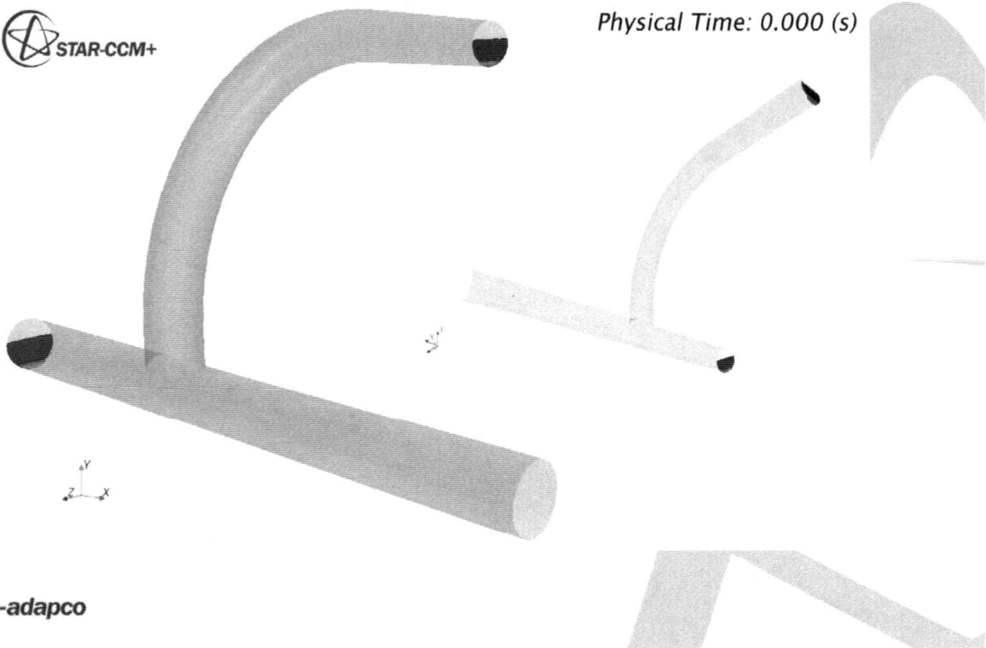

CD-adapco

Example: Multi-point couple

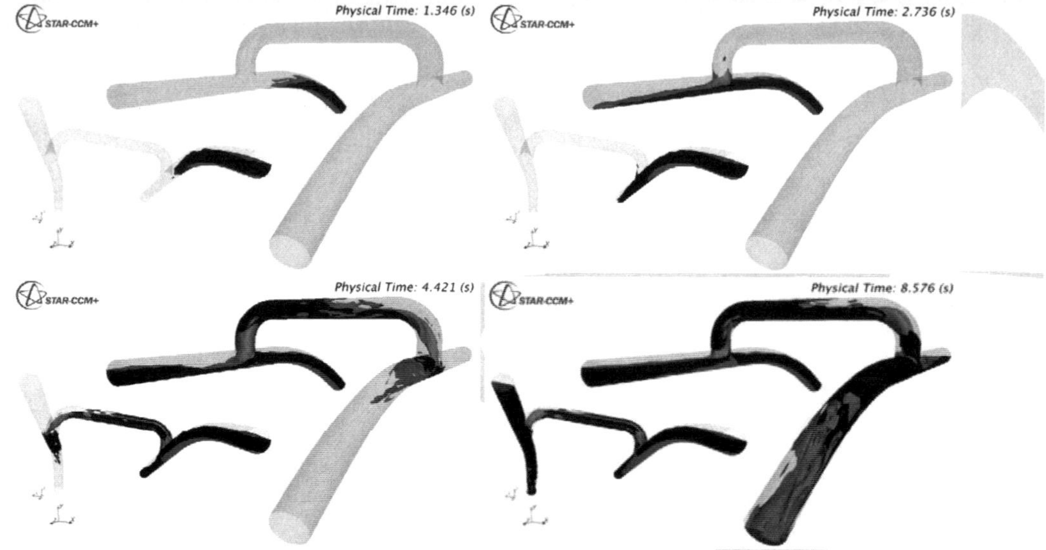

CD-adapco

17

Example: EXXON-MOBIL SLUG CATCHER

- Tanks: 45 m long / 3 m diameter
- Time scale large (2 hours)
- CFD Model: 2.2 Million cells, Time Step of 0.005 Sec

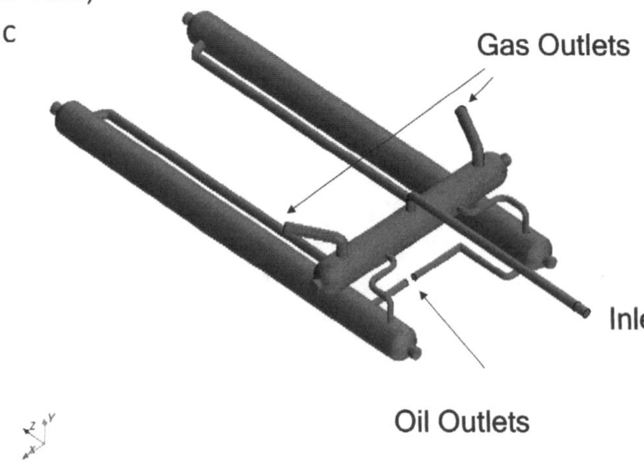

CD-adapco

EXXON-MOBIL SLUG CATCHER

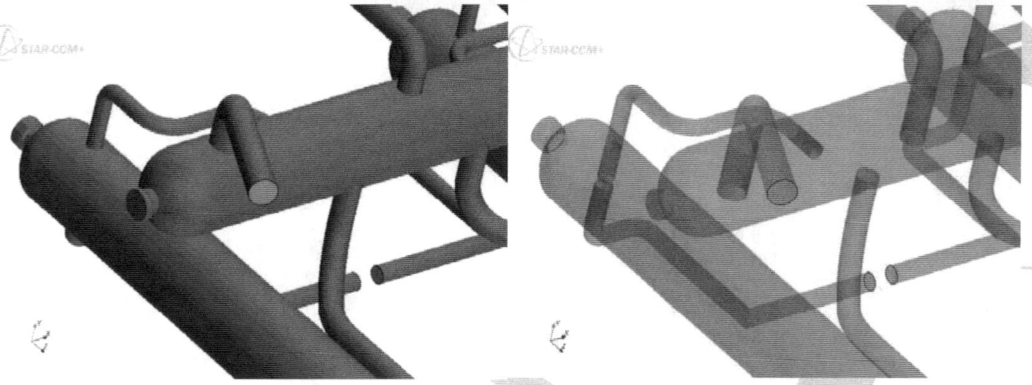

Geometry detail, showing how pipes extend into tanks

CD-adapco

EXXON-MOBIL SLUG CATCHER (Upstream Olga Model)

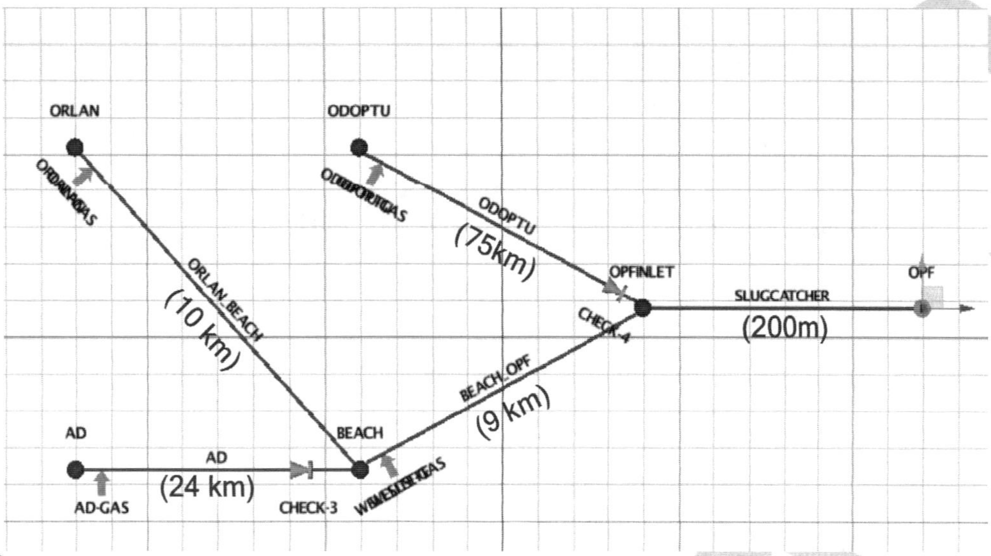

CD-adapco

EXXON-MOBIL SLUG CATCHER: Oil Surface

Olga Oil Flow Rate Blue graph below

Black:gas

Red: water

(BC to CFD Model)

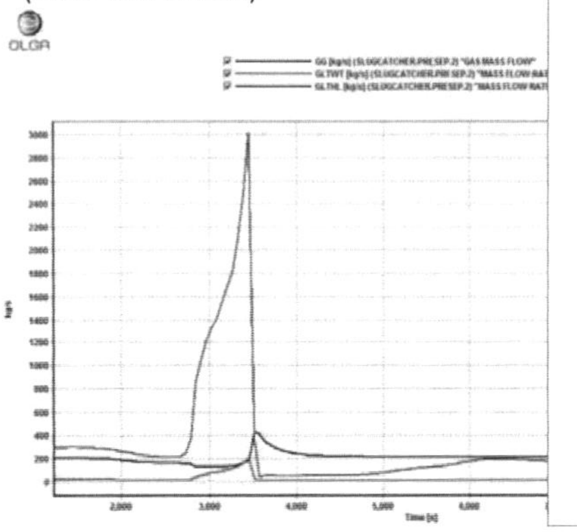

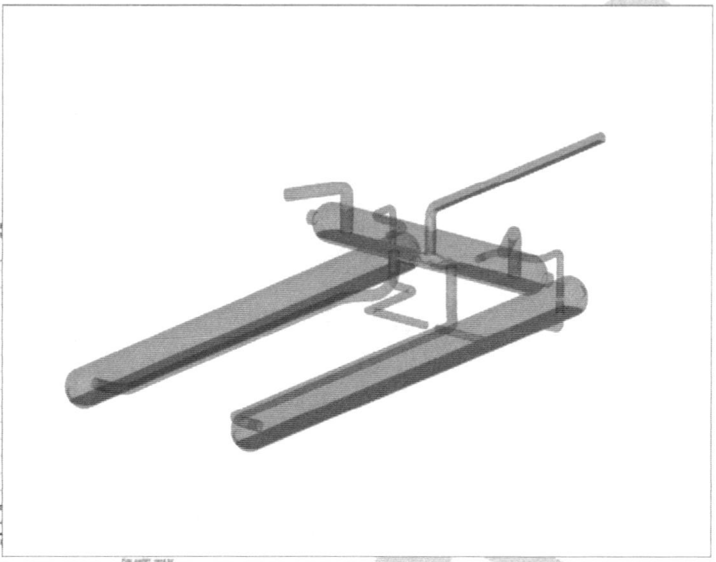

CD-adapco

Summary

- CD-adapco – SPT partnership and collaboration in development of Computation Flow Assurance tools
- Close working relationship between the two companies
- Successful development of coupling between OLGA and STAR-CCM+ for 1D analysis of long pipeline with detailed 3D CFD to study local effects
- One- and two-way, single- and multi-point coupling between OLGA and STAR-CCM+
- Successful applications of the STAR-OLGA model

CD-adapco

One-dimensional multiphase flow models on a coarse grid – theoretical and practical issues

By
J. Nossen, C. Lawrence, B Hu, R. Skartlien and P. Andersson
(IFE, Institute for Energy Technology)
and
D. Biberg
(SPT Group)

IFE

Acknowledgments

Part of this work was carried out in the HORIZON 2 project sponsored by Chevron, ConocoPhillips, ENI, ExxonMobil, Shell and Statoil

10/4/2011

IFE

Fundamental modeling challenge:

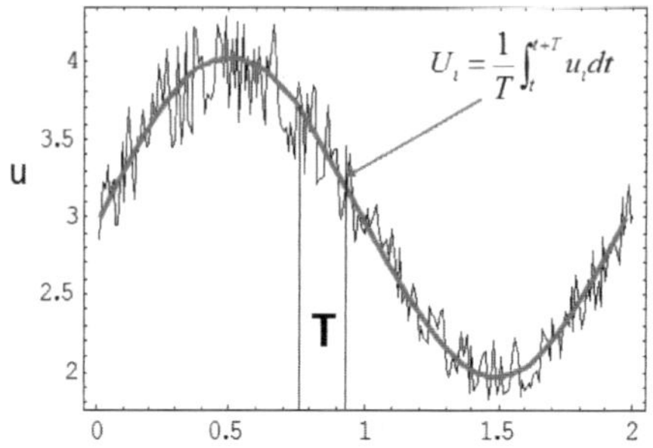

The closure problem of turbulent flow models
(more unknowns than equations)
makes heuristic closures based on data necessary

IFE

The frictions are the main unknowns in the integrated (1D) conservation equations

$$\rho_g\left(\frac{\partial A_g U_g}{\partial t} + \frac{\partial \Gamma_g A_g U_g^2}{\partial x}\right) + A_g\left(\frac{\partial p}{\partial z} + \rho_g g\left(\sin\theta + \cos\theta \frac{\partial h_l}{\partial x}\right)\right) = -S_g \tau_g - S_i \tau_i$$

$$\rho_o\left(\frac{\partial A_o U_o}{\partial t} + \frac{\partial \Gamma_o A_o U_o^2}{\partial x}\right) + A_o\left(\frac{\partial p}{\partial z} + \rho_o g\left(\sin\theta + \cos\theta \frac{\partial h_l}{\partial x}\right)\right) = -S_o \tau_o + S_i \tau_i - S_j \tau_j$$

$$\rho_w\left(\frac{\partial A_w U_w}{\partial t} + \frac{\partial \Gamma_w A_w U_w^2}{\partial x}\right) + A_w\left(\frac{\partial p}{\partial z} + \rho_w g\left(\sin\theta + \cos\theta \frac{\partial h_w}{\partial x}\right)\right) = -S_w \tau_w + S_j \tau_j$$

- The frictions are manifestations of the full 3D velocity distribution
- They are represented by single phase flow analogies or pure empirical correlations in most flow simulators
- OLGA applies 13 correlations for wall and interface frictions
- It has become increasingly difficult to tune the OLGA correlations to the constantly increasing data basis

IFE

What about field data?

- 13 parameters determine holdup(s) and pressure drop in three phase pipe flow
- We thus need e.g. $5^{13} \sim 10^{9}$ data points for 5 point resolution in each parameter
- We have ~ 200 field data points at present

- It is clearly impossible to cover the parameter space of three phase pipe flow with data

IFE

Conclusion: we need mechanistic models that extrapolate beyond lab data

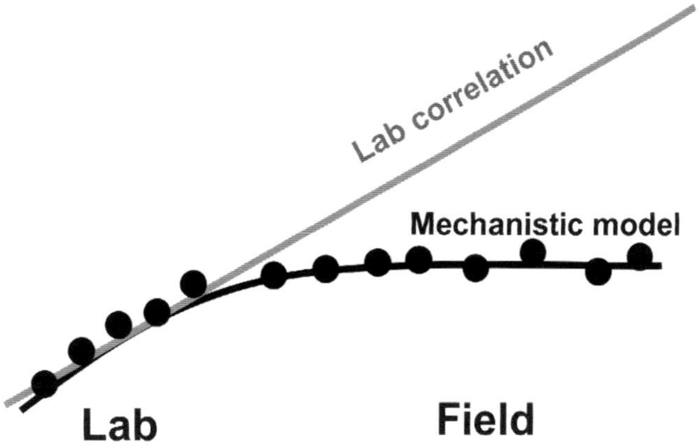

IFE

Computer limitations dictate the use of 1-D models on a coarse grid

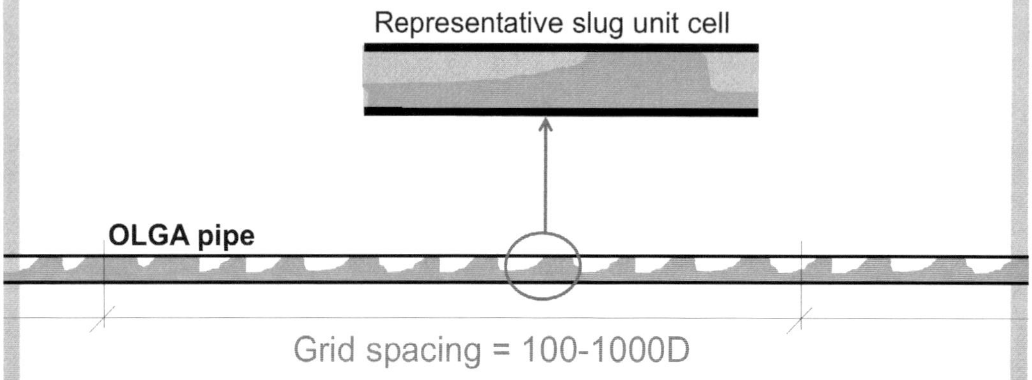

We want to preserve 3D effects in the 1D equations

Stratified Flow

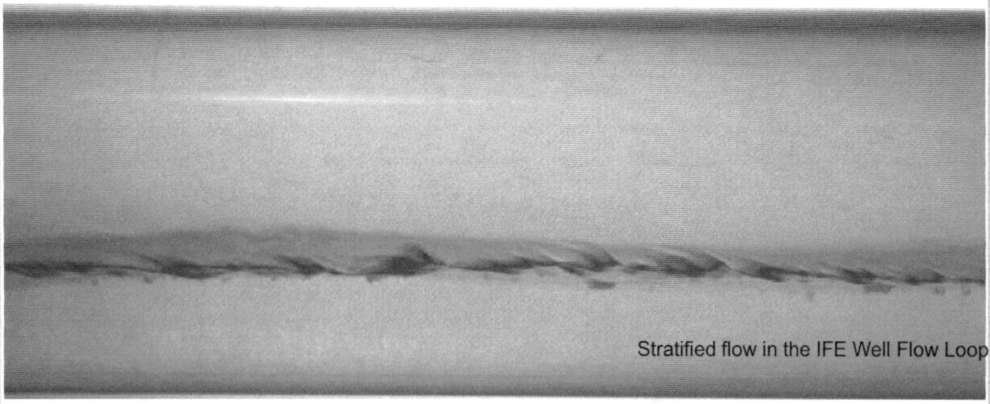

Stratified flow in the IFE Well Flow Loop

The HD stratified flow model computes friction and velocity in a slowly evolving 3-phase flow

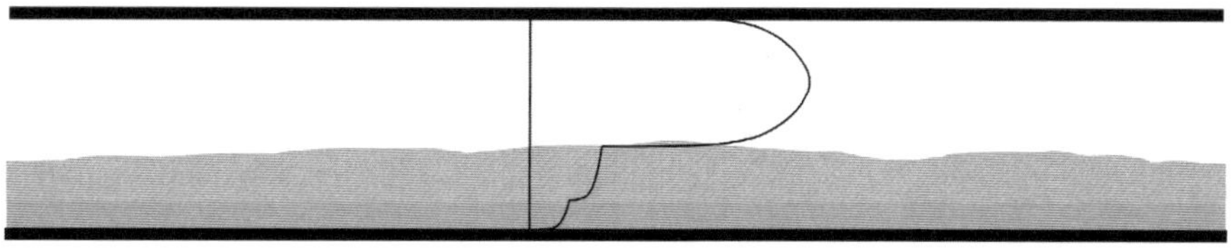

It is also applied within the slug model

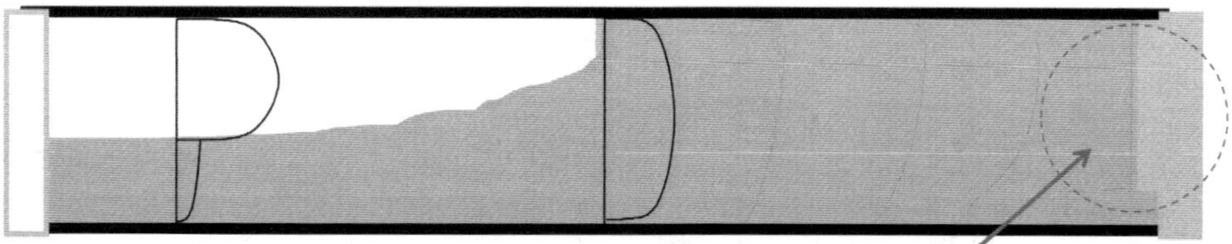

Mixing zones (e.g. slug front) require special treatment

The HD stratified flow model adds 2 dimensions to the 1-D flow simulator OLGA

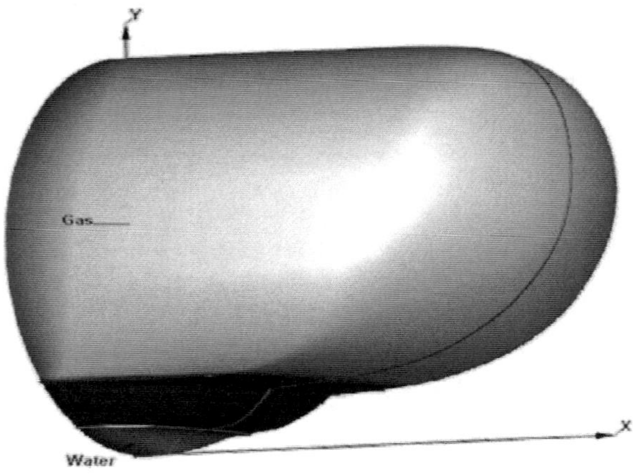

a 3D model with 1D computation time

Analytical description for flow in each layer

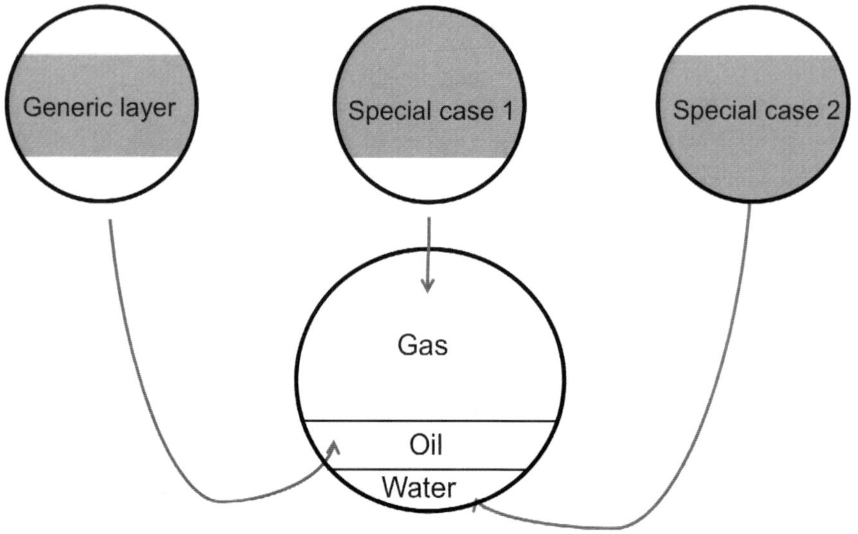

Coupling the layers yields the 3-phase model

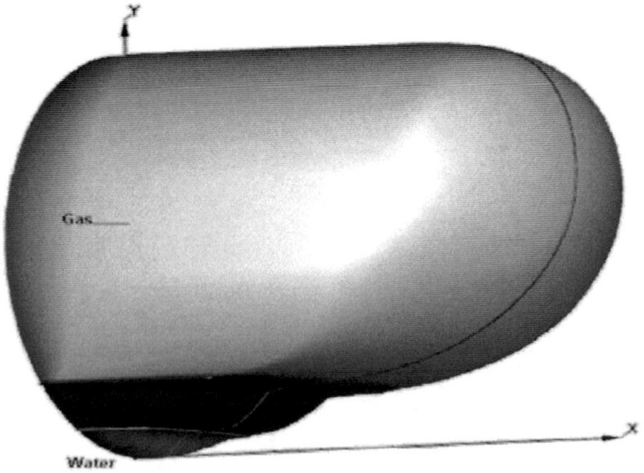

and the friction for the 1D equations

Need closures for the turbulent fields at the interface

Smooth interface
- Isolated turbulent fields
- Interface may be modeled as moving wall

Wavy interface
- Turbulence interaction
- Closure reflects turbulence level at interface associated with waves

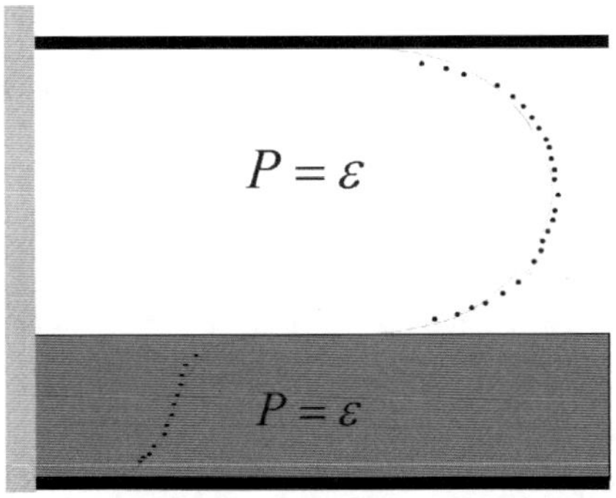

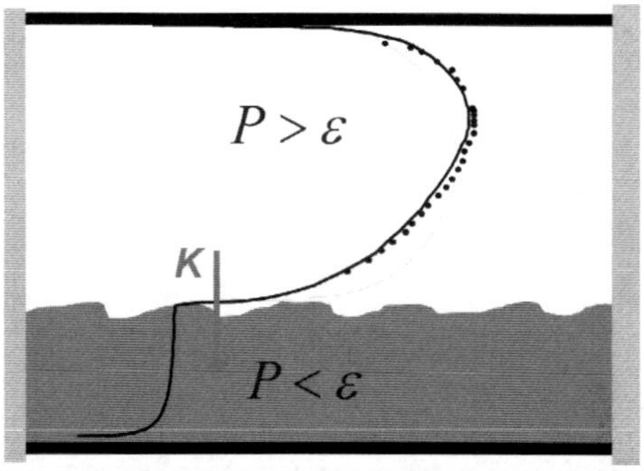

Most CFD models have closures of this kind!

IFE

The model agrees well with CFD

Centerline velocity vs. CFD for a smooth wall-like interface

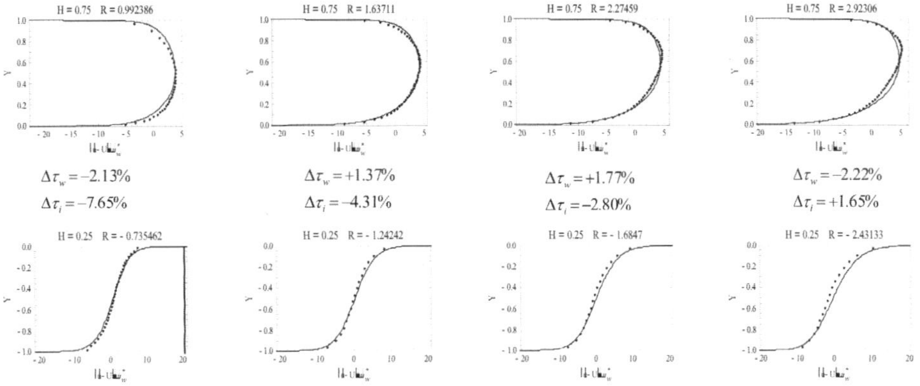

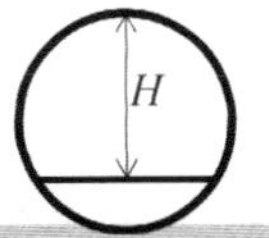

$$R = \frac{\tau_i}{\tau_w}$$

—— Horizon Model
..... Star CD

IFE

The model has excellent scaling properties

Tuning the model to low pressure small ID air/water data
yields improved predictions for high pressure large ID Tiller data !

OLGA Pressure drop

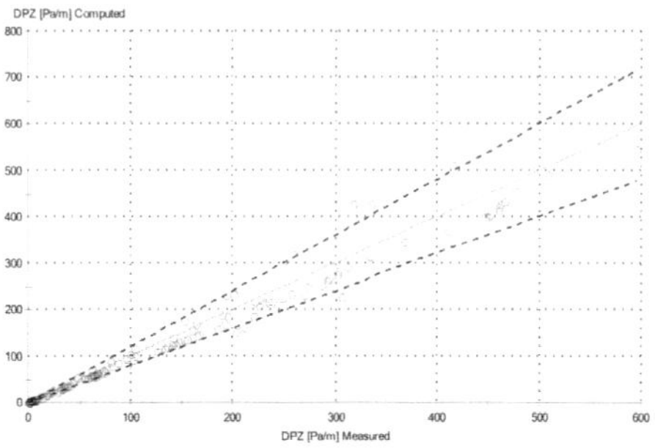

HORIZON Pressure drop

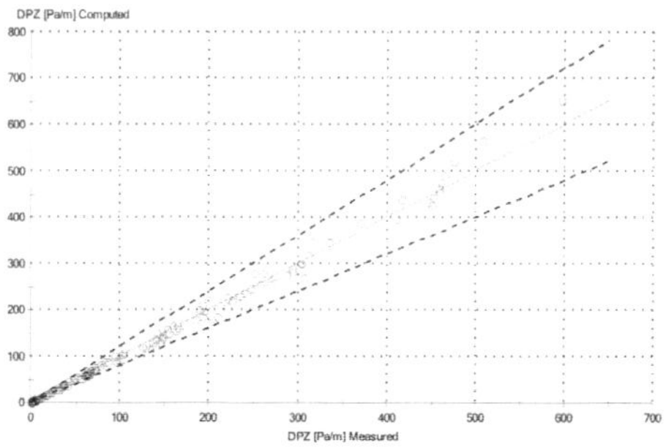

Data: Tiller Nitrogen diesel 8" ID, 20, 45 and 90 bar

IFE

Increasing the interfacial turbulence captures the large wave effect

Stratified flow model
Pressure drop vs. large wave data

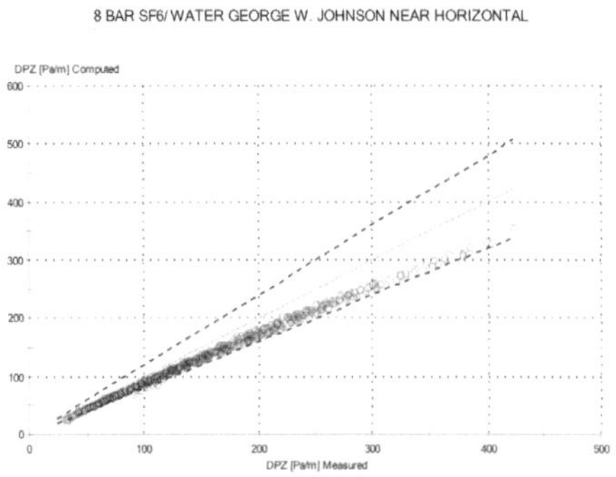

MODEL:Horizon NEXP = 578 ERR = -11.21% STD = 3.32 FRAC = 0.99 Error Bars = 20 %

Stratified flow model
with increased interfacial turbulence
Pressure drop vs. large wave data

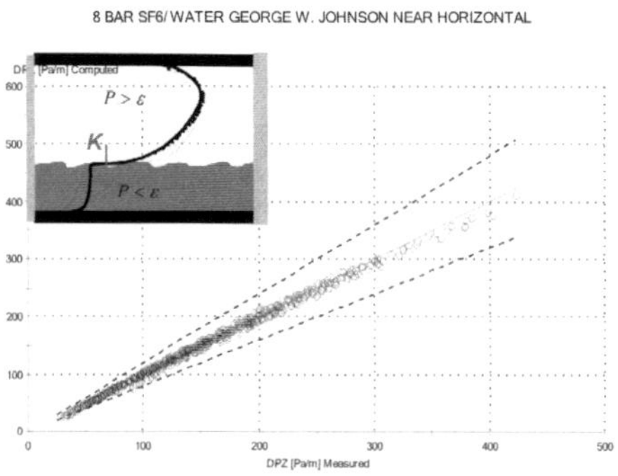

MODEL:Horizon NEXP = 578 ERR = 0.02% STD = 3.95 FRAC = 1.00 Error Bars = 20 %

G. Johnson 8 bar SF6/Water 578 large wave data points in IFE Well Flow Loop

IFE

Slug flow

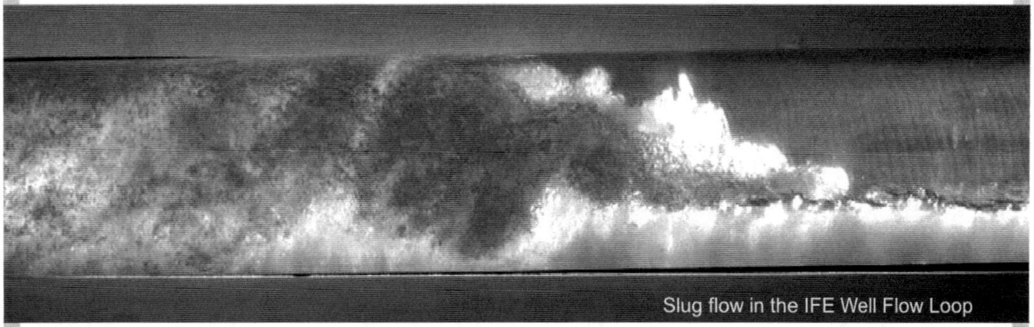

The slug flow model computes the average friction in slug flow

for an OLGA grid cell

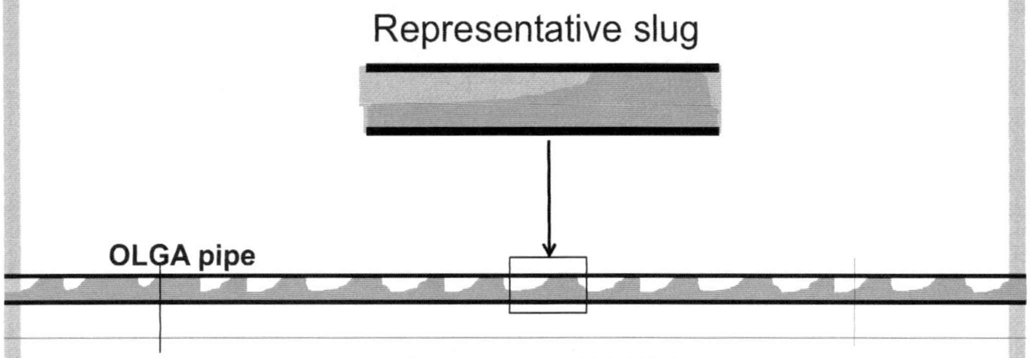

The HD stratified flow model is used locally for velocity and friction

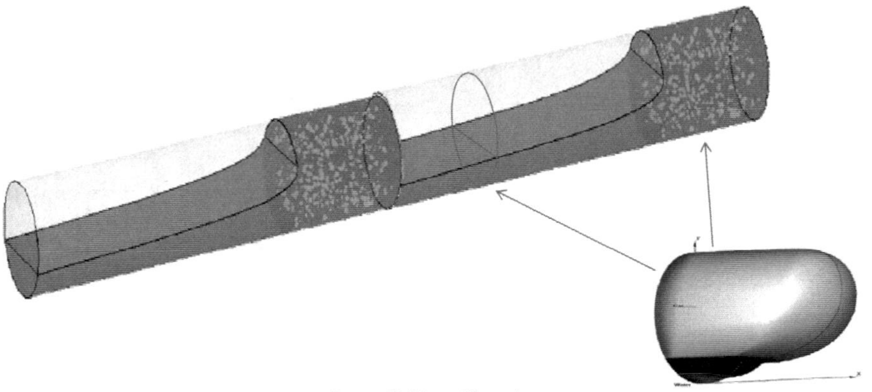

preserving 3D effects

IFE

A special model is required to capture the additional friction in the slug front

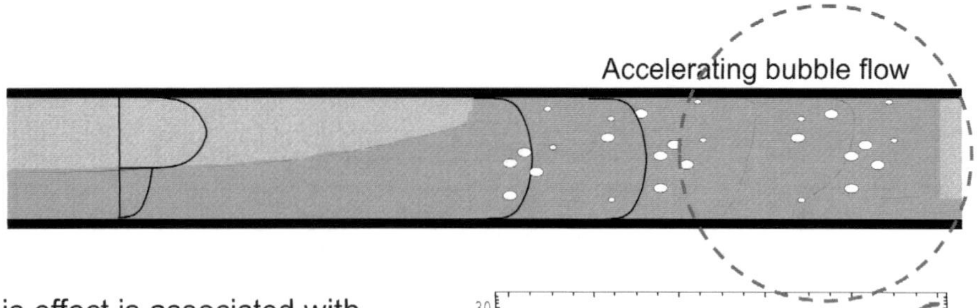

This effect is associated with turbulence generation and momentum mixing

NOT "acceleration pressure drop"

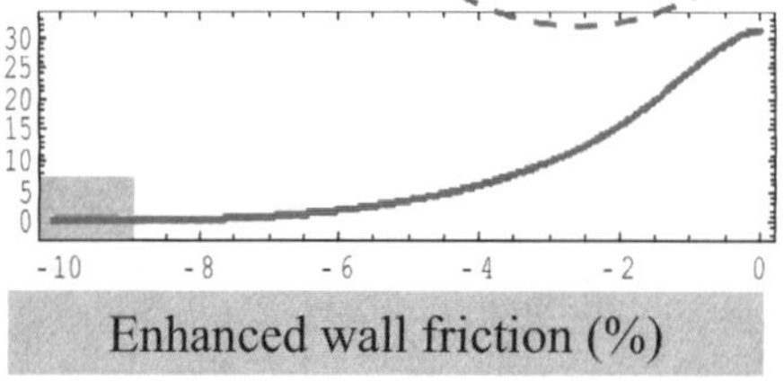

IFE

Effect of extra friction

- Data of Gokcal (2008) – oil viscosity 180-600 cP
- Void in slug correlation of Kora (2011)

Without extra friction

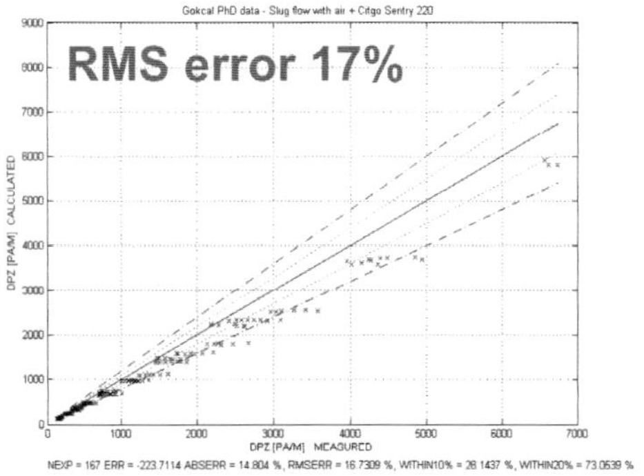

With extra friction

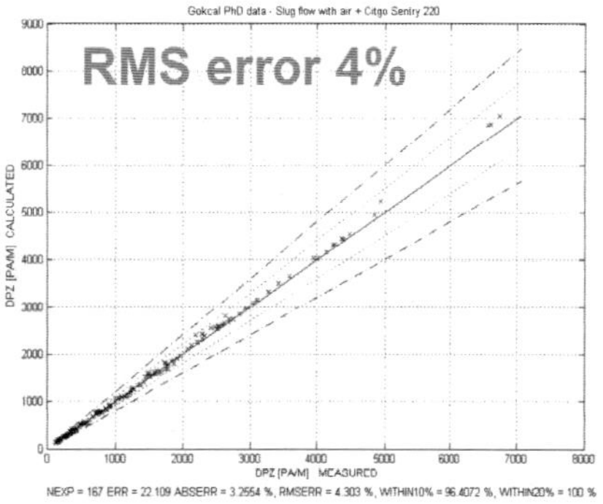

04/10/2011

Dynamic unit cell model yields dynamic evolution of slug length and frequency

(Ongoing work)

- Steady state spatially developing flow (Ujang)

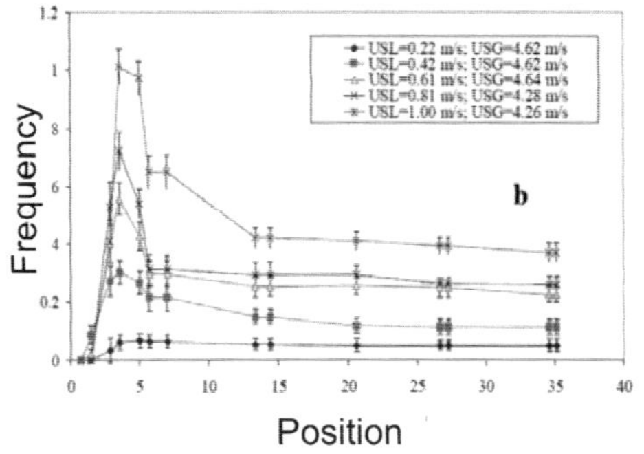

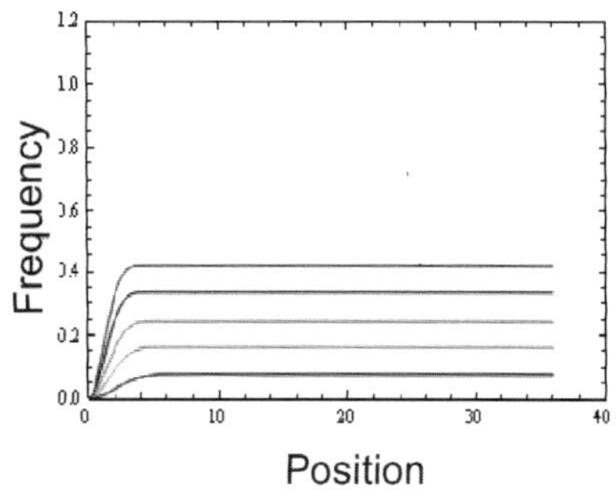

Modelling of gas entrainment at slug front

A

B

- A: Small Usg: **At slug foot.** (Inclination=1°Usl=1.5m/sUsg=**0.5m/s**)
- B: Larger Usg: **Ahead of slug front.** (Inclination=1°Usl=1.5m/sUsg=**2m/s**)

- **High speed videos by Peter Andersson IFE**

23

Gas entrainment processes

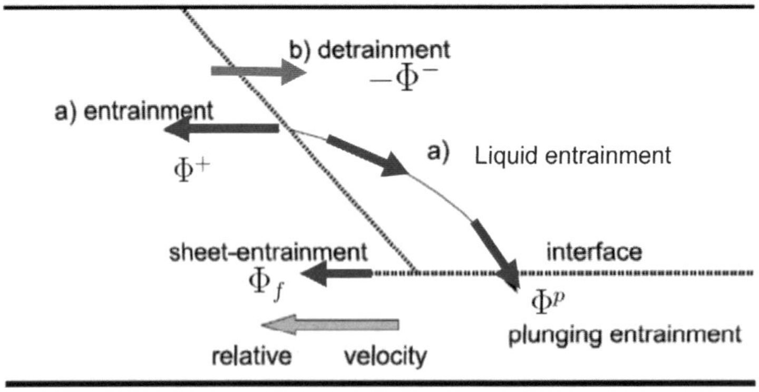

24

Gas entrainment compares well with data

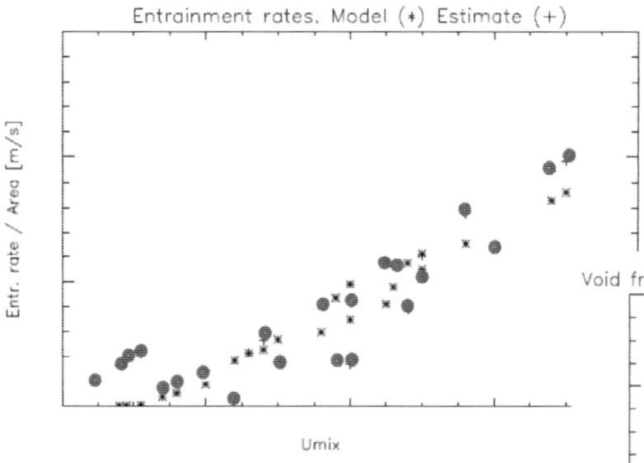

- ● Data
- * Model

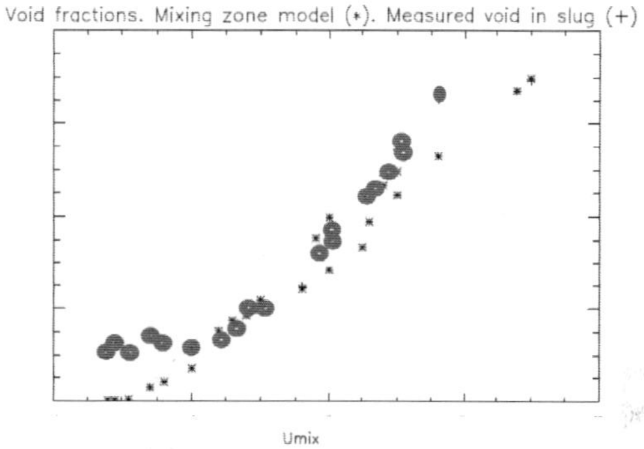

- This is the first attempt!
- <u>No fine tuning!</u>
- 20, 45, 90 bar works well (same parameters)

25

IFE

Conclusion

- We develop mechanistic flow models in order to improve scaling from lab to field conditions
- The approach is based on a "pre-integrated" flow description designed to preserve 3D effects in 1D models
- The work continues in HORIZON II 2008 – 2013

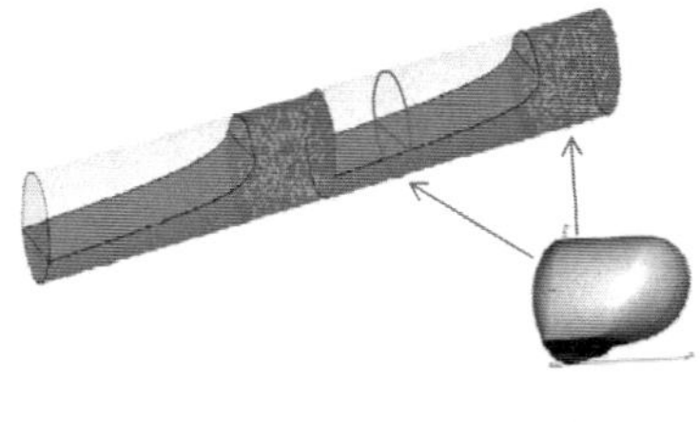

IFE

Applicability of Slug Flow Models to Heavy Oils

Challenges in Flow Assurance

September 26-28, 2011

C. E. "Rusty" Lacy
Shell Global Solutions (US) Inc.

1

DEFINITIONS AND CAUTIONARY NOTE

- The companies in which Royal Dutch Shell plc directly and indirectly owns investments are separate entities. In this presentation "Shell", "Shell group" and "Royal Dutch Shell" are sometimes used for convenience where references are made to Royal Dutch Shell plc and its subsidiaries in general. Likewise, the words "we", "us" and "our" are also used to refer to subsidiaries in general or to those who work for them. These expressions are also used where no useful purpose is served by identifying the particular company or companies. "Subsidiaries", "Shell subsidiaries" and "Shell companies" as used in this presentation refer to companies in which Royal Dutch Shell either directly or indirectly has control, by having either a majority of the voting rights or the right to exercise a controlling influence. The companies in which Shell has significant influence but not control are referred to as "associated companies" or "associates" and companies in which Shell has joint control are referred to as "jointly controlled entities". In this presentation, associates and jointly controlled entities are also referred to as "equity-accounted investments". The term "Shell interest" is used for convenience to indicate the direct and/or indirect (for example, through our 24% shareholding in Woodside Petroleum Ltd.) ownership interest held by Shell in a venture, partnership or company, after exclusion of all third-party interest.
-
- This presentation contains forward-looking statements concerning the financial condition, results of operations and businesses of Royal Dutch Shell. All statements other than statements of historical fact are, or may be deemed to be, forward-looking statements. Forward-looking statements are statements of future expectations that are based on management's current expectations and assumptions and involve known and unknown risks and uncertainties that could cause actual results, performance or events to differ materially from those expressed or implied in these statements. Forward-looking statements include, among other things, statements concerning the potential exposure of Royal Dutch Shell to market risks and statements expressing management's expectations, beliefs, estimates, forecasts, projections and assumptions. These forward-looking statements are identified by their use of terms and phrases such as "anticipate", "believe", "could", "estimate", "expect", "intend", "may", "plan", "objectives", "outlook", "probably", "project", "will", "seek", "target", "risks", "goals", "should" and similar terms and phrases. There are a number of factors that could affect the future operations of Royal Dutch Shell and could cause those results to differ materially from those expressed in the forward-looking statements included in this presentation, including (without limitation): (a) price fluctuations in crude oil and natural gas; (b) changes in demand for Shell's products; (c) currency fluctuations; (d) drilling and production results; (e) reserves estimates; (f) loss of market share and industry competition; (g) environmental and physical risks; (h) risks associated with the identification of suitable potential acquisition properties and targets, and successful negotiation and completion of such transactions; (i) the risk of doing business in developing countries and countries subject to international sanctions; (j) legislative, fiscal and regulatory developments including potential litigation and regulatory measures as a result of climate changes; (k) economic and financial market conditions in various countries and regions; (l) political risks, including the risks of expropriation and renegotiation of the terms of contracts with governmental entities, delays or advancements in the approval of projects and delays in the reimbursement for shared costs; and (m) changes in trading conditions. All forward-looking statements contained in this presentation are expressly qualified in their entirety by the cautionary statements contained or referred to in this section. Readers should not place undue reliance on forward-looking statements. Additional factors that may affect future results are contained in Royal Dutch Shell's 20-F for the year ended 31 December, 2010 (available at www.shell.com/investor and www.sec.gov). These factors also should be considered by the reader. Each forward-looking statement speaks only as of the date of this presentation, 28 September 2011. Neither Royal Dutch Shell nor any of its subsidiaries undertake any obligation to publicly update or revise any forward-looking statement as a result of new information, future events or other information. In light of these risks, results could differ materially from those stated, implied or inferred from the forward-looking statements contained in this presentation. There can be no assurance that dividend payments will match or exceed those set out in this presentation in the future, or that they will be made at all.
-
- We use certain terms in this presentation, such as discovery potential, that the United States Securities and Exchange Commission (SEC) guidelines strictly prohibit us from including in filings with the SEC. U.S. Investors are urged to consider closely the disclosure in our Form 20-F, File No 1-32575, available on the SEC website www.sec.gov. You can also obtain these forms from the SEC by calling 1-800-SEC-0330.

Outline

- Background
- Flow Pattern Prediction
- Closure Relationships
- Laboratory Data
- Summary

3

Background

- Large quantities of heavy oil to be produced now and in the future
  - Viscosities 0.1 to 10 Ns/m^2 (100 – 10,000 cp)
- Multiphase flow heavy oil models needed in general
- Slug Flow models needed specifically
  - Slug flow pattern can dominate operability window
  - Annular and Stratified flow may be outside of operability window
- Issues
  - Modeling – What needs to be changed?
  - Data – Do we have enough information to determine the changes?
  - Scaleup – Will the changes work for actual design predictions?
    - Laminar at lab scale
    - But laminar, transitional, or weakly turbulent at field scales

4

Flow Patterns - Horizontal

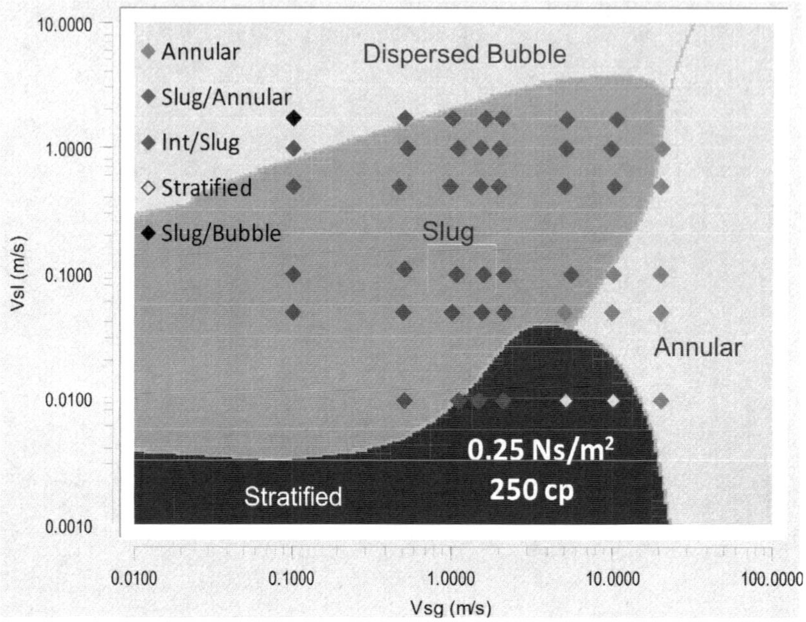

B. Gokcal, MS Thesis, Petroleum Engineering, The University of Tulsa, 2006.

Flow Pattern – Vertical

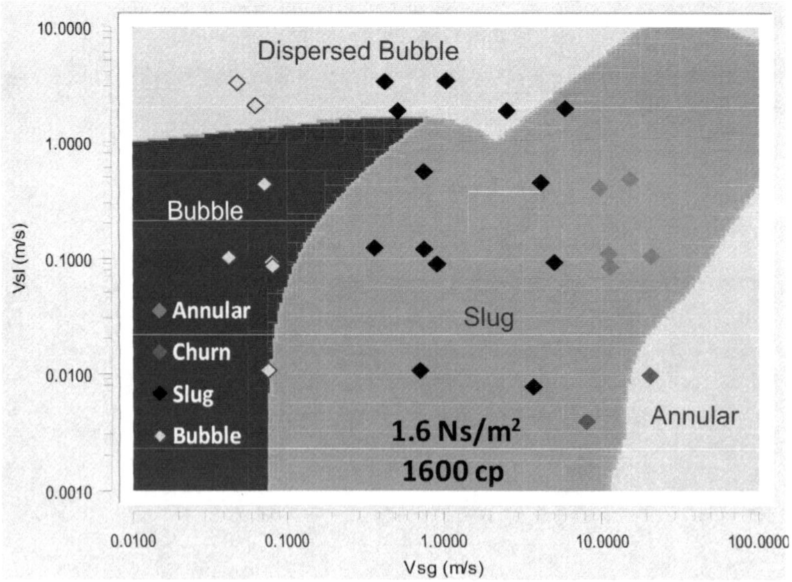

Schmidt et al., Int. J. Multiphase Flow, Vol. 34, pg. 363, 2008.

Unit Cell Model for Slug Flow

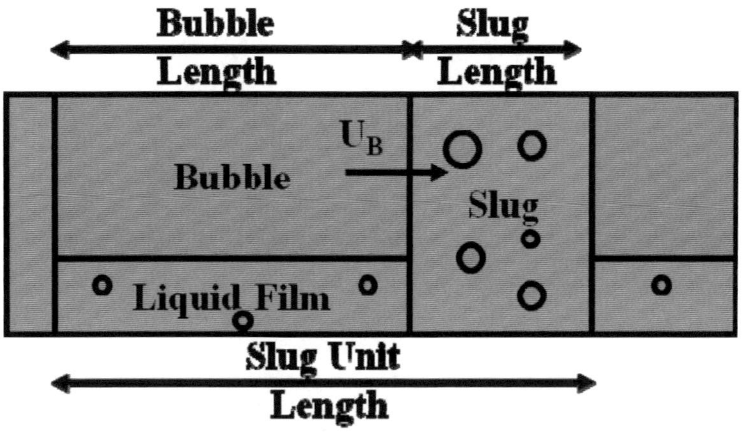

Unit cell model for two-phase flow of heavy oil and gas

7

Closure Relationships

- Translational velocity of the bubble

$$U_B = C_0 U_{Mix} + U_{Drift}$$

- Distribution Parameter, C_0

$$C_0 = \frac{Lam}{1+x^2} + \frac{Turb.}{1+x^{-2}}$$

$$x = Re_L / Re_c$$

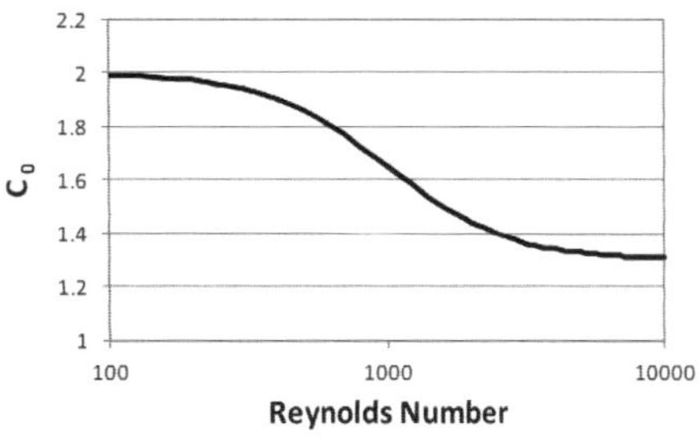

- Drift Velocity, U_{Drift}
  - Both physical property and angle dependence
  - Vertical – Viana et al.
  - Horizontal – Ben-Mansour et al.

R. Ben-Mansour, et al., 7th North American Conf. Multiphase Technology, Banff, Canada, 2010.

F. Viana etal.. Journal of Fluid Mechanics, Vol. 494, pp. 379-398, 2003.

8

Closure Relationships -2

- Slug characteristics
  - Voidage in liquid slug – Kora, 2010 (horizontal)
    - Common models only a function of U_{mix}
  - Voidage in liquid film (lumped into bubble)
  - Slug length / frequency
    - Is physics "frozen in"?
    - For results here, Shell's model (Zabaras, SPE 65093, 2000)

- Sufficient data??
  - Can we distinguish between different closure alternatives?
  - How important are these distinctions?

C. Kora, MS Thesis, Petroleum Engineering, The University of Tulsa, 2010.

9

Vertical Data – Extra Heavy Oil Holdup

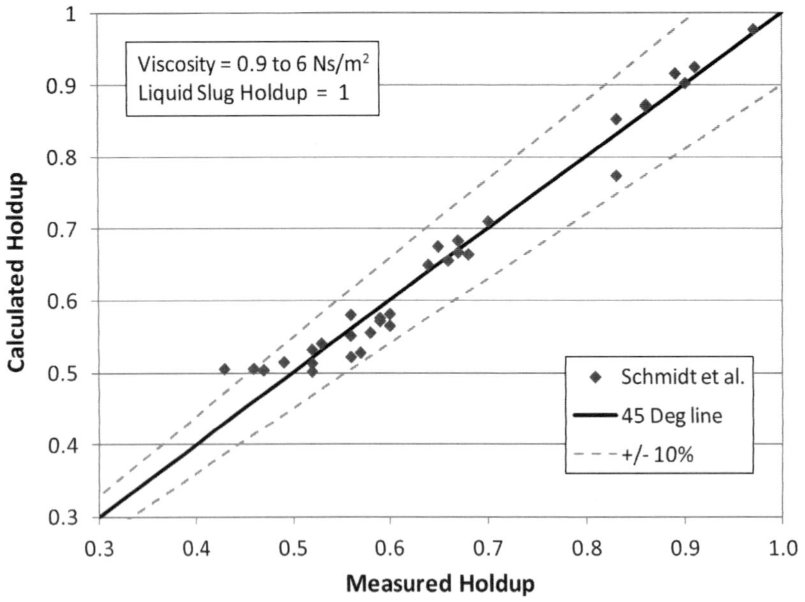

Schmidt et al., Int. J. Multiphase Flow, Vol. 34, pg. 363, 2008.

Horizontal Data – Heavy Oil Pressure Gradient

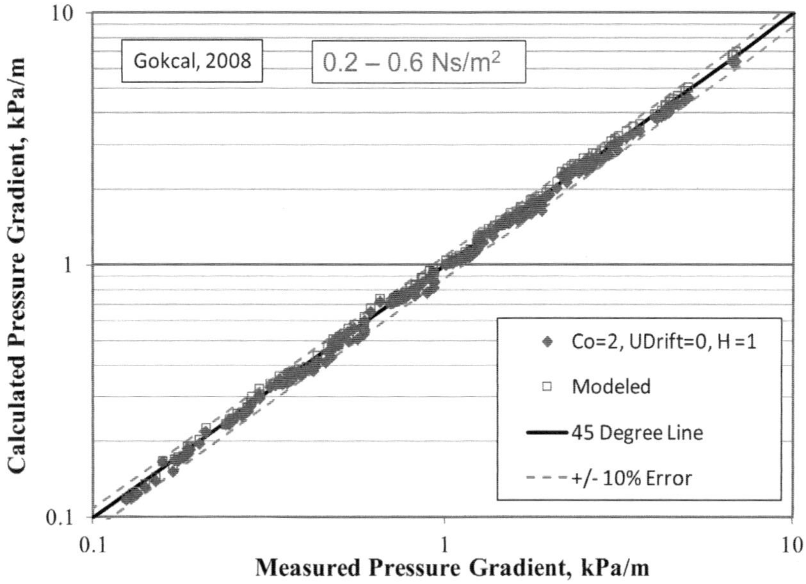

B. Gokcal, PhD Thesis, Petroleum Engineering, The University of Tulsa, 2008.

Horizontal Data – Heavy Oil Holdup

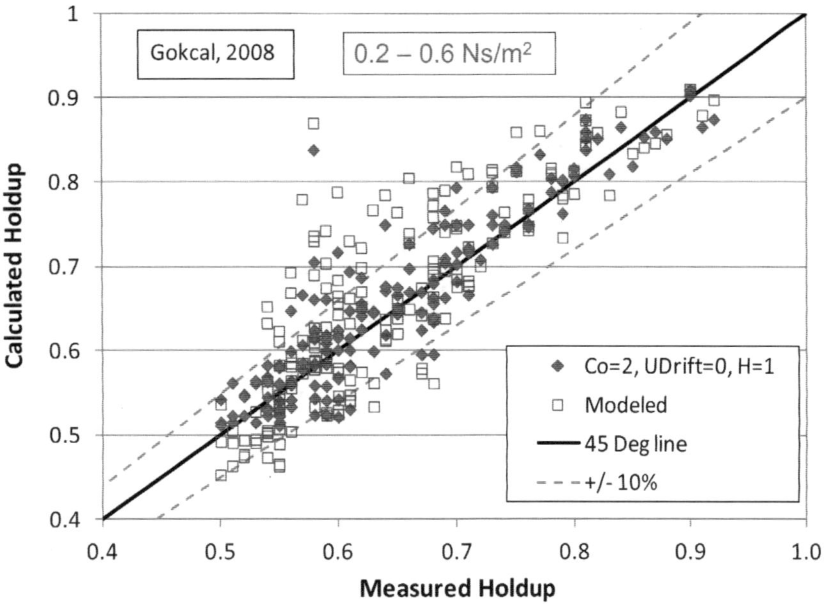

B. Gokcal, PhD Thesis, Petroleum Engineering, The University of Tulsa, 2008.

12

Horizontal Data – Comparison of Closure Relationships

Conditions			% dP/dx Error		% Holdup Error	
C_0	U_{Drift}	α^S_L	Avg	Abs Avg	Avg	Abs Avg
2	0	1	-2.0	4.4	1.8	5.1
Model	Model	Model	1.6	5.5	2.5	7.9

- Data is reasonably predicted independent of modeling choice
  - Though good quality, cannot distinguish between some modeling choices
  - Drift velocity and liquid slug holdup developed under similar conditions but can make results worse.
- Why does simplest choice work well?
  - C_0 is incorporating part or all of U_{Drift}.

R. Ben-Mansour, et al., 7th North American Conf. Multiphase Technology, Banff, Canada, 2010.

C. Kora, MS Thesis, Petroleum Engineering, The University of Tulsa, 2010.

13

Vertical Data – Heavy Oil Holdup

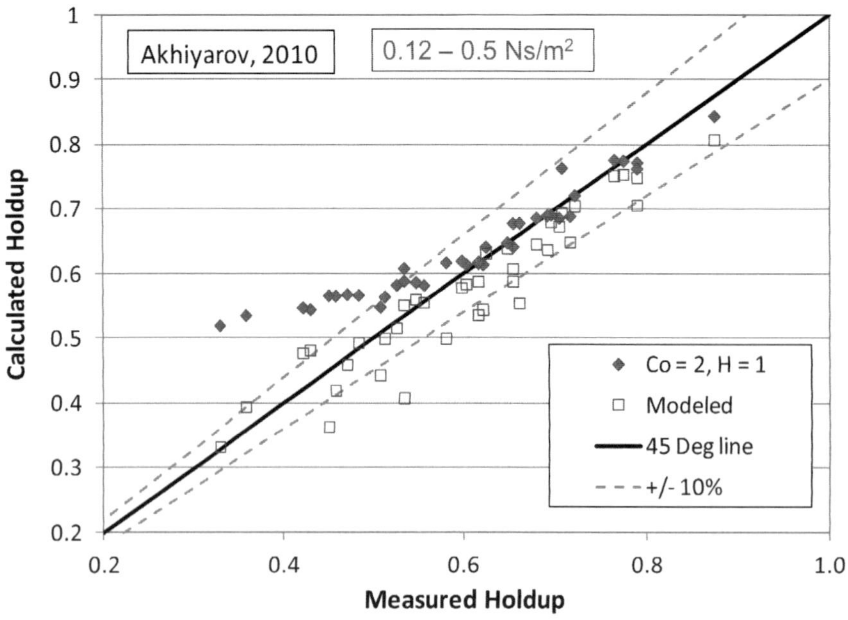

D. Akhiyarov, MS. Thesis, Petroleum Engineering, The University of Tulsa, 2010

Vertical Data – Heavy Oil Pressure Gradient

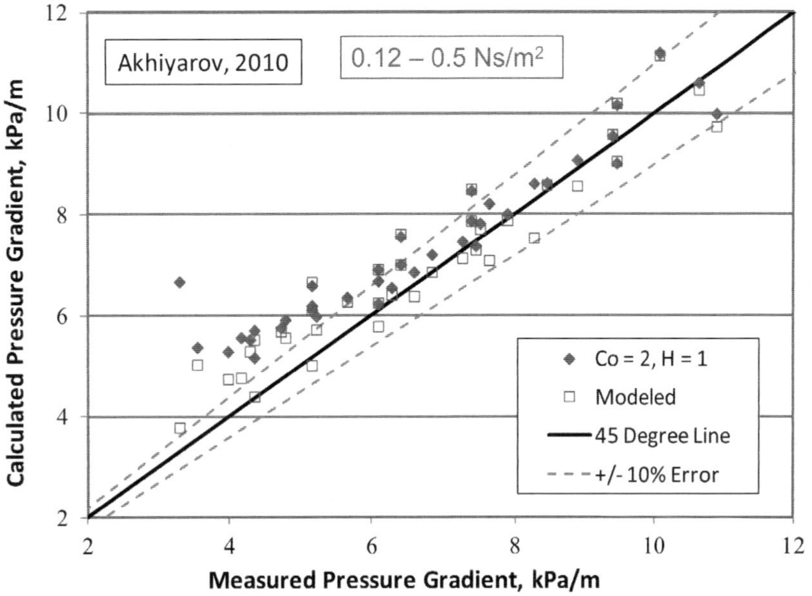

D. Akhiyarov, MS. Thesis, Petroleum Engineering, The University of Tulsa, 2010

Laminar/Turbulent Transition – Medium Oils

- Industrial systems
  - Can operate in transitional, or weakly turbulent flow.
  - This requires medium oils of 0.03 - 0.1 Ns/m^2 (30 – 100 cp)
    - Re_L = 900 – 3100 equals a 6" ID pipe with 300 cp oil at 2 – 7 m/s mixture velocity

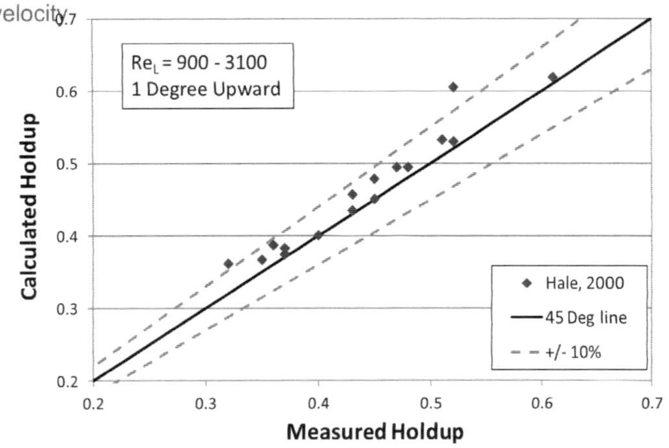

C.P. Hale, PhD Thesis, Imperial College, London, 2000.

16

Transition to Weakly Turbulent – Medium Oils

- On example – Manolis horizontal flow data
  - Re_L = 2400 -12,000
  - Analysis here shows the importance of slug holdup relationship
  - Optimizing both holdup and pressure gradient difficult however

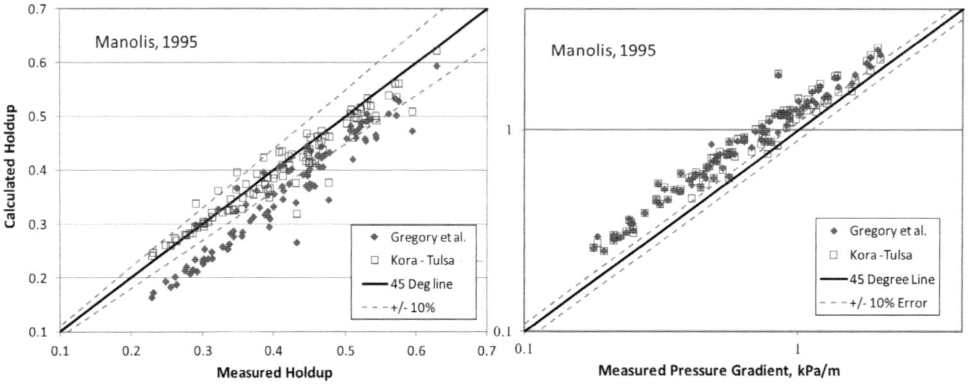

G.A. Gregory, Int. J. Multiphase Flow, Vol. 4, pp. 33-39, 1978.
I.G. Manolis, PhD Thesis, Imperial College, London, 1995.

17

Summary

- Flow pattern prediction appears ok.
  - More data needed to verify boundaries however
- Current slug flow modeling can explain much of the data available
  - More work is definitely needed
  - Simple closure relationships can often give reasonable predictions
- Issues remain
  - Optimize laminar/turbulent transition and weakly turbulent flows
  - Inclination effects only beginning to be addressed
  - Limited data
    - Distinguishing between model deficiencies and experimental errors
    - Verify closure relationships
  - "Extrapolation" from lab to field potentially tougher for heavy oil
    - Deal with laminar to turbulent differences instead of different levels of turbulence

18

The 6th AIChE/SPE Joint Workshop - Challenges in Flow Assurance

Flow Assurance Engineering -

The Role in Projects and the New Challenges

Y. Doreen Chin

Upstream Major Projects
Shell International Exploration and Production

Houston, TX, USA
September 27-28, 2011

Date 17/10/2011

Outline

1) Introduction.
2) Flow Assurance Engineering building blocks .
3) The integrated role of Flow Assurance engineer in development projects
4) Flow Assurance Engineering in various phases of the projects
5) Flow Assurances Engineering facing new challenges
6) Summary

1. Introduction

- Flow Assurance gained its name from "*Garantia de Escoamento*" ⇒ "Guarantee the Flow" – by Petrobras in the early 1990's.
- It started long ago, from
  - Correlation based pipeline hydraulic/thermal calculations,
  - Solid deposition and pipeline blockage avoidance,
  - Chemical treatment.
- Over one and half decades, it has evolved into an integrated engineering discipline in petroleum industry.
- The objectives of this presentation to present
  - An integrated picture of Flow Assurance Engineering.
  - Key facts contributing to Flow Assurance Engineering success.

3 of 20

1. Introduction

- A famous old question: "What is Flow Assurance"?
- **Flow Assurance** is the technologies and sciences that help the industry to ensure the production of hydrocarbon fluids from reservoir to end point,
  - Safely,
  - Economically ,
  - Over the life of a field, and
  - In any environments.
- **Flow Assurance Engineering** is an emerging discipline in the industry that applies the science and technology of Flow Assurance into project, and is responsible for the continuity of flow in an asset by defining
  - Key mechanical parameters,
  - Operating conditions and strategies, and
  - Chemicals and deliver systems required to maintain production and asset integrity.

4 of 20

2. Flow Assurance Engineering Building Blocks

- On project environment, FA engineer need to *deliver*/to *help deliver* the project.
- The resource involved is project dependent.

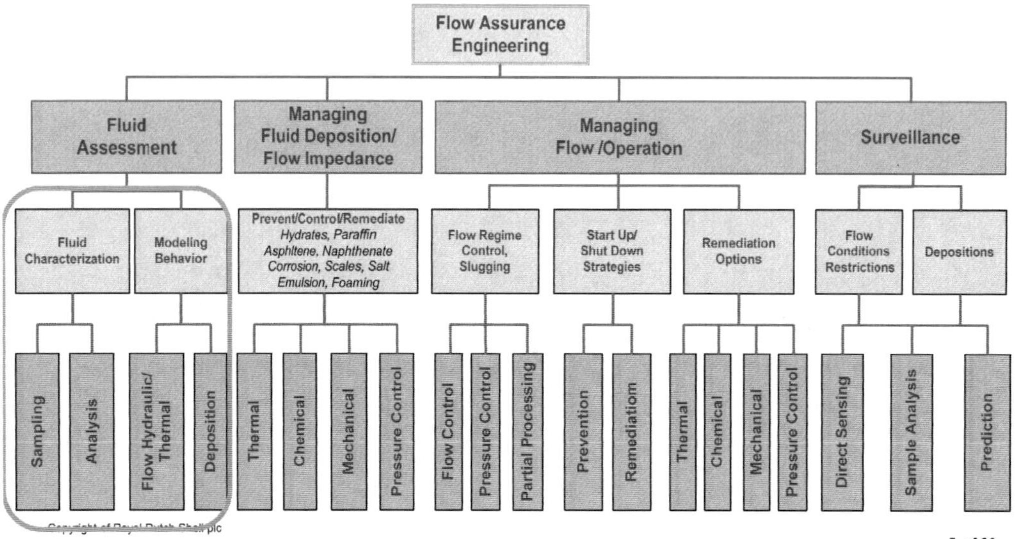

Fluid & Flow Modeling & Flow Assurance Engineering

- **Fluid and Flow Modeling** describe the fundamentals and applications of transport of gas, liquid, and solids (either as single phase or multiphase flow) in wells, pipelines, and facilities. It includes,
  - —Fluid analysis and simulations
    - —Fluid characterization, PVT properties, rheology,
    - —Organic solids properties, deposition rate.
  - —Flow modelling
    - —Single phase flow, multiphase flow (2, 3 and 4 phases);
    - —Heat and mass transfer at interfaces;
    - —Flow system dynamic behaviours;
    - —Operation scenarios, upset scenario,s and
    - —Fluid-structure interaction, Computational Fluid Dynamics
  - —Fluid and flow simulation combined
    - —Solid deposition rates and locations.
- Fluid and Flow Modelings are the most critical block of Flow Assurance Engineering, and impact every phase of the project.

Fluid & Flow Modeling & Flow Assurance Engineering

- **Flow Assurance Engineering** applies the outcome of fluid & flow analyses to
  - Design of components and systems
    - Sizing of pipelines and flowlines,
    - Integrated design of the subsea system (trees, manifolds, umbilicals, controls, and flowlines).
  - Development of strategies to ensure optimal, uninterrupted production from the wellbore to the processing facility, both onshore and offshore.
  - Determination of subsea production system performances, and
  - Surveillance during operation, particularly during start-ups and shutdowns.

Fluid & Flow Modeling & Flow Assurance Engineering

- Multiphase: Thermo-Hydraulics & Fluid-Related Effects.

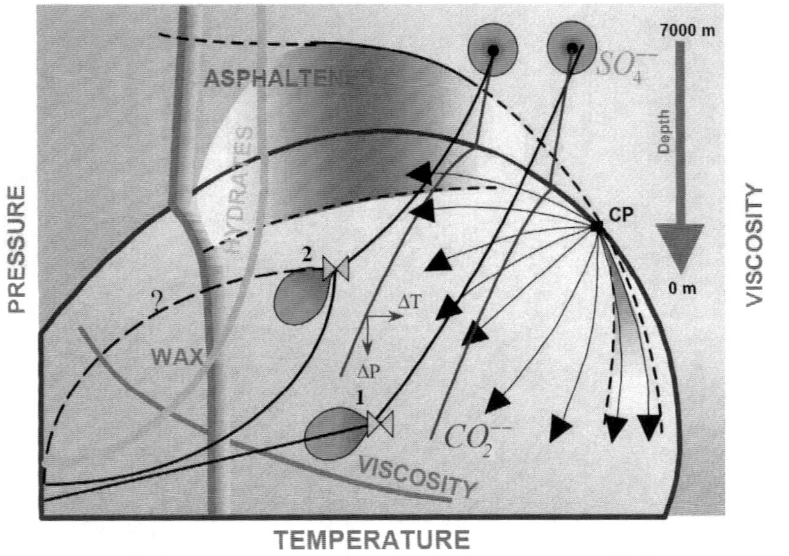

1: shallow water **2: deep water offshore**

3. Flow Assurance Engineering in Deepwater Projects

- Flow Assurance Engineering is a transverse and multidisciplinary approach/process.

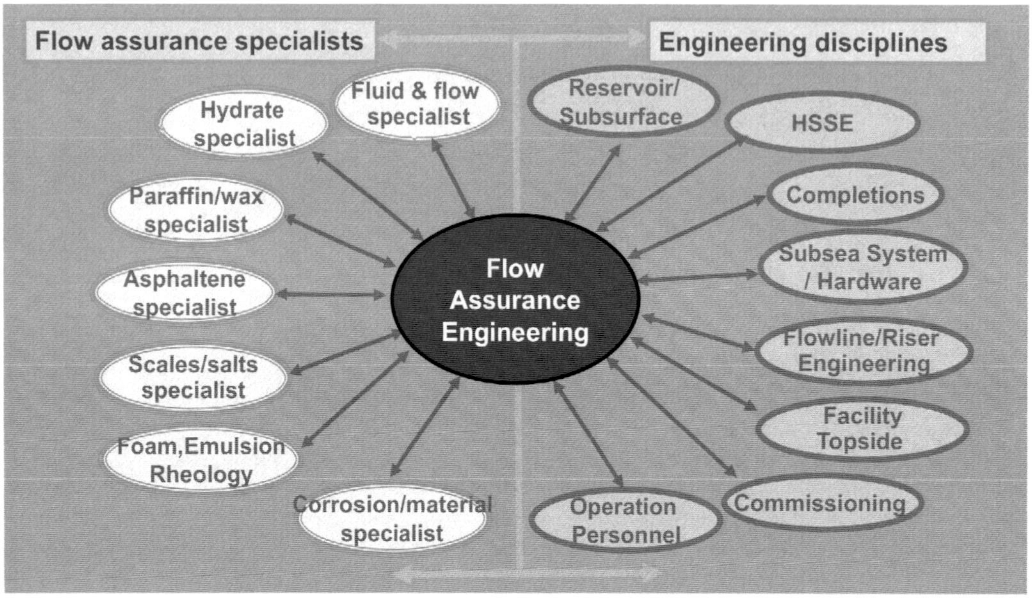

Flow Assurance Engineering in Deepwater Projects

- Flow Assurance Engineering drives system selection, design concept, and operation strategy of deepwater fields.

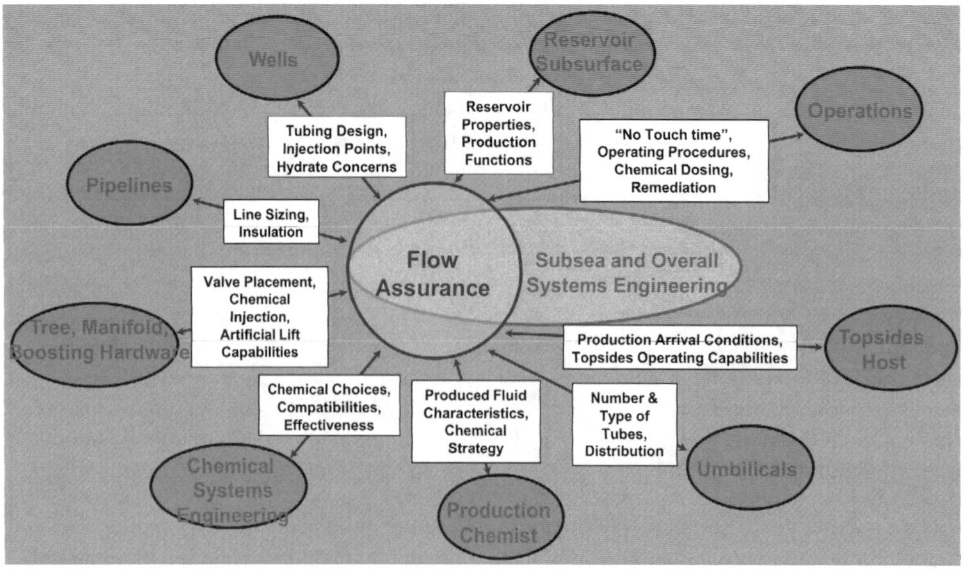

Flow Assurance Engineering in Deepwater Projects

- Flow Assurance Engineering impacts on topside design
- Integration of Flow Assurance Engineering and Topside engineering is critical.

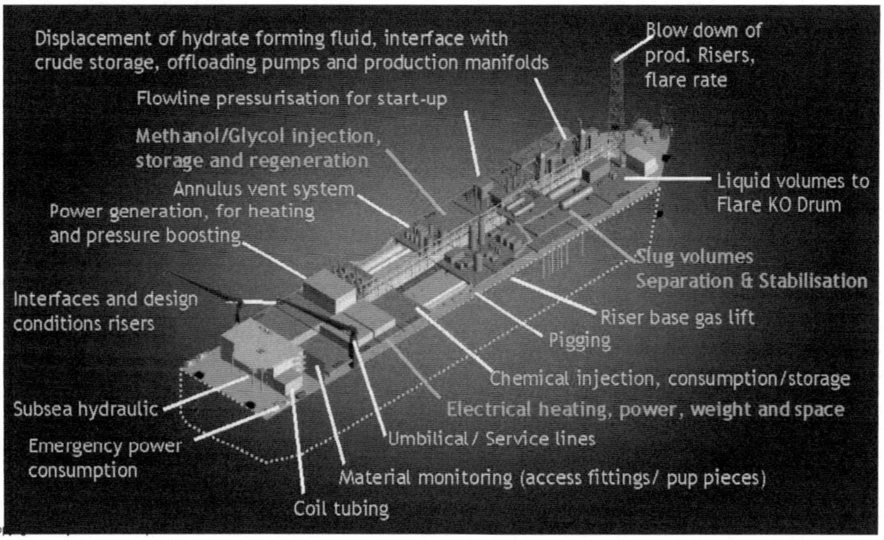

Flow Assurance Engineering in Deepwater Projects

- Flow Assurance Engineering impacts on costs
- Reduce CAPEX:
  - Elimination/Simplification/Standardization/Automation

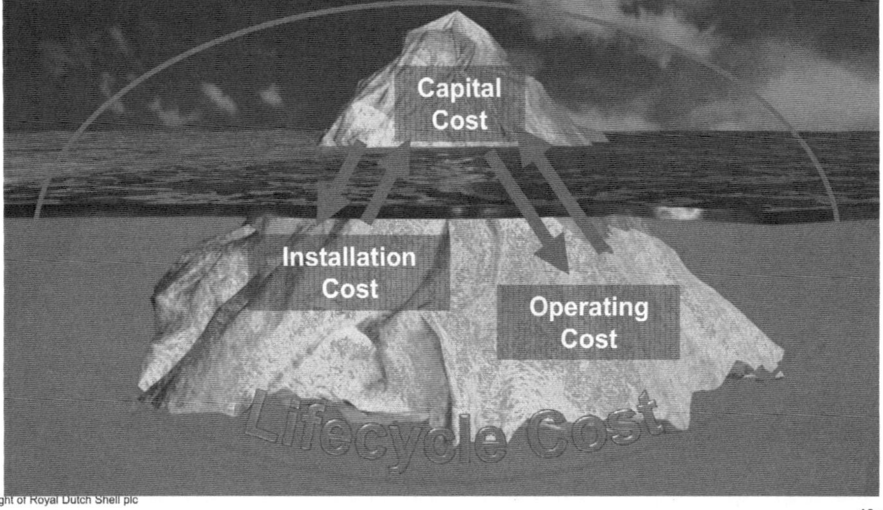

4. Flow Assurance Engineering in Project Phases

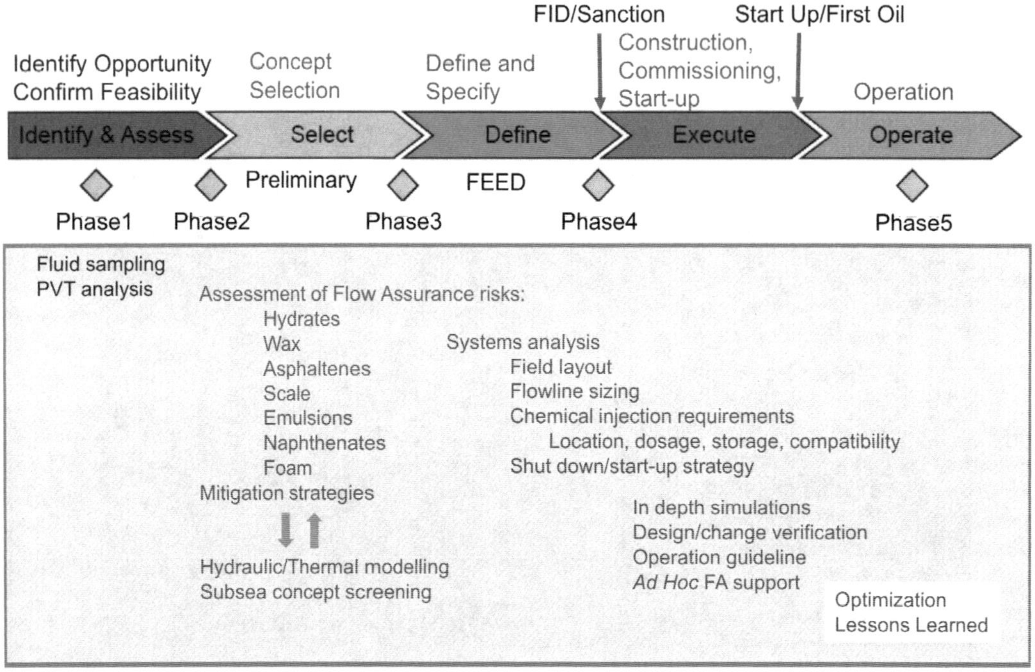

Flow Assurance Engineering in Project Phases

Keys for Flow Assurance Engineering Success

- Characterization of fluids and fluid-related risks identified at an early stage for production & injection.
  - The higher the uncertainty, the
    - more risks,
    - more built-in flexibilities.
    - higher capital cost.
- Transverse and multidisciplinary approach including specialists and operations, coordination of inputs from
  - Flow assurance specialists,
  - Many disciplines, from reservoir engineer to operations.
- Accuracy of prediction depends on quality of subsurface work.
- All issues fully addressed in concept selection and design.
- All technical solutions in place to deal with uncertainties in operation phase.
- Contingency plan.

5. Flow Assurance Facing New Challenges Today

- Two deepwater projects with successful subsea processing technology deployment, the enablers.
- Artificial Lift needed for economic development, 30%-50% Increase in Reserves.
- Matured technology – Caisson ESP system.

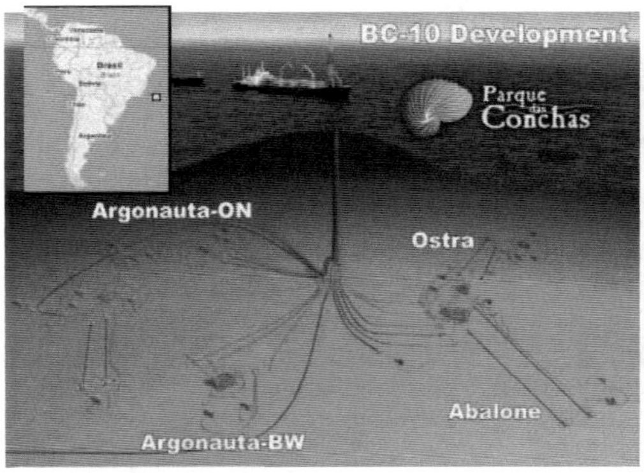

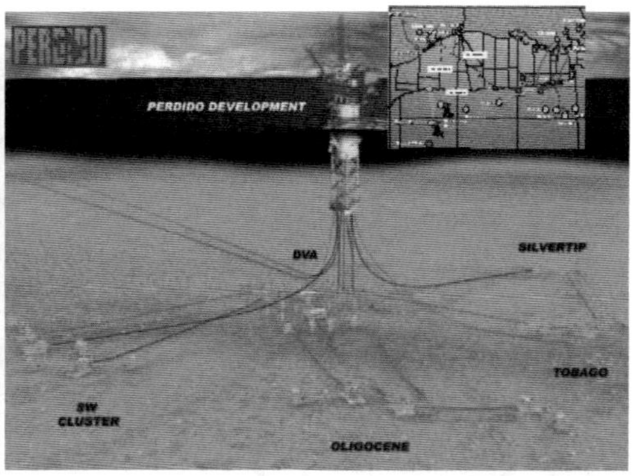

BC-10 Brazil

- Large Reserves
- ~6000 ft water depth
- Low Pressure Reservoirs
- Heavier Oils

Perdido GoM

- Large Reserves
- ~8000 ft water depth
- Low Pressure Reservoirs
- Variable Fluid Properties
- No Production Analog

Flow Assurance Facing New Challenges Today

Subsea processing

- Has been an emerging technology in the industry.
- Is becoming field proven technology (some concepts).
- Can increase productivity by using:
  - Single or multiphase pumping systems,
  - Two-phase separation,
  - Three-phase separation.
- Has expanded Flow Assurance Engineering to new territory
  - Examples: BC-10 and Perdido – Caisson ESP systems
    - Subsea separation
      - Flow pattern entering caisson
      - Foaming
      - Liquid carry over into gas line
      - Gas carry under into pumps
    - Subsea multiphase pumping
      - Gas –Volume – Factor (GVF) in to pump
      - Rapid gas decompression during start up

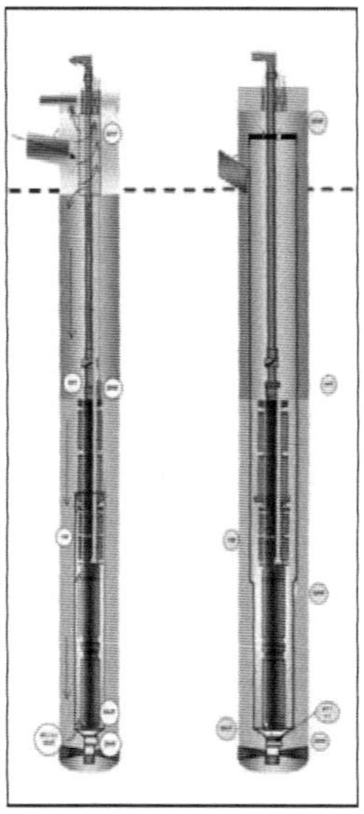

FAE Support in Subsea Processing/Boosting System Design

- Flow Assurance Engineering supports in BC-10 Caisson ESP system design and subsea system performance determination.

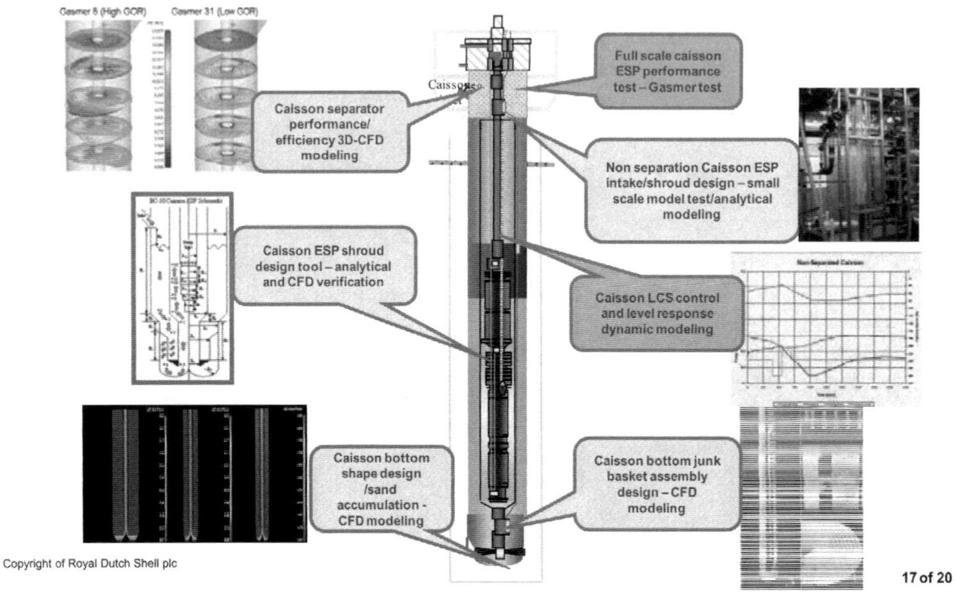

17 of 20

Subsea Production System Operating and Performance Envelopes

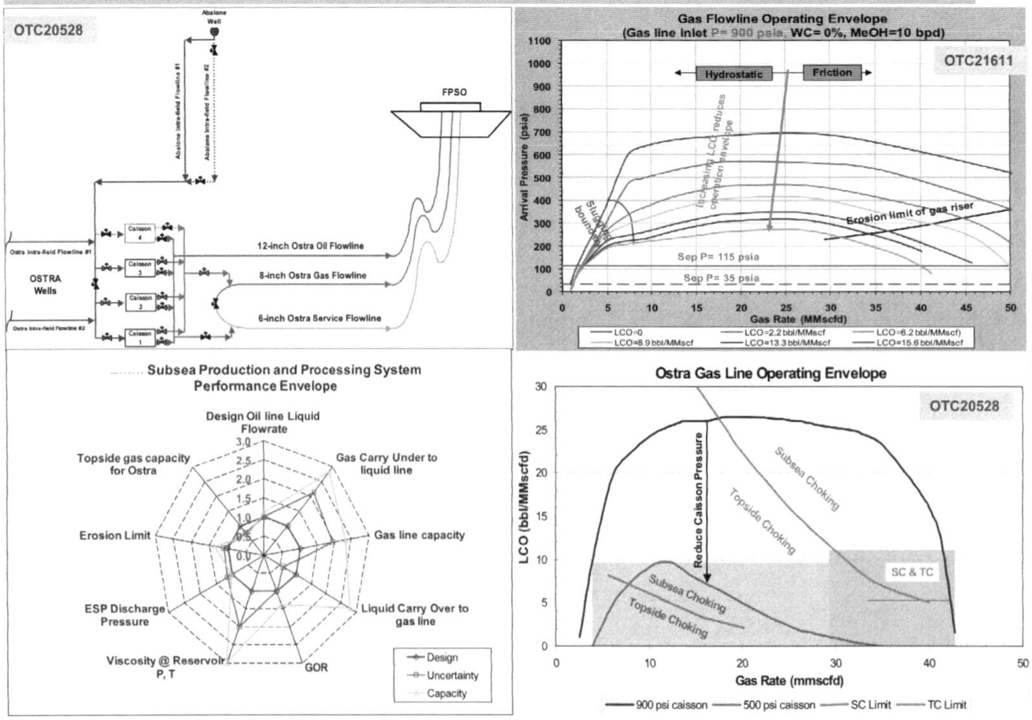

FA Engineering- Conventional and New Territory

- Example of FAE at one major green field development (BC-10) – from Selection to First Oil.

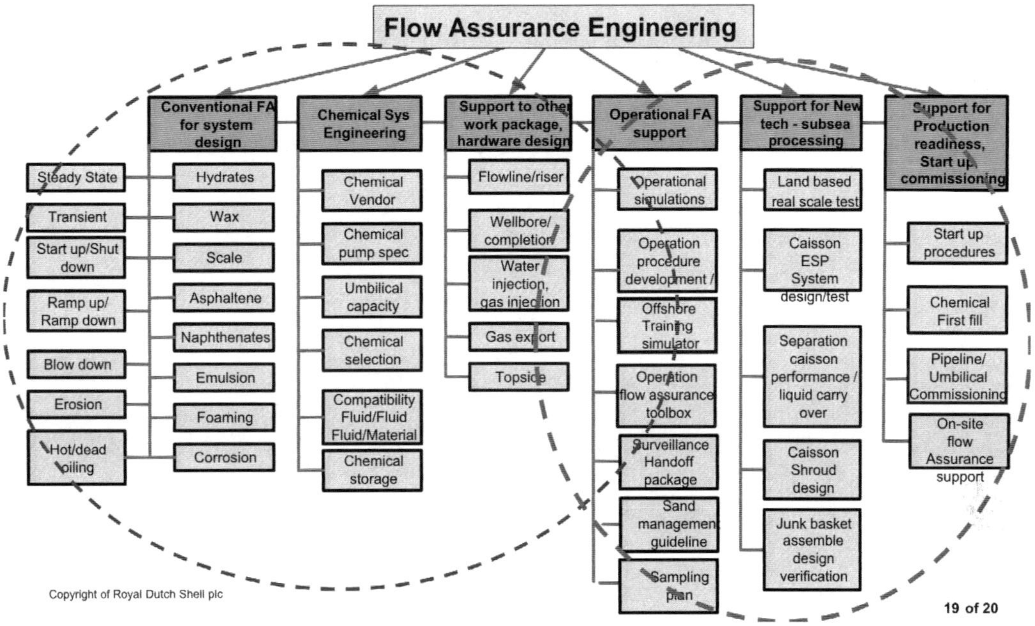

19 of 20

6. Summary

- Flow Assurance project engineer integrates all the key flow assurance building blocks into to projects,
  - Prediction and interpretation of the fluid characteristics,
  - Flow hydraulics and thermal behaviors,
  - Prevention strategies of solids deposition
  - Remediation methods
  - Determination of production system performance, and
  - Guidance of operation strategies.
- Transverse and multidisciplinary approach is a key to successful Flow Assurance Engineering.
- Flow Assurance research advances in academia is helping the industry progressing in FA risk management.
- Subsea processing technology deployment in deepwater development have raised new challenges for Flow Assurance Engineering, and expanded the flow assurance domain.

20 of 20

Thank You!

Subsea Oil Spill Flow Assurance

Hatton, G., Shell

Macondo and Ixtoc were two different large-volume oil-spill events. To improve response to possible future events, the industry is preparing for the broad domain of possible oil spill scenarios–from a first set of spill scenarios for which metal-to-metal, mechanical connection systems are suited to a second set of other scenarios with disparate spill sources not amenable to metal-to-metal seal solutions. For this second set, better insight and scientific understanding of the flow assurance challenges are important in developing robust response and capture systems. This paper discusses some of the flow assurance issues and challenges associated with capture systems for the second set of scenarios.

AUTHOR INDEX